景山南立面图
South elevation of Jingshan Park

寿皇殿建筑群外院南立面图

South elevation of the outer court of Hall of Imperial Longevity complex

寿皇殿建筑群内院南立面图
South elevation of the inner court of Hall of Imperial Longevity complex

寿皇殿建筑群纵剖面图
Longitudinal section of Hall of Imperial Longevity complex

寿皇殿建筑群横剖面图
Transverse section of Hall of Imperial Longevity complex

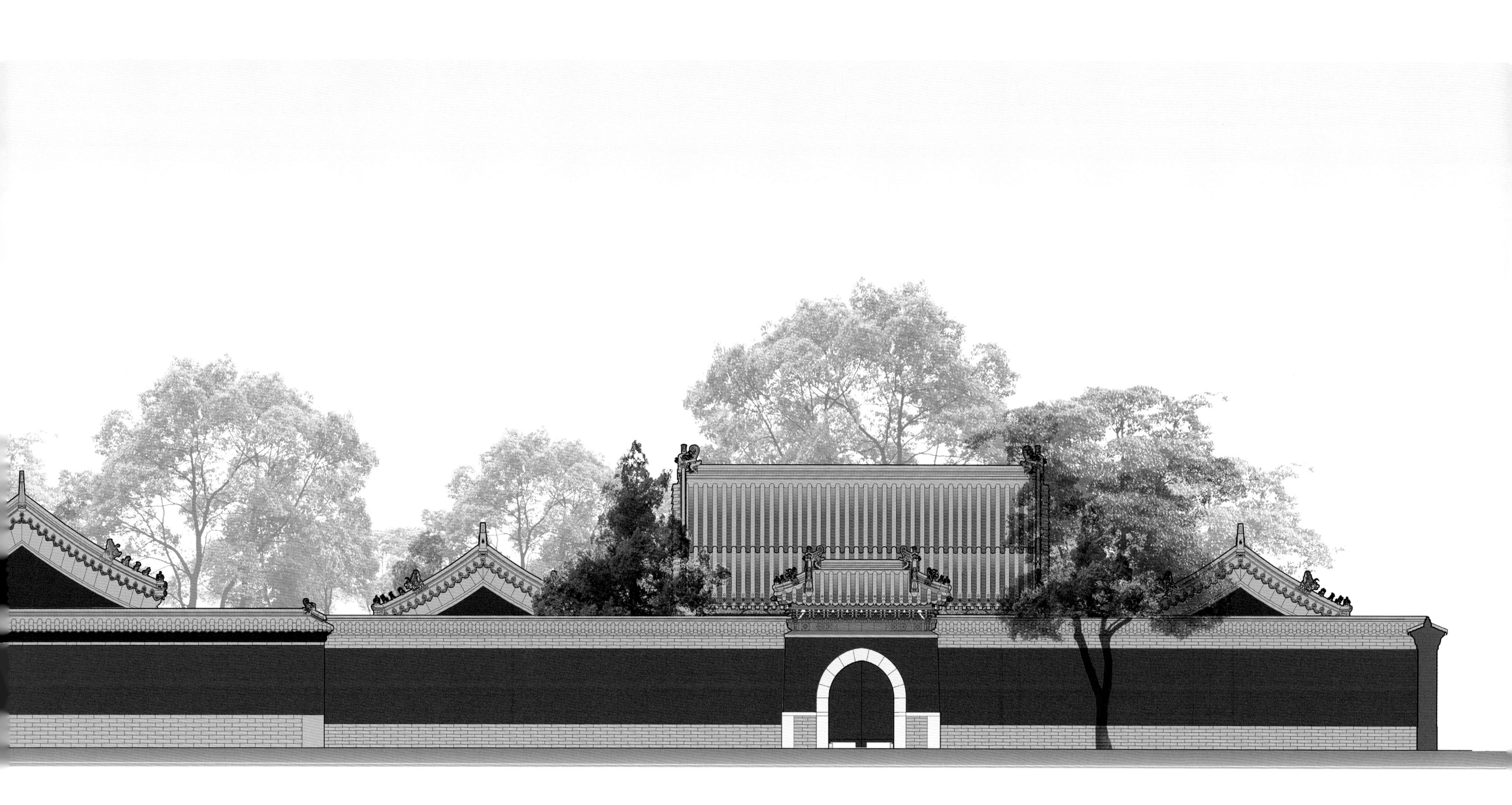

观德殿与关帝庙建筑群南立面图

South elevation of the complex of Guande Hall and the Temple of Guandi

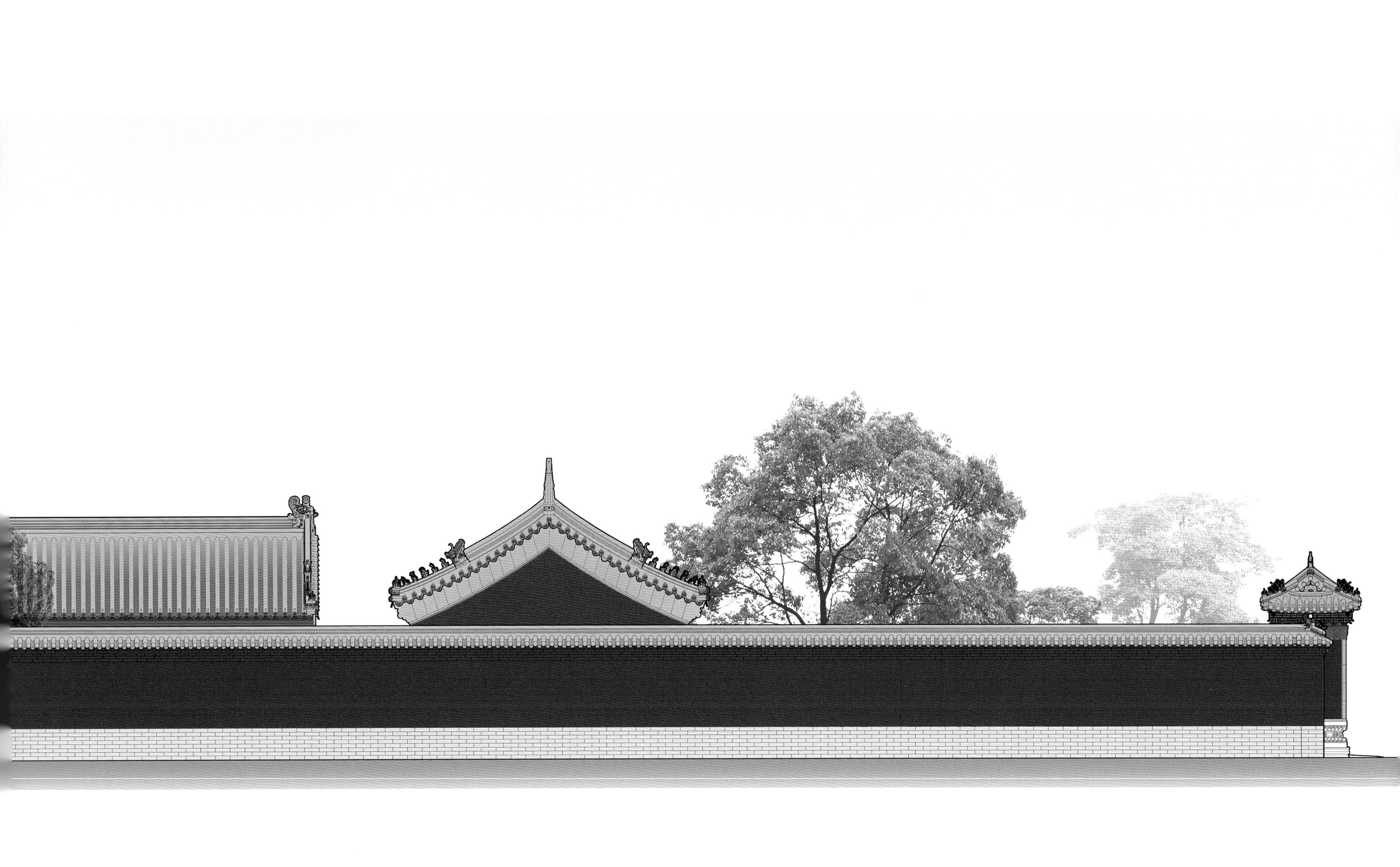

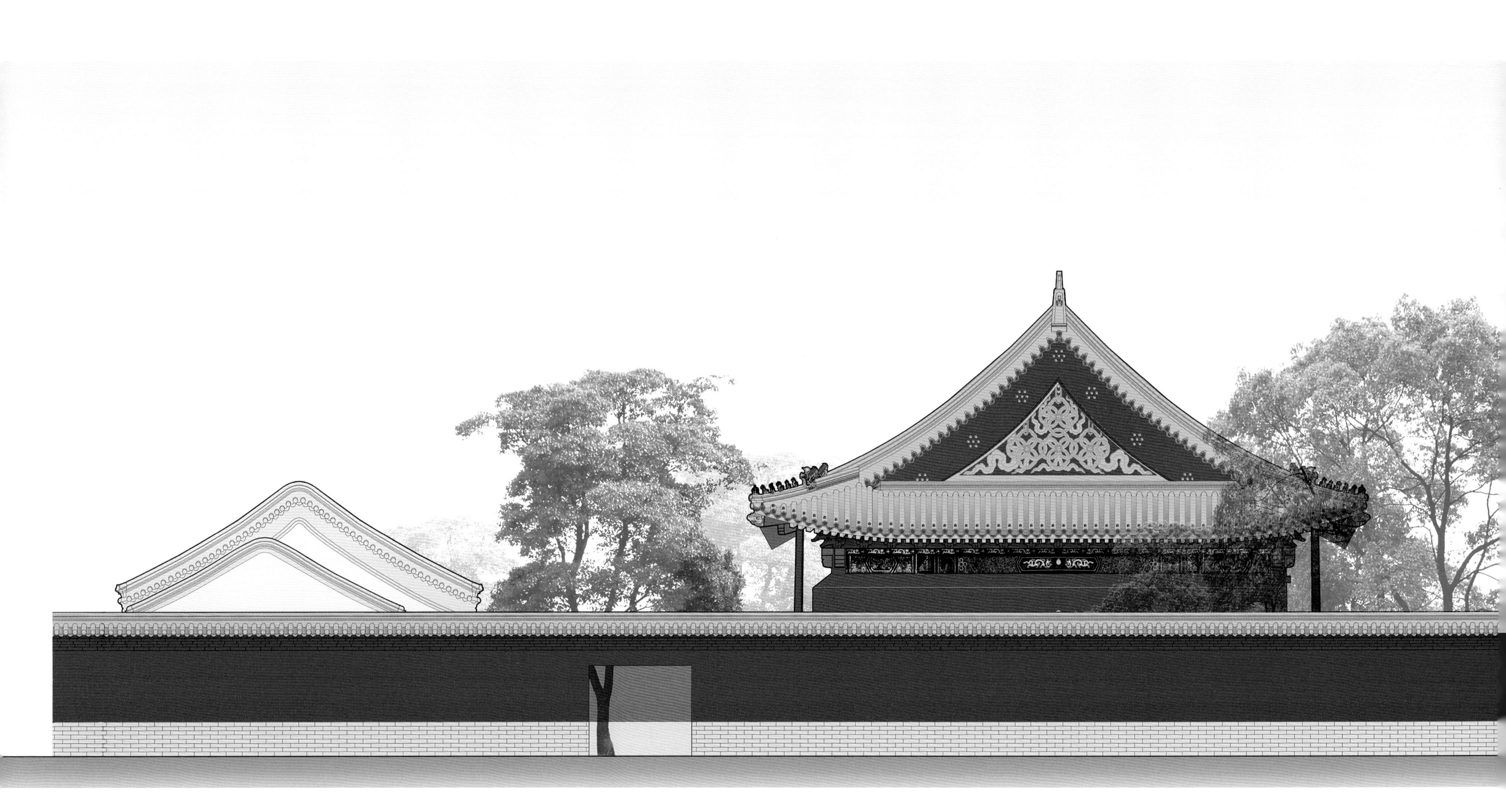

观德殿建筑群西立面图
West elevation of the Temple of Guande complex

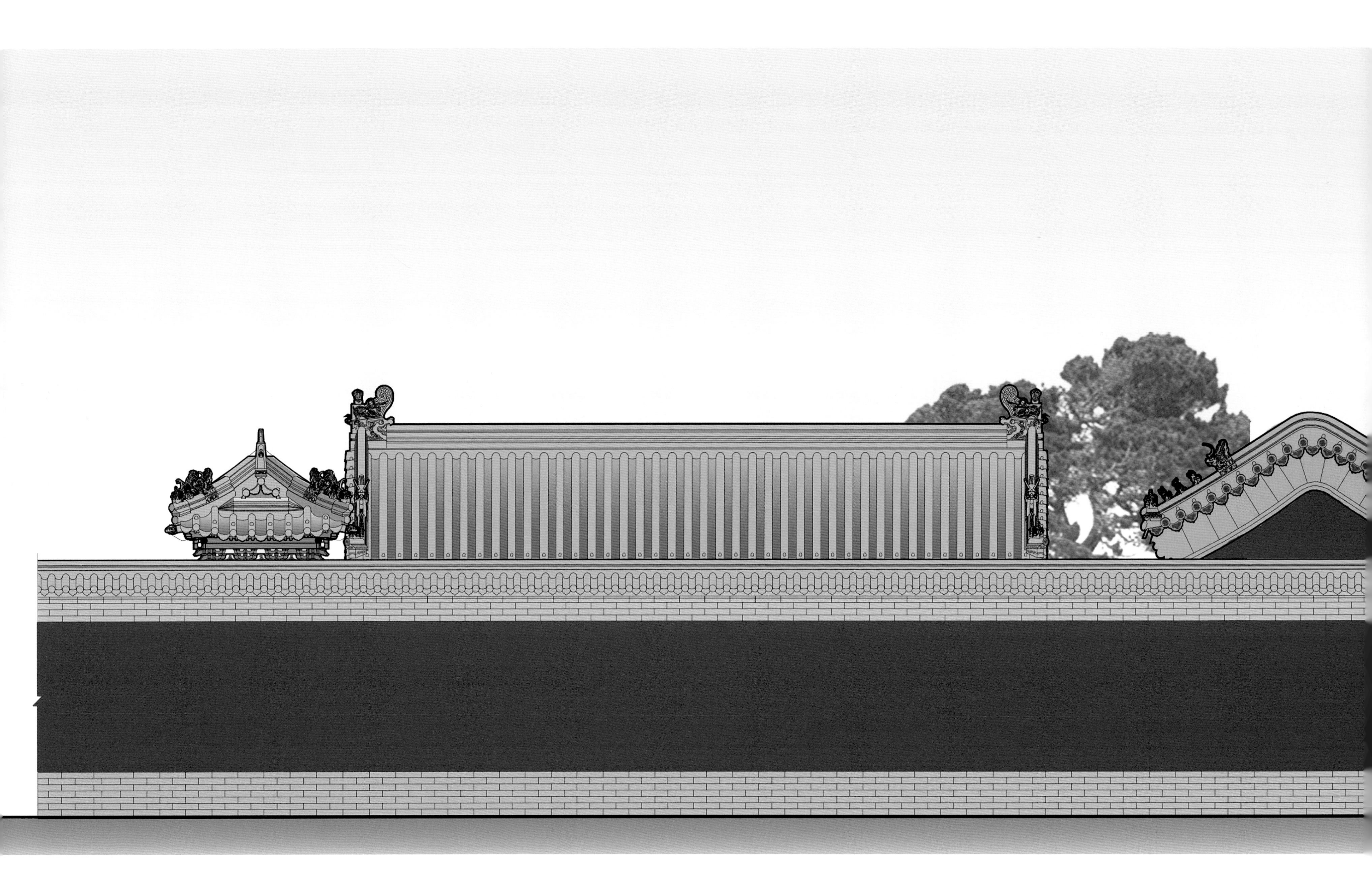

关帝庙建筑群东立面图
East elevation of the Temple of Guandi complex

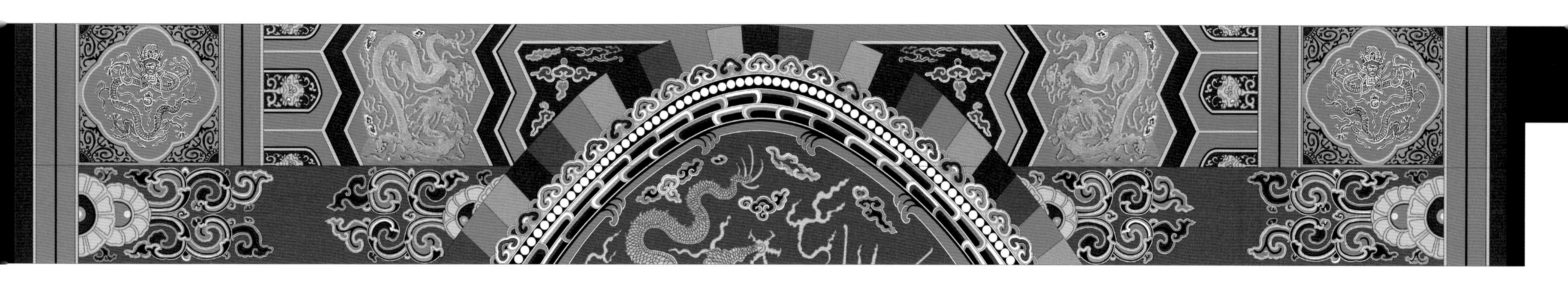

寿皇殿大殿明间脊檩彩画
Detail drawing of colored pattern of ridge purloin of Hall of Imperial Longevity

国家出版基金项目

『十二五』国家重点图书出版规划项目

中国古建筑测绘大系·园林建筑

景山

天津大学建筑学院　北京市景山公园管理处　合作编写
王其亨　主编
张凤梧　杨菁　编著

中国建筑工业出版社

Traditional Chinese Architecture Surveying and Mapping Series:
Garden Architecture

JINGSHAN PARK

Compiled by School of Architecture, Tianjin University &
Beijing Municipal Administration Office of Jingshan Park
Chief Edited by WANG Qiheng
Edited by ZHANG Fengwu , YANG Jing

China Architecture & Building Press

Contents

目录

地　　址：北京紫禁城北方

始建年代：元代

占地面积：二二点七公顷

主管单位：北京市景山公园管理处

测绘单位：天津大学建筑学院

测绘时间：二〇一四—二〇一七年

Location of Jingshan Park: The north of Forbidden City in Beijing

The beginning of the construction: The Yuan Dynasty

Occupied area: 22.7 hectares

Governing body: Beijing Municipal Administration Office of Jingshan Park

Surveying and mapping unit: School of Architecture, Tianjin University

Time of surveying and mapping: 2014-2017

Introduction

Jingshan, or Prospect Hill, is located in the center of the former imperial city (*huangcheng*) of Beijing, north of the Forbidden City (formerly palace city or *gongcheng*), south of the Drum and Bell Towers, east of Beihai Park. Jingshan Park covers an area of 22.7 ha. The 'five pavilions' (*wuting*), each located at one of Jingshan's five summits, are landmarks of Old Beijing and the best places to overlook the city. The park contains all kinds of different building types (*diange, tingxie, paifang*), in total more than forty structures that stand independently or grouped in a cluster. They can be subdivided into distinct zones, including a front-hill palace or entrance area, the five-hilltop pavilion (*wuding wuting*) area, and on the northern hill slope, the building clusters of Shouhuangdian (Imperial Longevity Hall), Guandedian (Observing Virtue Hall), Guandimiao (Emperor Guan [legendary figure of Guan Yu] Temple), Yongsidian (Perpetual Reflecting Hall), Jixiangge (Condensed Auspiciousness Pavilion), and Xingqingge (Exalted Felicity Pavilion).

Probably because of the beautiful scenery created by the old towering trees reaching high into the sky, there was building activity at Jingshan as early as the Jin lasting until the Yuan period, when it served as a garden for the imperial family. Construction intensified during the Ming and Qing dynasties, and Jingshan became an imperial park with grand buildings and picturesque vistas full of profound cultural connotations. Jingshan also performed multiple socio-cultural and recreational functions as an ancestral temple of the Qing imperial family once stood here. Additionally, the park corresponded to the geometric central point of the north-south city axis and was a topographical marker of Beijing urban landscape.

Jingshan embodies the great wisdom of harmonious co-existence inherent in traditional Chinese multi-ethnic culture and deeply ingrained within the Chinese tradition of cultural tolerance. It is the largest garden ensemble with the longest history and the most complete layout situated along the culturally significant central axis of Beijing city. Not only did it witness garden construction and development in the imperial city, it also actively affected the urban transformation of thte city and even the formation of the imperial city itself. In 1928, Jingshan was opened to the public. In 1957, it was declared a Beijing Priority Protected Site on the city level in the first

导言

景山位于北京皇城的中心，南临紫禁城，北望钟、鼓楼，西邻北海。全园占地二十二点七公顷，以景山五亭为地标，是俯瞰北京城市风貌的最佳地点，现存各式殿阁、亭楼、牌坊等共四十余座，或独立点景，或组群布置，围绕景山山体，划分为前山宫门区，山顶五亭，山北寿皇殿建筑群、观德殿建筑群、关帝庙建筑群和永思殿建筑群以及集祥阁、兴庆阁等。

因风景优美，古木参天，金代在这里已有建设活动，至迟在元代，景山便已成为皇宫后苑，又经明、清两朝的持续建设，成为一处建筑恢弘、风景绮丽、内涵深厚、功能独特的皇家宫苑，曾是清朝皇家祭祖家庙，更是北京城南北中轴线的中心点和制高点。

景山凝聚了古代中国多民族文化和谐共生的大智慧，承续了中华文明有容乃大的深厚传统，是北京中轴线上规模最大、延续时间最长、格局最完整的古代皇家宫苑杰作，这里不仅见证了北京皇城宫苑的建设与更迭，甚至影响了城址的变迁乃至皇城的架构。一九二八年，景山开放为公园，一九五七年列为北京市第一批重点文物保护单位，二〇〇一年列为全国重点文物保护单位，二〇一二年列入我国世界文化遗产备选名录。

batch of cultural heritage designations. In 2001, it was listed as a National Priority Protected Site on the state level and included in the World Cultural Heritage List in 2012.

1. Beginnings in the Yuan Period

Following the construction of Taining Palace in the Three Lakes district in the northeastern part of the Jin-dynasty city of Beijing (then known as central capital or Zhongdu), Jingshan was used as farmland as it did not have any buildings yet. In 1215, the third year of the Zhenyou reign period of Jin, the Mongolian army seized Zhongdu and destroyed Taining Palace. In 1271, the eighth year of the Zhiyuan reign period of Yuan, Kublai Khan founded the Yuan dynasty and established Beijing as the Great Capital (Dadu, also known as Khanbaliq), where he built a new imperial palace on the site of the former (Jin-period) Taining Palace. Jingshan was incorporated into the imperial-city area as a piece of land where the emperor could learn about planting crops, irrigation techniques, and other agricultural activities.

2. Expansion in the Ming Period

In 1368, the twenty-eighth year of the Zhizheng reign period of Yuan, emperor Hongwu, personal name ZHU Yuanzhang, the founder and first emperor of the Ming dynasty, established his capital in Nanjing, soon after declaring the new Hongwu reign period of Ming. His fourth son, the later Yongle emperor also known under his personal name ZHU Di, pushed pacification and re-conquered the territories still under Yuan rule, which allowed him to move the capital northward back to 'Beiping' (which he renamed Beijing). It then took him thirteen years to build a new imperial palace and garden on top of the razed Yuan capital while simultaneously designing the urban landscape of Beijing from scratch. To the north of the imperial-city, he built Nanhai (Southern Sea) as an extension of Taiye (Great Liquid) Pond. At the site of the former Yuan-dynasty Yanchunge, he piled up the excavated soil and debris into a tall artificial mound named Zhenshan (Repression Hill). This hill served as a central landmark of Beijing and was also called Wansuishan (Ten-thousand-years Hill). It borrowed from the scenery of Qionghuadao (Jade Flower Island) situated in present-day Beihai Park, as both sites were part of the new imperial pleasure grounds known as Xiyuan (West Park). Thereafter, the Ming dynasty successively built platforms, pavilions, and halls around Wansuishan, the then popular name of Jingshan, and planted many trees. For example, they planted a fruit forest in front of the hill and a large number of pine and cypress trees on the hill slopes. North of the hill, they built a multi-functional "green backyard" for roaming the landscape, archery, keeping birds, and planting flowers.

一、元代的初创

金代在中都城东北部湖泊一带建设太宁宫，景山区域虽无建筑，但已有农事活动。金代贞祐三年（一二一五年），蒙古军队攻占金中都，攻克太宁宫。至元八年（一二七一年），忽必烈建立了元帝国，并设立北京为『元大都』，在太宁宫的基础上建造皇宫，并将景山一带纳入皇城范围，作为皇帝从事耕种、灌溉等农事活动的场所。

二、明代的扩建

元至正二十八年（一三六八年），明太祖朱元璋定都南京，后燕王朱棣发动『靖难之役』，将都城迁至北平，改名为『北京』。

永乐年间，永乐皇帝用十三年时间在元朝都城的基础上整修皇城宫苑，重新规划和建设北京城市格局，在皇城正北用挑挖南海和营建新宫的渣土堆在元代延春阁旧址上，筑起高大的『镇山』，将之作为北京城的中心地标和制高点，后称之为『万岁山』，与西苑北海琼华岛景致相互因借。

此后，明代各朝陆续围绕万岁山前后建设亭台殿宇，并在山前种植了百果林，在山上和山后种植了大量的松柏树，营造出一处兼具游赏、校射，以及饲养禽鸟、种植花草等多重功能的皇家『后苑』。

弘治十一年（一四九八年），在万岁山下建毓秀亭；嘉靖十三年（一五三四年），万岁山后建玩芳亭；万历十三年（一五八五年），在万岁山北建寿皇殿、左毓秀馆、右育芳亭、后万福阁、臻禄

In 1498, the eleventh reign year of the Hongzhi emperor, Yuxiuting was built at the foot of Wansuishan; in 1534, the thirteenth reign year of the Jiajing emperor, Wanfangting was built behind Wansuishan; and in 1585, the thirteenth year of the Wanli emperor, Shouhuangdian was built to the north of Wansuishan, Yuxiuguan to its left, Yufangting to its right, and behind it were built Wanfuge, Zhenlutang, Yongxige, Juxianshi, Yanningge, and Jixianshi. In the fifth month of 1600, the twenty-eighth year of the Wanli emperor, Guande (Observing Virtue) Hall was added and named after the cultural practice of watching the archery training (*guande* i.e. "observing the virtue") that took place on the open space on the left side of the hill. In the same year, Wanfangting was first renamed Wanjingting and in 1601, Wanli's twenty-ninth reign year, again renamed Yuxiuting. In the second month of 1602, Wanli's thirtieth reign year, Guanhuadian was added and peonies were planted. In the second month of 1603, Wanli's thirty-first reign year, the two small pavilions of Shouchunting and Jifangting were built. On the twenty-eight day of the fourth month in 1613, Wanli's forty-first reign year, Xingqingge was renamed Wanchunlou. Under emperor Wanli, the entrance to the garden was renamed Wansuimen (or Ten-thousand-year Gate). As the garden area was now entered through a north gate located at the southern end and enclosed by walls on the east, west, and north sides, Jingshan essentially followed the standard design pattern for palace construction. At that time, it contained the following buildings and sites: Wansuimen, the inner gates on the left and right hill-sides, the left and right hill-side gates, Shouhuangdian, Guandimiao, Guandedian, Yong'anting, Shouanshi, Jingmingguan, Tudimiao, Longwangmiao. Additionally, the area included a deer park, pheasant house, chicken house, and waterwheel house. A tragic event happened at the end of the Ming dynasty, as the last Ming emperor Chongzhen hanged himself on the eastern slope of Jingshan.

3. Enhancement in the Qing Period

The architectural layout of the garden did not change much in the early Qing period, only the name of the garden changed: formerly known as Wansuishan that was renamed Jingshan. Both the Shunzhi and Kangxi largely followed the form, function, layout of architectural and landscape elements of Wansuishan as established during the Ming dynasty. Here they still enjoyed the beautiful scenery, practiced archery, and kept coffins temporarily before burial (*ting ling*). Kangxi's grandson, the fifth Qing emperor Qianlong, then launched a large-scale construction project. He had not only five Buddhist statues cast and five pavilions (*wuting*) on top of the five summits erected, but he also made substantial adjustments to the overall design. In 1744, Qianlong's ninth reign year, Guandimiao was renovated, and the directional

堂、永禧阁、聚仙室、延宁阁、集仙室。万历二十八年（一六〇〇年）五月，添盖观德殿，以山左空旷处作为射箭场所，故名『观德』。同年，玩芳亭更名『玩景亭』，万历二十九年（一六〇一年）更名『毓秀亭』；万历三十年（一六〇二年），闰二月，添盖观花殿，植牡丹、芍药。万历三十一年（一六〇三年）二月，添盖寿春亭、辑芳亭。万历四十一年（一六一三年）四月二十八日，兴庆阁更名『玩春楼』，万历年间后苑正门更名『万岁门』。至此，景山一区基本形成以南端北上门为起点，东、西、北三面宫墙环绕的完整宫苑格局，包括万岁门、山里左门、山里右门、山左门、山右门、寿皇殿、关帝庙、观德殿、永安亭、寿安室、景明馆，以及鹰房、鹿苑、野鸡房、家鸡房、土地庙、龙王庙等为主的庞大建筑群。明末，崇祯皇帝自缢于景山山体东侧。

三、清代的恢弘

清初，自顺治皇帝入关，直至康熙朝，景山格局并未发生改变，仅将万岁山更名为『景山』，清代两朝皇帝延续明代万岁山的宫苑格局和功能，在此赏景、校射、停灵。迨至乾隆朝，乾隆皇帝对景山进行了大规模修建，不仅在山顶修建五亭并铸造五方佛像，更对山后进行大幅度调整：乾隆九年（一七四四年），重修关帝庙，将东西向调整为南北向；乾隆十五年（一七五〇年）仿照太庙的规制、圆明园安佑宫的格局，将东偏的寿皇殿建筑群重新修建于中轴线上，作为皇家祭祖家庙；挪盖后的观德殿与关帝庙并置，山前添建绮望楼，山后添修集祥阁，与兴庆阁对称。

至此，景山已形成集皇帝后妃游乐、祭祖、登高、校射以及官炮制作等多重功能，汇集亭楼、殿阁等各式建筑的重要场所。直至清末，各朝未改变格局，仅对景山内古建筑进行维修。

一九〇〇年辑芳亭被八国联军烧毁，一九〇二年修复，观德殿及永思殿内外檐遭拆毁。至中华人民共和国成立时，永思殿正殿大部分已塌毁，仅存永思门、朝房和配殿。山顶东西四亭内佛像被劫走，万春亭佛像虽幸免于难，但也在『文化大革命』期间遭破坏（图一～图四）。

四、近现代的保护实践

自清末至今，景山古建筑群历经多次管理权与使用权的变革，作为中轴线上的重要古建筑群，始终备受关注，针对文物建筑的保护维修实践时断时续。

一九二八年，景山开放为城市公园，并由故宫博物院管理至一九四五年。一九五四年十二月一日，北京市人民政府将景山公园移交北京市园林局管理。一九五五年三月一日园林局正式接管，并成立景山公园管理处。

一九五七年二月，北京市园林局撤销景山公园管理处，成立北海景山公园管理处，统一领导两园的管理工作。二〇〇三年七月，景山公园从北海景山公园管理处中独立出来，成立景山公园管理处。

（一）民国期间

一九四一年至一九四五年，由张镈先生组织的北京中轴线测绘项目，囊括了景山建筑群轴线建筑，

layout of the garden was changed from east-west to north-south. In 1750, Qianlong's fifteenth reign year, the official rules and regulations drafted for the Imperial Ancestral Temple (Taimiao) and Anyougong in Yuanmingyuan, the new imperial garden complex in the western outskirts of Beijing, were also implemented at Jingshan. Shouhuangdian, originally situated at the eastern garden side, was relocated to a position along the central axis where the hall was rebuilt and rededicated as a temple for ancestral worship of the imperial family. Guandimiao was now juxtaposed with Guandedian, it enhanced the symmetry of the garden layout even more so since Qiwanglou was added in front of the hill and the Jixiang and Xingqing pavilions behind the hill.

The variegated and diverse architecture at Jingshan served multiple functions. The garden was used for recreation by emperors and their wives and concubines who amused themselves there. Members of the imperial court would worship their ancestors in the complex, while male figures would practiced shooting with bows (archery), and guns (artillery). This pattern was continued until late imperial times, and succeeding Qing emperors ordered only maintenance and repair of Jingshan's historical architecture and landscape.

At the turn to the 20th century, Jifangting was burnt down by the soldiers of the Eight-Nation Alliance but repaired in 1902. Additionally, the architectural clusters of Guandedian and Yongsidian were partly demolished. At the time of the establishment of the People's Republic of China, the main hall at Yongsidian had already collapsed and only the entrance gate to the complex and several auxiliary buildings (*chaofang*, *peidian*) were still standing. The Buddha statue at Wanchunting survived until the Cultural Revolution (when it was destroyed), but the statues of the other hilltop pavilions were stolen during this period (Fig.1-Fig.4).

4. Monument Protection in Modern Times

Since the late Qing period, the Jingshan architectural ensemble has undergone tremendous change in management and usage rights. As a prominent group of historical monuments standing on the central axis of Beijing, it has always attracted much attention and concern regarding cultural heritage protection. Therefore, maintenance and repair were continuously carried out over the years with only some interruptions.

In 1928 Jingshan was finally opened to the public, and the new urban park was managed by the Palace Museum until 1945. On December 1st, 1954, the Beijing Municipal

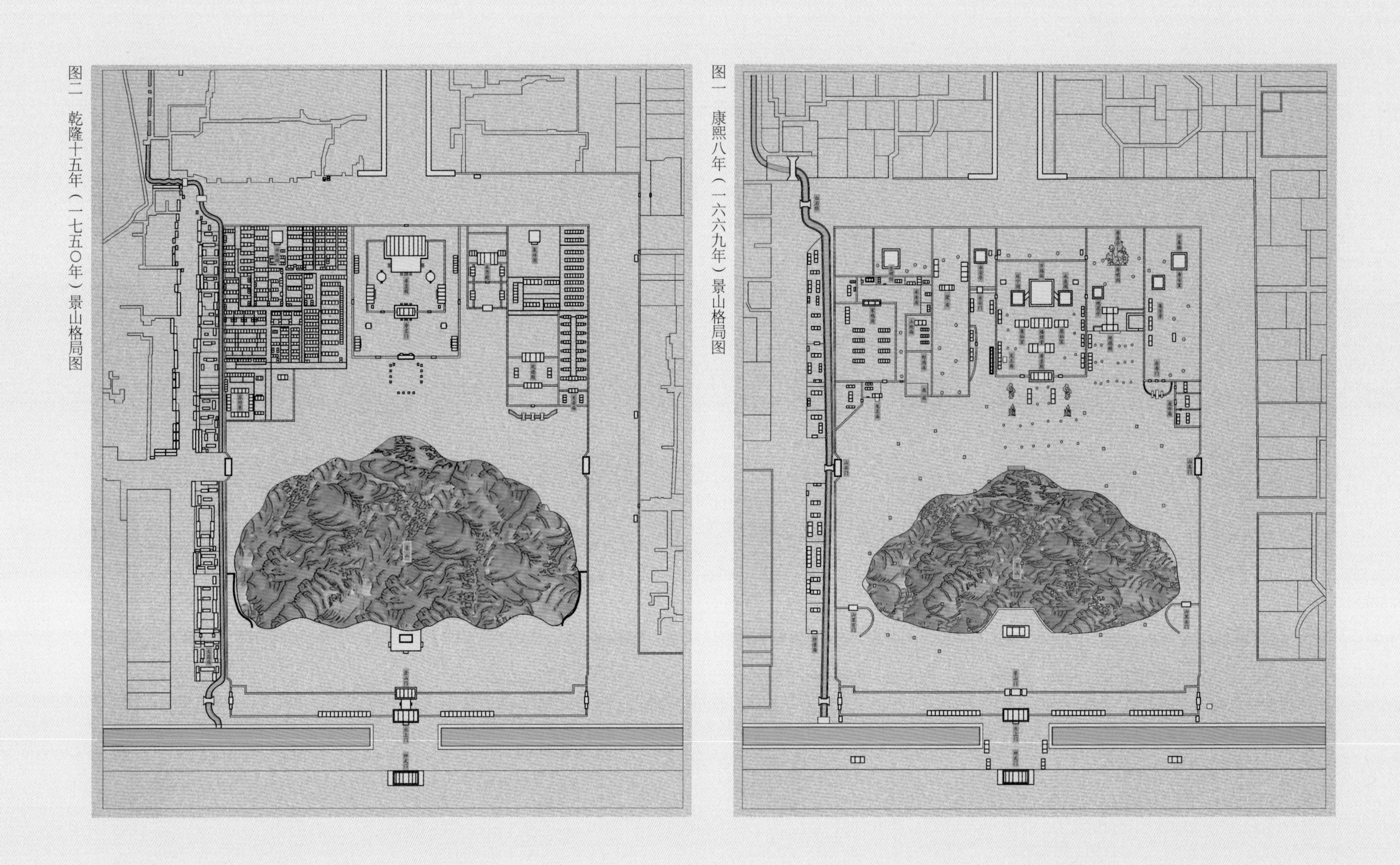

图二　乾隆十五年（一七五〇年）景山格局图

图一　康熙八年（一六六九年）景山格局图

Fig.1 Site plan of Jingshan from the eighty reign year of emperor Kangxi (1669)
Fig.2 Site plan of Jingshan from the fifteenth reign year of emperor Qianlong (1750)

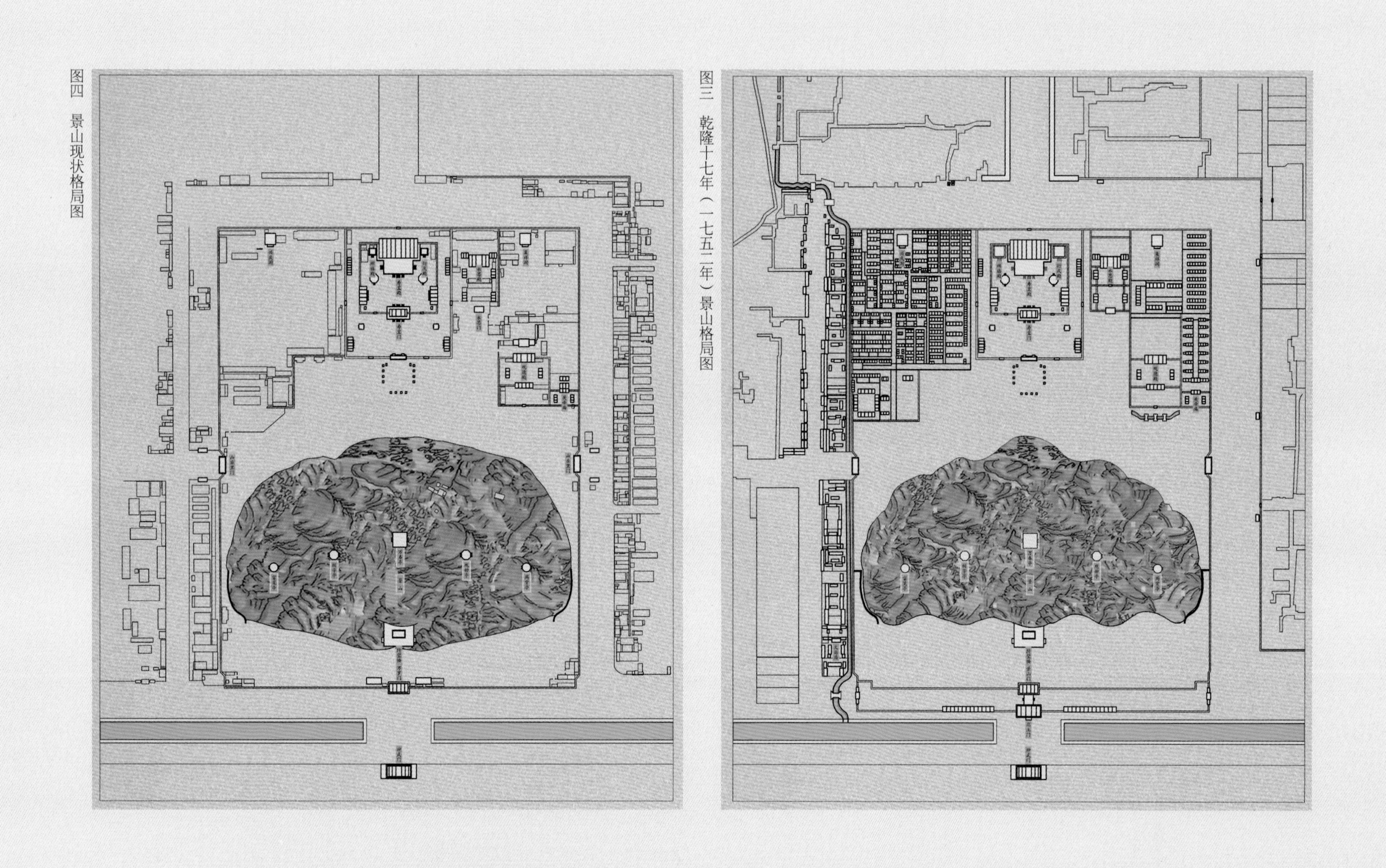

图四　景山现状格局图

图三　乾隆十七年（一七五二年）景山格局图

Fig.3 Site plan of Jingshan from the seventeenth reign year of emperor Qianlong (1752)
Fig.4 Site plan of Jingshan, present-day condition

Government transferred the administration of Jingshan Park to the Beijing Municipal Garden Bureau, which became effective a few months later on March 1st, 1955, when the bureau officially established the Jingshan Park Management Office.

In February 1957, the Beijing Municipal Garden Bureau revoked the Jingshan Park Management Office and established instead the Beihai-Jingshan Park Management Office, with the aim to unify the two gardens and simplify park administration. However, in July 2003, Jingshan Park became independent again as the Jingshan Park Management Office was reestablished.

(1) Republican Era

From 1941 to 1945, an important surveying and mapping project of architecture along Beijing's central axis was carried out under supervision of ZHANG Bo. In the case of Jingshan, this included all the monuments aligned along the central park axis, in total thirteen structures represented in eighty-one drawings, namely: Qiwangmen (today Wansuimen), Qiwanglou, the inner gate on the right hill-side, three small pavilions named Guanmiaoting, Zhoushangting, and Wanchunting, the jade stoneway (*fang*) at Shouhuangdian, a glazed-tile gate, Shouhuangmen, a silk burning furnace (*fenbolu*), and two auxiliary buildings (*peidian* and *dongwu*) (Fig.5, Fig.6).

Until 1945, Jingshan Park was managed by the Imperial Palace Museum. The Jingshan Administration Office carried out building renovation on several occasions. For example, in 1918, the seventh year of the People's Republic of China, Guandimiao was rebuilt after it was destroyed in a fire. In 1934-1939, the Imperial Palace Museum asked LIANG Sicheng and LIU Dunzhen from the Society for Research in Chinese Architecture (Zhongguo Yingzao Xueshe) to develop a recovery plan for the five pavilions (*wuting*). Work started on April 1st, 1935, when the Wanchunting complex was renovated under supervision of LIU Dunzhen. Unfortunately, the architecture was struck by lightning in 1938, although in the next year, the ceiling (*baoding*), wooden pillars, glazed brick building(s), and decorative windows (*linghua chuanghu*) were repaired again. Afterwards, in 1942, Guandemiao was repaired and in 1947, the wooden pillars and architraves of the three nine-bay-wide archways (*pailou*) in front of Shouhuangdian were rebuilt in concrete (but the other parts remained wood)(Fig.7).

During this period, some buildings were damaged and even collapsed due to limited funds for restoration. In 1941,Yongsidian fell into a state of disrepair and collapsed.

包括绮望门（今万岁门）、绮望楼、山右里门、观妙亭、周赏亭、万春亭、寿皇殿宝坊、琉璃门、寿皇门、焚箔炉、配殿、东庑、碑亭等共十三座建筑，绘制图纸八十一张（图五，图六）。

至一九四五年前，景山公园由故宫博物院管理，设景山管理处，对部分文物建筑进行了整修，如：民国七年（一九一八年），关帝庙失火烧毁窗棂，后修复。民国二十三年（一九三四年）至二十八年（一九三九年），故宫博物院委托中国营造学社梁思成、刘敦桢等代拟设计修理五座亭子计划，于民国二十四年（一九三五年）四月一日开工，修理万春亭，刘敦桢为修亭工程师。民国二十七年（一九三八年）万春亭遭雷击并于民国二十八年（一九三九年）对宝顶、木柱、琉璃砖楼及菱花窗户进行过修理。此后，民国三十一年（一九四二年）修理观德殿，民国三十六年（一九四七年）寿皇殿建筑群前部三座九楼牌楼被改为钢筋混凝土柱子、额枋，其他部位仍用木料，撤去戗木（图七）。

在此期间，由于经费紧张，部分建筑出现破损，甚至年久失修，倒塌损毁，如，民国三十年（一九四一年）永思殿年久失修，倒塌残毁。民国三十七年（一九四八年），经勘查，寿皇殿上层檐西北角梁后部翘起，琉璃脊筒臌出两节，内裂大缝。近该角处左右翘飞及翼角下垂，且殿顶积土生草，陇灰脱落之处甚多。查得观德殿琉璃门内东部须弥座石及两门墩稍有沉陷，以致东北角由上部闪裂严重，且门框、门龙均错离部位，殊有塌落之虞，似应修整，以免损伤琉璃瓦件。

（二）一九四九年后

随着一九四九年中华人民共和国成立后，北京城市建设的需要以及功能使用的需求，对文物建筑

图五 一九四〇年代景山测绘图

Fig.5 Survey map of Jingshan from 1940

图六 一九四四年（民国三十三年）景山寿皇殿纵剖面图

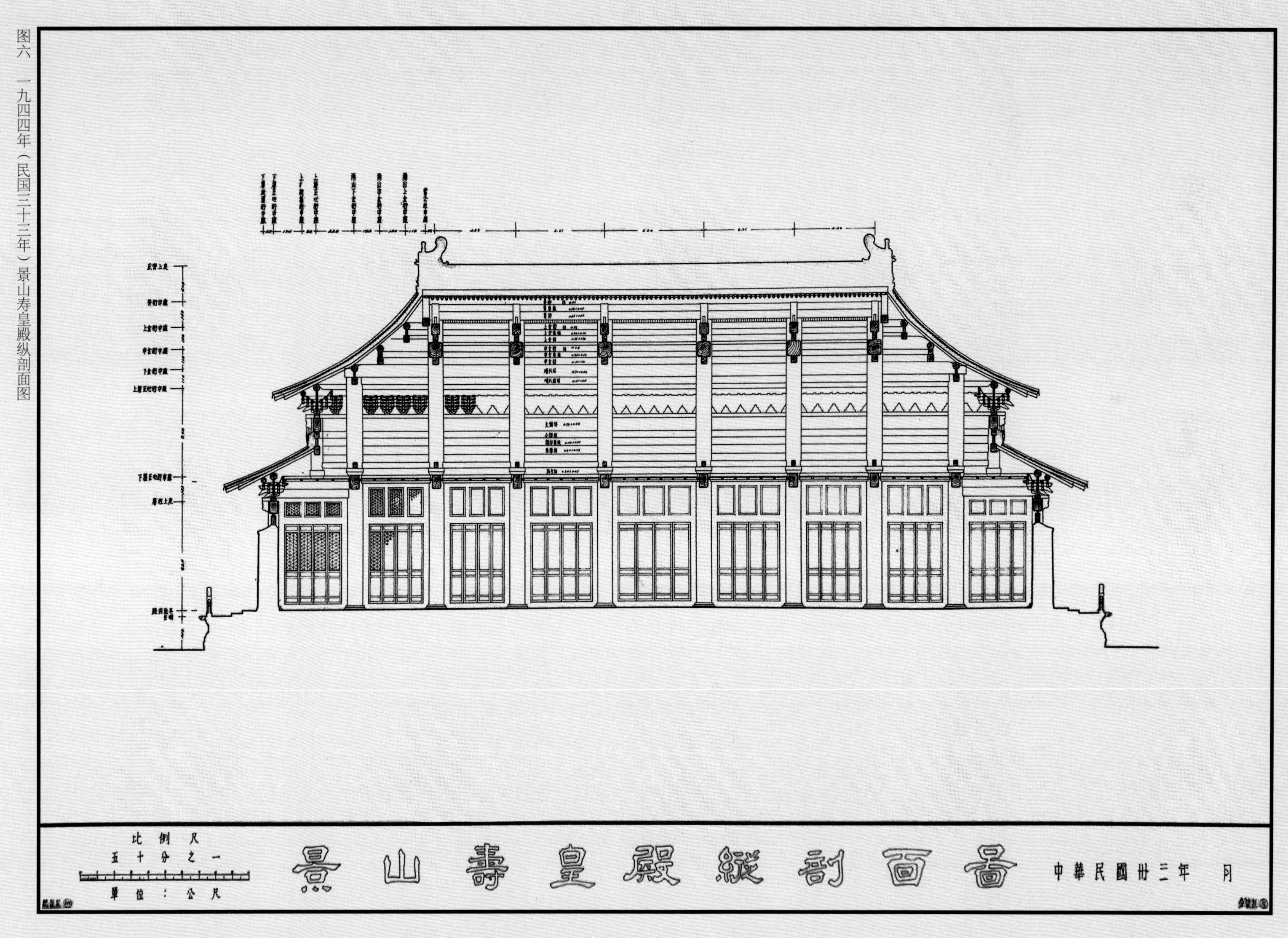

Fig.6 Longitudinal section of Shouhuang Hall of Jingshan from 1944

图七 五亭修缮图

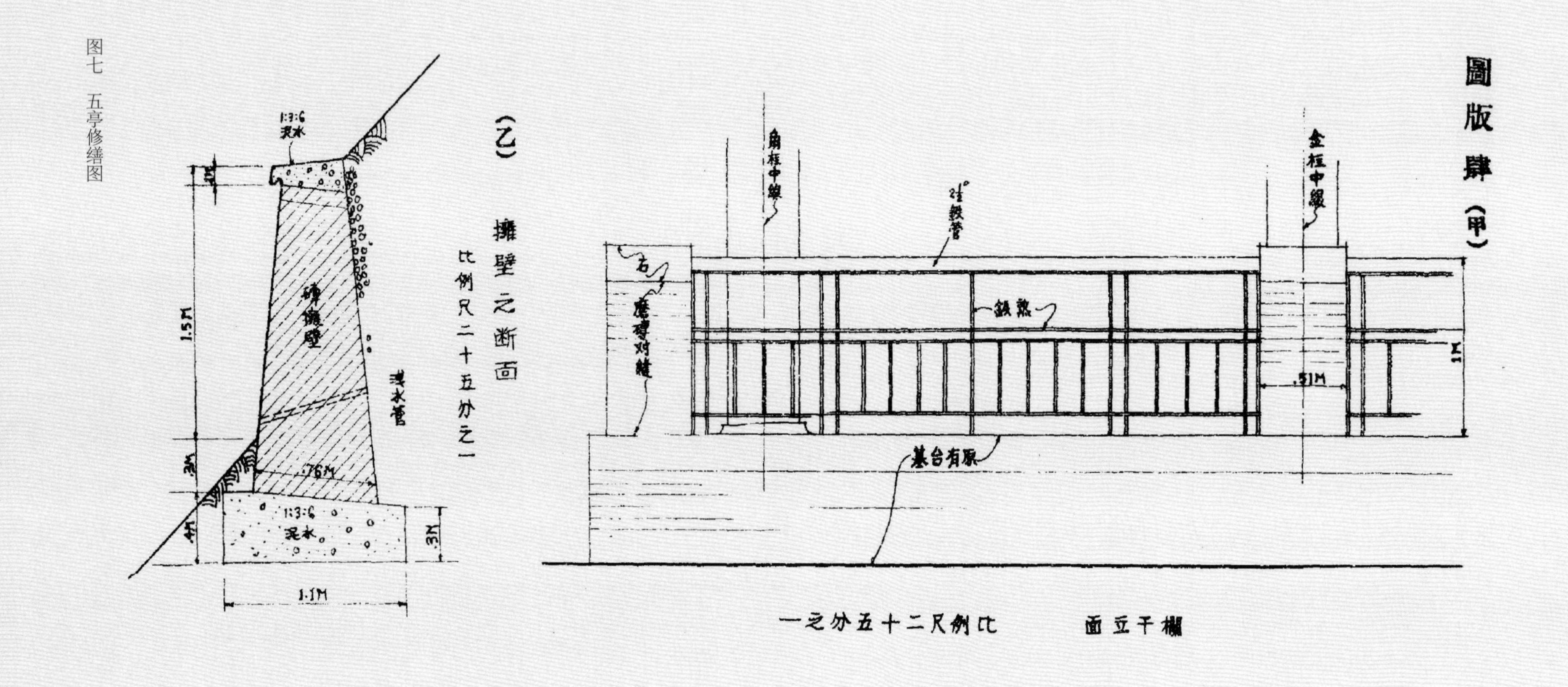

Fig.7 A recovery plan for five pavilions

The 1948 site survey of Shouhuangdian drew attention to the alarming state of the hall: the northwestern corner beam of the upper eaves was tilted; (two rows of) the semi-circular glazed ridge tiles were bulging outwards; and a large crack had appeared in the hall inside. The nearby roof sections were almost dropping, grass was growing all over the roof, and plaster was falling off many places.

The glazed-tile gate at the Guandedian complex was also in a precarious state. Several stones positioned at the east-side inner section of the *Xumizuo* (Mount Sumeru shaped podium) and the gate pier stones were slightly sunken, causing a top-down crack at the northeastern corner and an off-axis shift of the doorframe. Without intervention, the structure would collapse and ruin the glazed tiles.

(2) After 1949

After 1949, many cultural relics at Jingshan were renovated in order to meet new needs resulting from Beijing's urban design renewal and the increased requirements for safe usage. For example, in 1950, a new staircase was built at Qiwanglou (installed in the eastern side bay of the building), and the wooden latticework (fences) on the left- and right-side doors was repaired. In 1954, the upper northern gate and the public school located in the two side halls (*peidian*) were demolished, and Yuanyang Bridge was buried. After maintenance and repair, Qiwanglou was transformed into an office space used by the park administration and for government-funded extra-curricular children activities (Children's Palace, *shaoniangong*). In 1955, Wansuimen, the five pavilions, and Qiwanglou were completely renovated.

On August 16th, 1955, the Cultural Relics Bureau of the Central Committee of the Ministry of Culture officially designated the Shouhuangdian, Guandedian, Yongsidian, and Guangdimiao building clusters as a public facility for children and charged the Beijing Children's Palace with the task of monument preservation. In 1965, Beijing Children's Palace tore down the main hall at Yongsidian, although the base on which the building was standing remained intact. In the same year, Guandedian was refurbished on a large scale. From 1966 to 1989, several gates (the inner gates on the left and right hill-sides and Wansuimen) were successively repaired and repainted.

During the "Cultural Revolution", several cultural relics were destroyed. In 1966, the large copper statue in Wanchunting was taken down and sold. As the latticed windows on all sides were also torn down, the building was transformed into an open-side pavilion (*Changting*).

进行了多次整修，如：一九五〇年，绮望楼新做楼梯（东次间）及左右门木栅栏修缮。一九五四年，拆除北上门及两侧配殿官学，鸳鸯桥被埋没；绮望楼作为景山公园筹备处及少年宫筹建处办公用房，进行维护性修缮。一九五五年整修万岁门、五亭、绮望楼。

一九五五年八月十六日文化部文物局正式将寿皇殿、观德殿、永思殿、关帝庙一区批给北京市少年宫使用，随后由少年宫负责保护整修。一九六五年北京市少年宫将永思殿正殿拆除，殿基完好无损。同年，对观德殿进行大规模整修。一九六六年至一九八九年期间，多次对山左里门、山右里门和万岁门进行修缮和油饰见新。

『文化大革命』期间，部分文物遭到破坏，最重要的事件是一九六六年，万春亭内的铜佛被拆除卖掉，四周菱花格隔扇窗一并拆掉，改为敞亭。

一九七三年为迎接景山闭园三年后重新开放，对景山三宫门与五峰亭、绮望楼等开放区古建筑进行修缮，油饰见新。

一九八一年四月十日晚，因北京市少年宫工作人员失误引起火灾，烧毁寿皇殿建筑群寿皇门及周围有数百年树龄的古柏。一九八三年由故宫博物院负责复原设计，一九八五年重建竣工。同时对寿皇殿进行挑顶整修，更换东北角梁。

在维修文物建筑破损的同时，为满足功能使用需求，公园也对古建筑进行了局部改建，如一九六六年绮望楼改作商业用房，将金步装修改至前檐，并将东次间楼梯改在西侧后廊步位置。一九七三年，在五座亭子的檐柱位置添配了坐凳，以便于游人休息。二〇〇六年和二〇〇八年，为迎接北京夏季奥运会，分别对景山五亭、绮望楼、三牌楼进行整修

In 1973, to prepare for the upcoming reopening of Jingshan Park three years later, the monuments located in the Sangongmen, *wuting*, and Qiwanglou zone that would be open to the public were repaired and freshly painted.

On the evening of April 10th, 1981, a mistake by a staff member of the Beijing Children's Palace caused a fire, and the buildings at Shouhuangmen and the surrounding ancient cypress trees were burnt to the ground. In 1983, the Imperial Palace Museum became responsible for the recovery design, and in 1985, the reconstruction was completed. At the same time, the roof at Shouhuangdian was mended and the northeastern corner beam replaced.

While aiming to safeguard the damaged cultural heritage site, several buildings were still modified in order to meet the new requirements caused by tourism and public usage. For example in 1966, Qiwanglou was made available for commercial use; some interior decoration was relocated to a position under the front eaves; and the staircase installed in the second east-side front bay was moved to the west-side back corridor. In 1973, benches were added between the columns of the five hilltop pavilions to allow visitors to rest.

In 2006 and 2008, for the Beijing Summer Olympics, Jingshan's five pavilions, Qiwanglou, and the three archways (*pailou*) were renovated and repainted. In 2009, 2011, and 2013 Jixiangge, Guandimiao and Guandedian were successively restored.

Following the 2013 relocation of the Beijing Children's Palace, Tianjin University performed surveying and mapping work at Jingshan in 2014, 2016, 2017, precisely of the Shouhuangdian complex, the five pavilions, the three gates (inner gates on the left and right hill-sides and Wansuimen), Guandedian, Guandimiao, and Jixiangge. As Project group also measured and mapped Qiwanglou during this period, Project group have essentially completed the field survey documentation of all the historical monuments and sites at Jingshan. In total, six hundred and thirty line drawings were produced, among which Project group chose two hundred and sixty-nine to be published in this volume of the *Traditional Chinese Architecture Surveying and Mapping Series*.

Since the renovation of Shouhuangdian was completed in 2017, today, the two tasks of data collection and physical repair of the historical architecture at Jingshan Park have been accomplished, the only exception being Yongsidian, which awaits future repair and renovation.

并油饰见新。二〇〇九年、二〇一一年、二〇一三年又陆续对集祥阁、关帝庙和观德殿进行整修。

二〇一三年，随着少年宫腾退搬迁，天津大学于二〇一四年、二〇一六年、二〇一七年先后三次分别对寿皇殿建筑群、五亭及三门（万岁门、山左里门、山右里门）、观德殿和关帝庙及集祥阁进行扫描测绘，在此期间结合文物本体勘察对绮望楼进行测绘，基本完成了景山所有古建筑的信息采集工作，共绘制二维线图六百三十张，从中挑选二百六十九张，作为本册图集的主要内容。

与此同时，寿皇殿启动修缮工程，二〇一七年竣工。至此，景山公园主要建筑群都已完成基础信息采集和残损整修工作，仅永思殿尚未进行修复和整修。

图版

Drawings

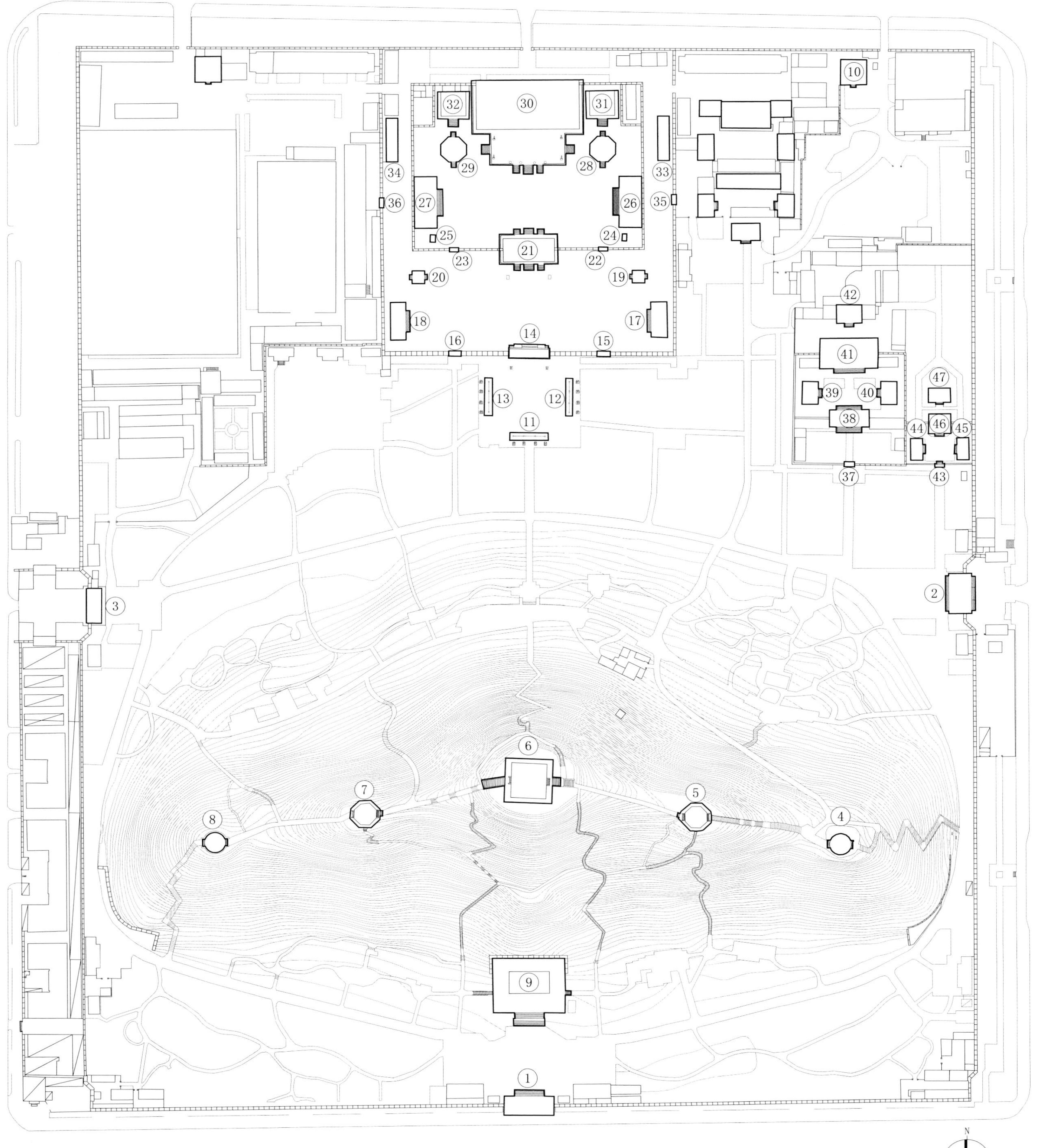

宫门 Gate

1. 万岁门 Longevity Gate
2. 山左里门 Inside Gate of east of Jing shan Park
3. 山右里门 Inside Gate of west of Jing shan Park

五峰亭 Wufeng Pavilion

4. 周赏亭 Zhoushang Pavilion
5. 观妙亭 Guanmiao Pavilion
6. 万春亭 Everlasting Spring Pavilion
7. 辑芳亭 Jifang Pavilion
8. 富览亭 Fulan Pavilion

楼阁 Building

9. 绮望楼 Qiwang Building
10. 集祥阁 Jixiang Building

寿皇殿建筑群 The Groups of Shouhuang Hall

11. 显承无斁昭格惟馨牌坊 Xianchengwuduzhaogeweixin Archway
12. 继序其皇绍闻祗遹牌坊 Jixuqihuangshaowendiyu Archway
13. 世德作求旧典时式牌坊 Shidezuoqiujiudianshishi Archway
14. 南砖城门 South Bricked Gate
15. 东南砖城门 Southeast Bricked Gate
16. 西南砖城门 Southwest Bricked Gate
17. 神库 Sacred Warehouse
18. 神厨 Sacred Kitchen
19. 东井亭 East Well Pavilion
20. 西井亭 West Well Pavilion
21. 寿皇门 Gate of Shouhuang
22. 东琉璃门 East Colored-glaze Gate
23. 西琉璃门 West Colored-glaze Gate
24. 东燎炉 East Flame Burner
25. 西燎炉 West Flame Burner
26. 东配殿 East Side Hall
27. 西配殿 West Side Hall
28. 东碑亭 East Tablet Pavilion
29. 西碑亭 West Tablet Pavilion
30. 寿皇殿 Shouhuang Hall
31. 衍庆殿 Yanqing Hall
32. 绵禧殿 Mianxi Hall
33. 东值房 East Duty Room
34. 西值房 West Duty Room
35. 东砖城门 East Bricked Gate
36. 西砖城门 West Bricked Gate

观德殿建筑群 The Groups of Guande Hall

37. 琉璃门 Colored-glaze Gate
38. 二宫门 Palace Gate
39. 西配殿 West Side Hall
40. 东配殿 East Side Hall
41. 观德殿 Guande Hall
42. 后殿 Back Hall

关帝庙建筑群 The Groups of Guandi Temple

43. 琉璃门 Colored-glaze Gate
44. 西配殿 West Side Hall
45. 东配殿 East Side Hall
46. 护国忠义庙 Huguozhongyi Temple
47. 真武殿 Zhenwu Hall

景山总平面图
Plan of Jingshan Park

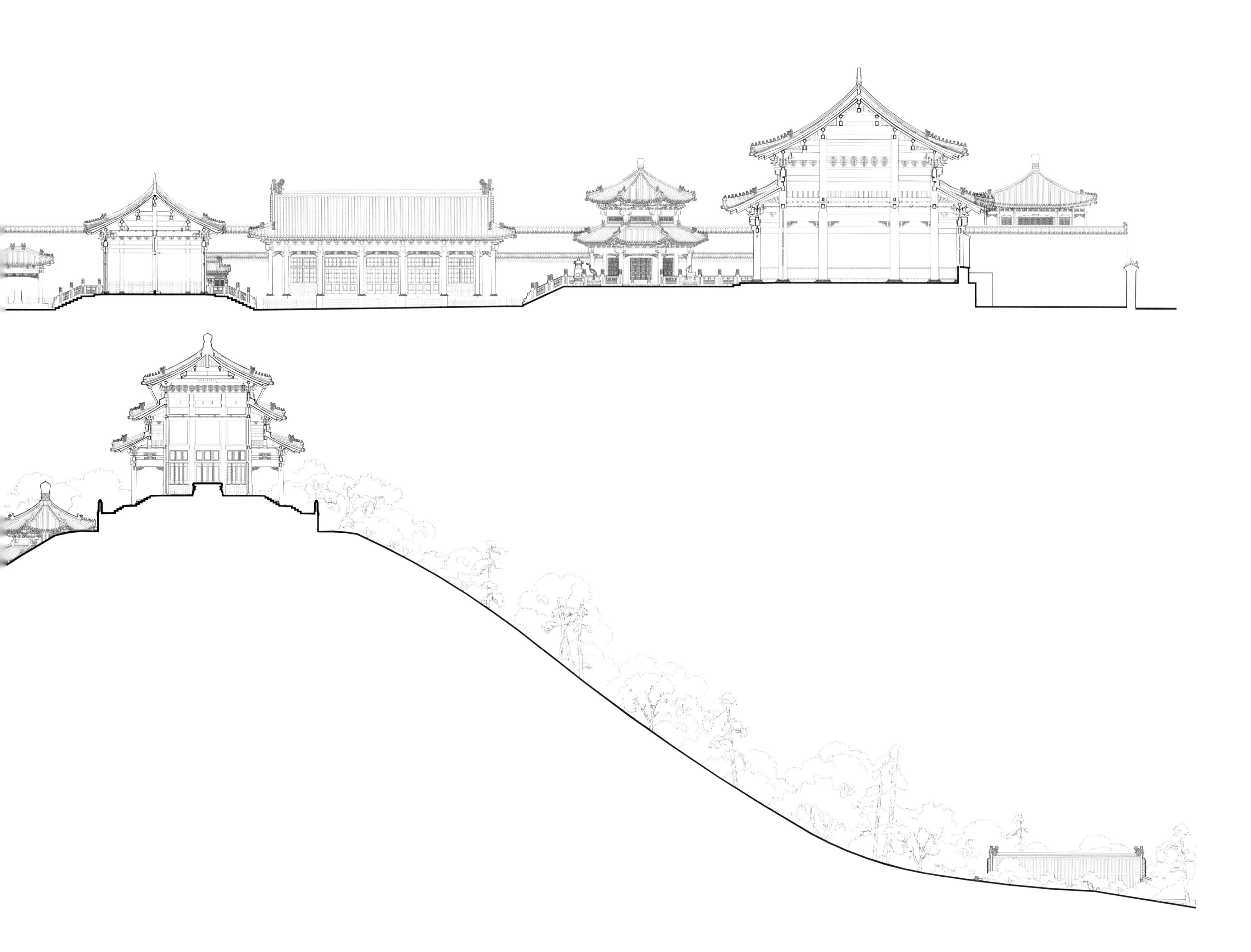

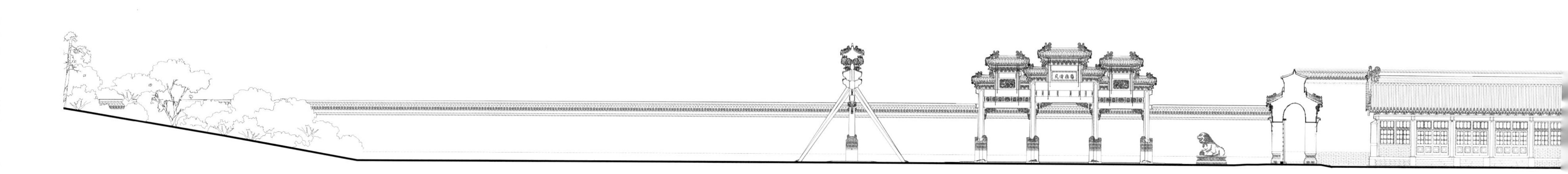

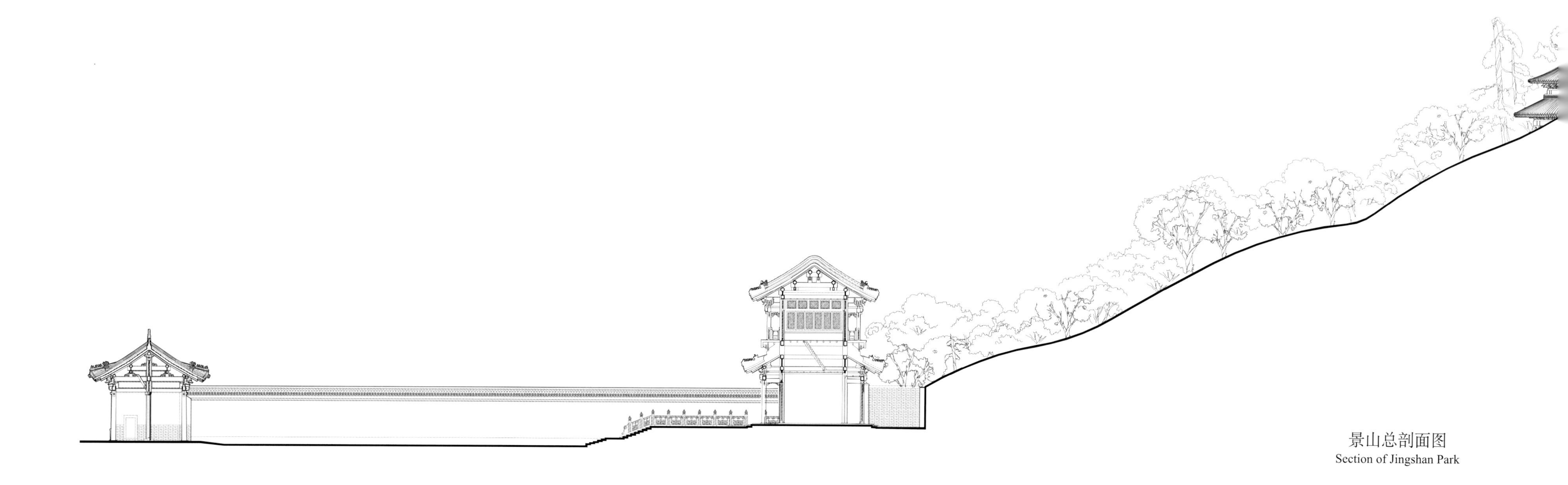

景山总剖面图
Section of Jingshan Park

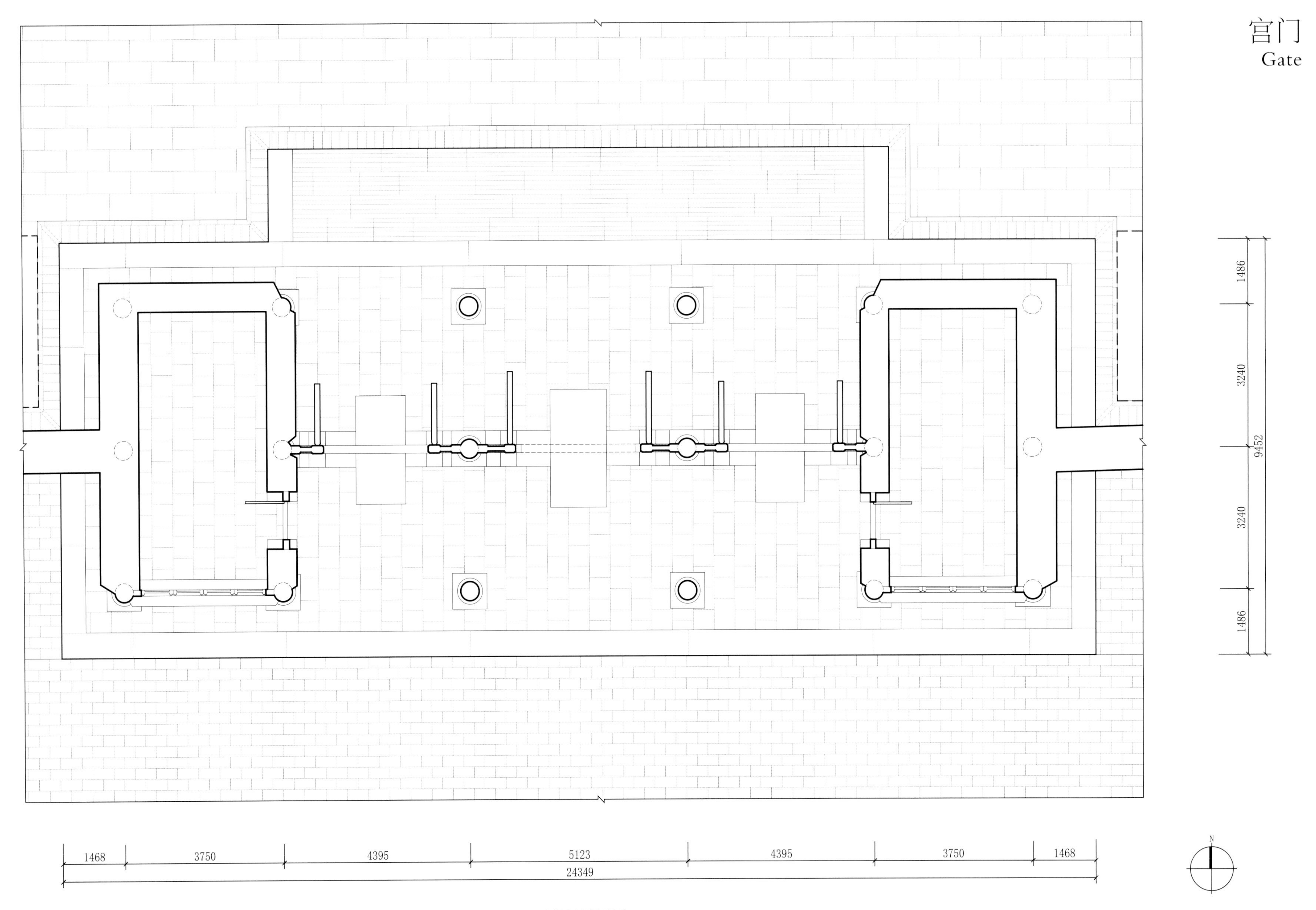

万岁门平面图
Plan of Longevity Gate

万岁门屋顶平面图
Roof Plan of Longevity Gate

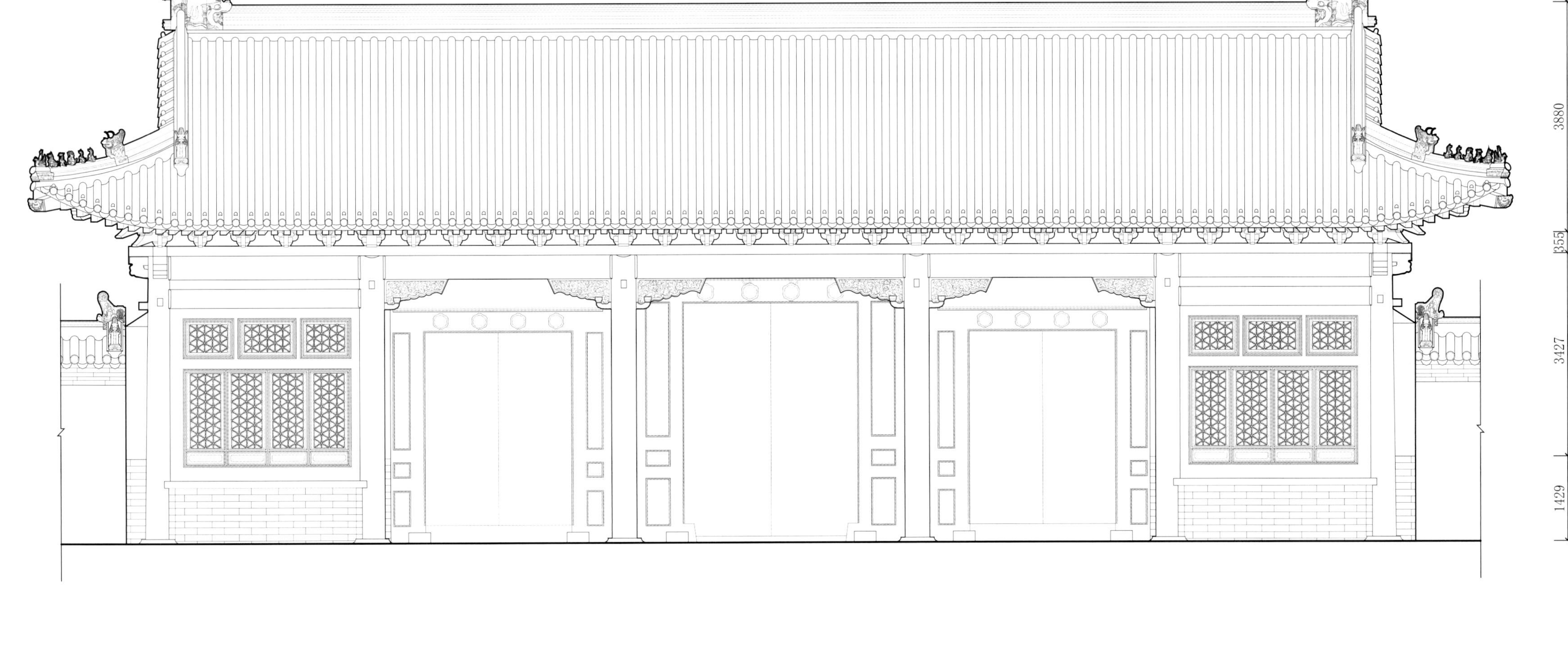

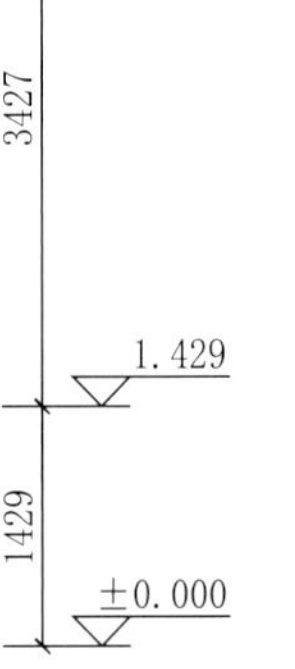

万岁门正立面图

Front elevation of Longevity Gate

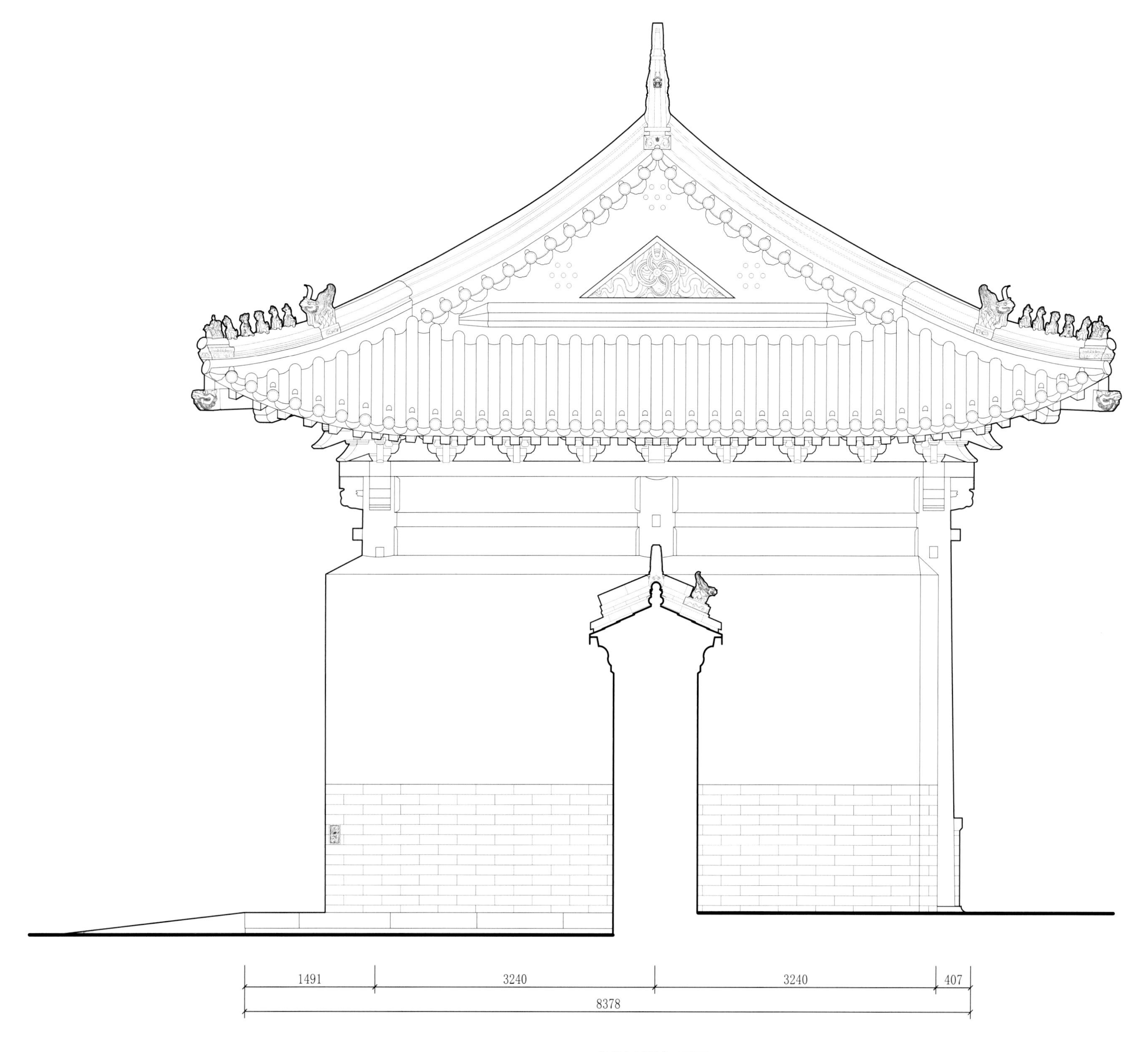

万岁门侧立面图

Side elevation of Longevity Gate

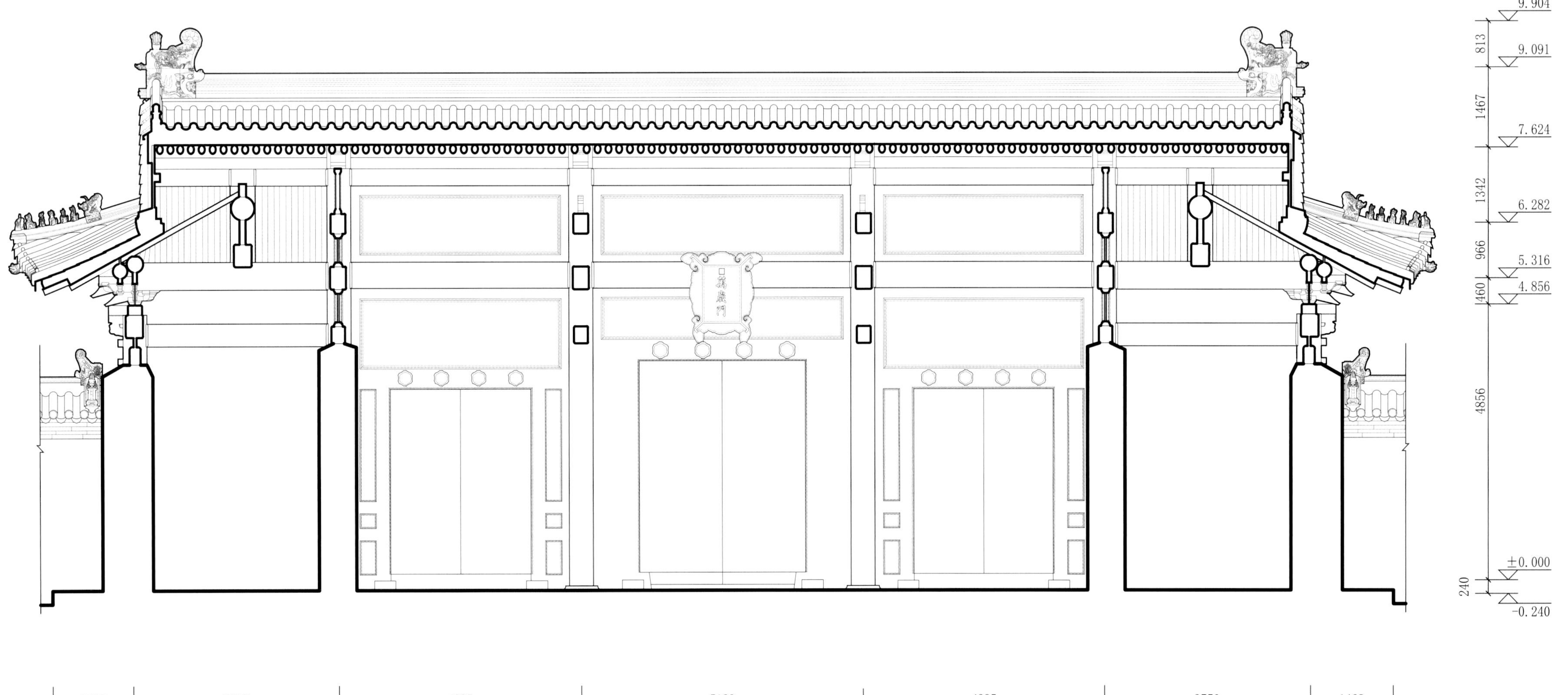

万岁门纵剖面图
Longitudinal section of Longevity Gate

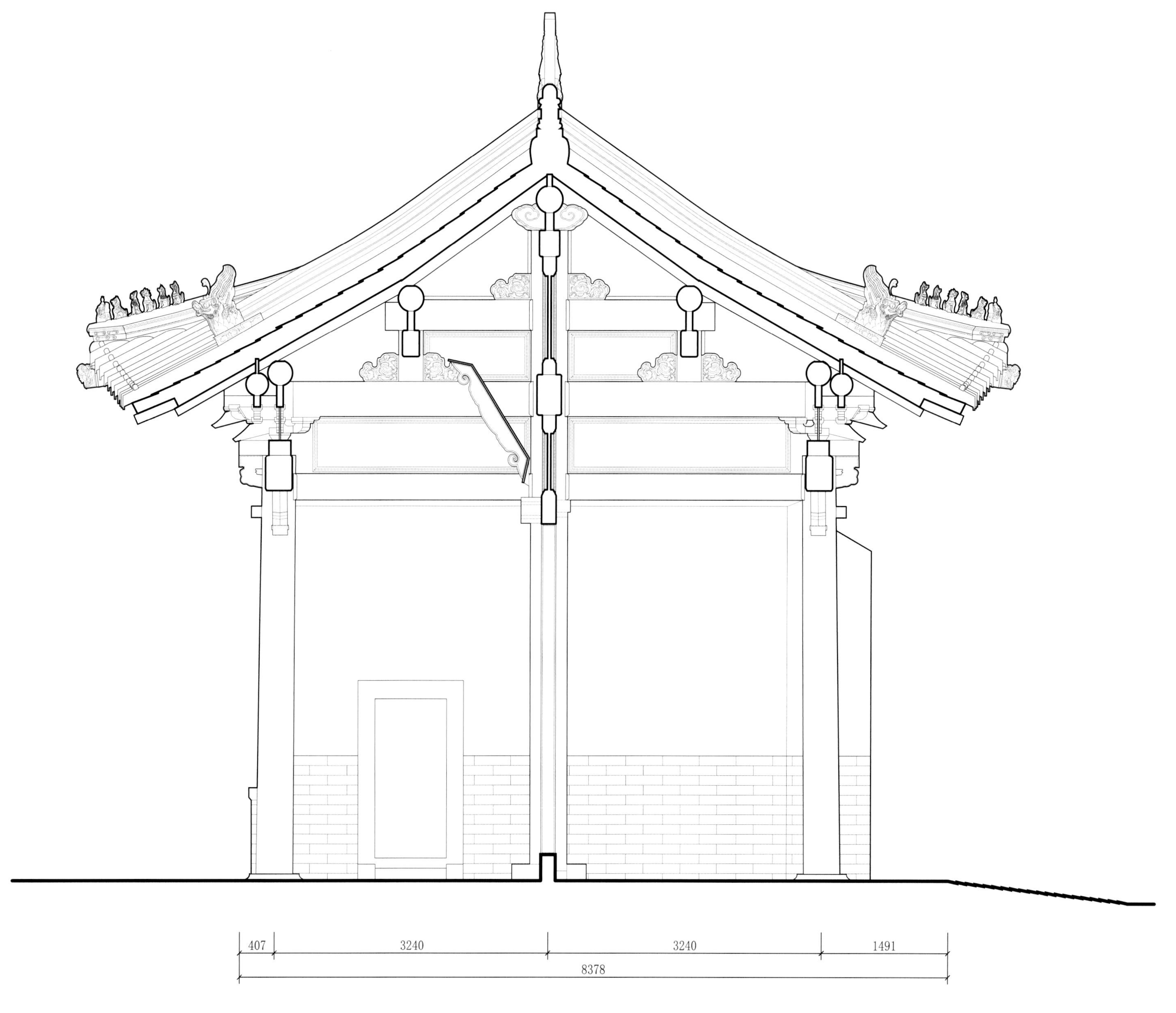

万岁门明间横剖面图
Central bay section of Longevity Gate

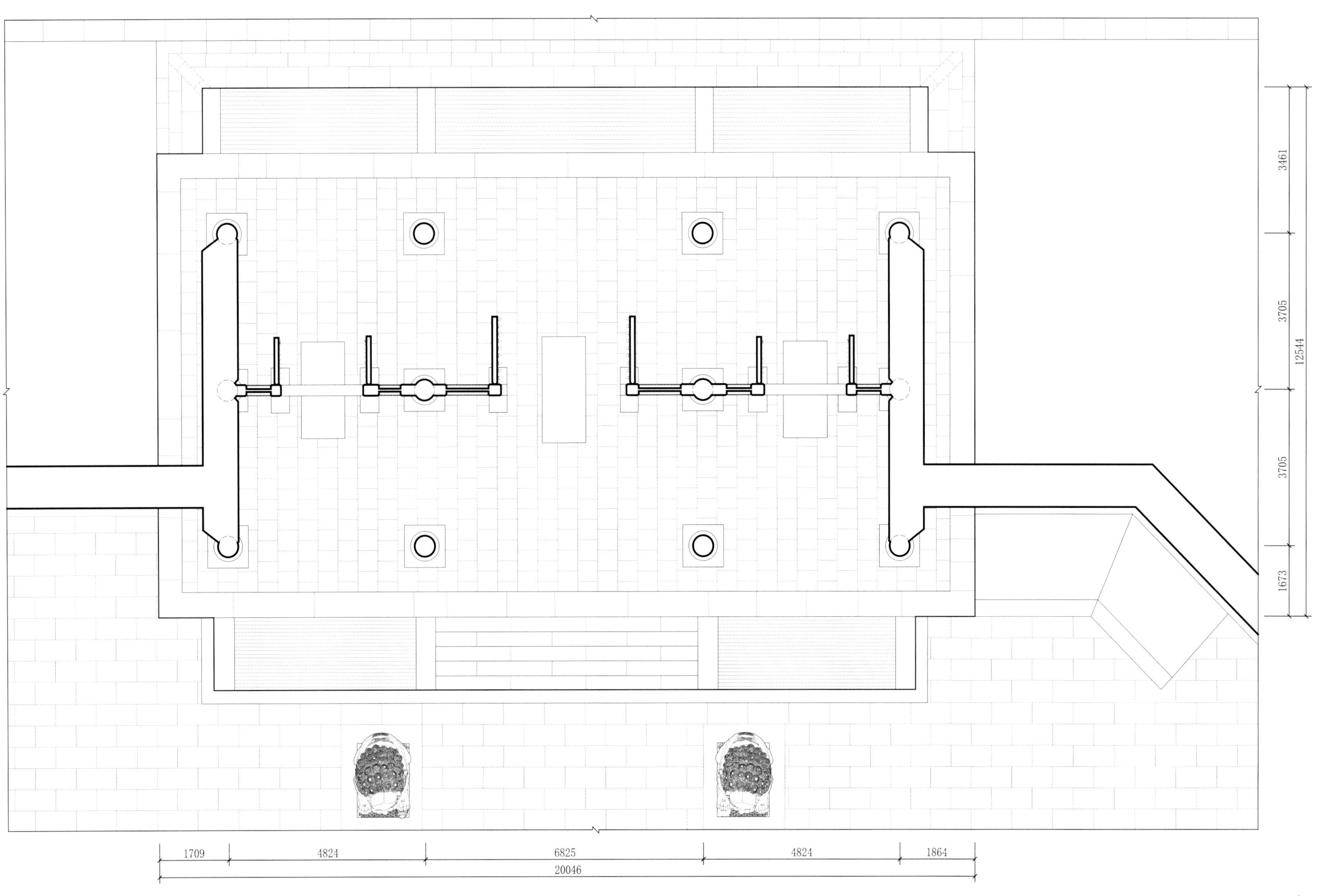

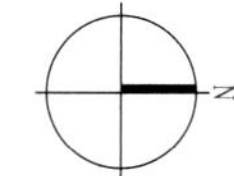

山左里门平面图
Plan of inside Gate of east of Jingshan Park

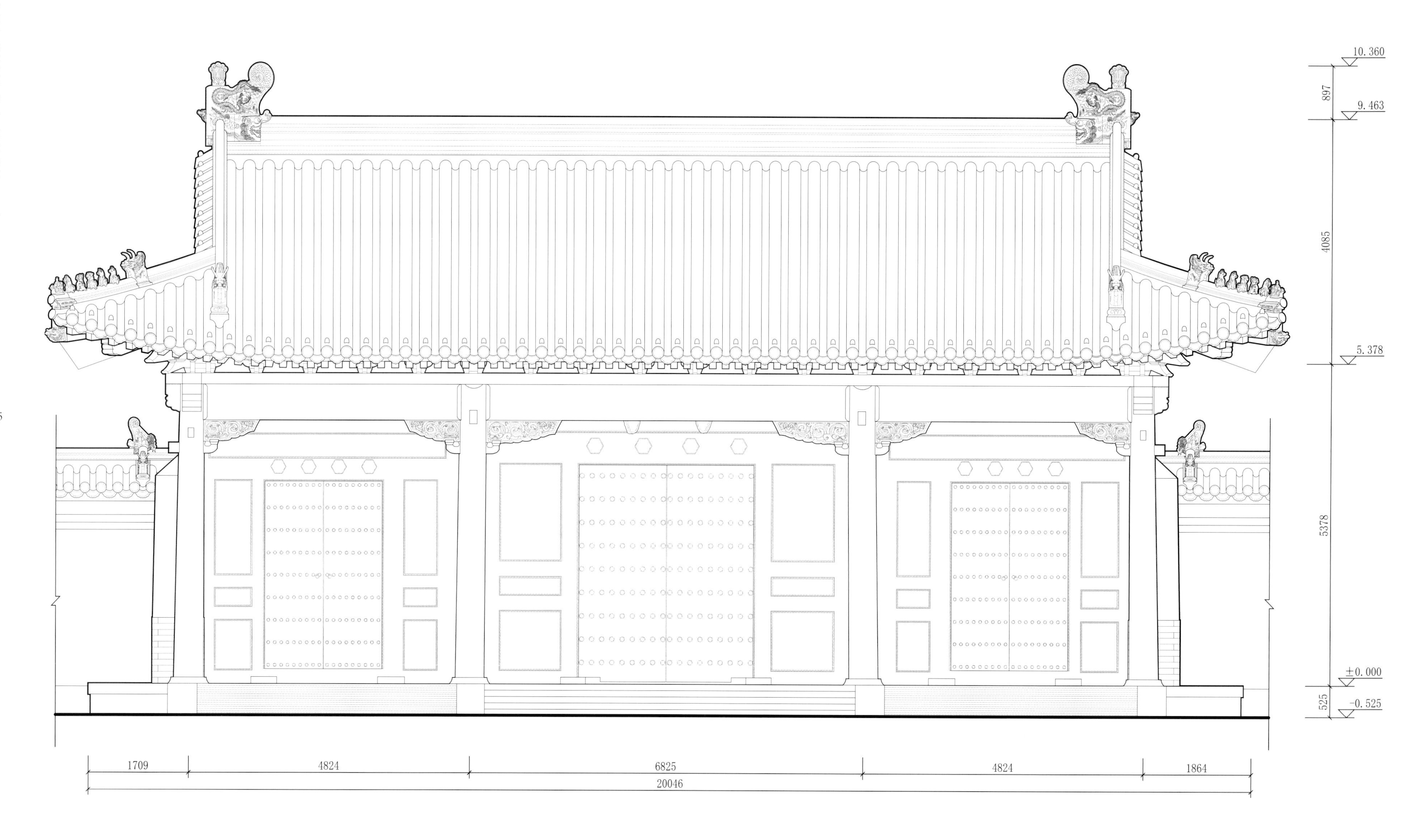

山左里门正立面图

Front elevation of inside Gate of east of Jingshan Park

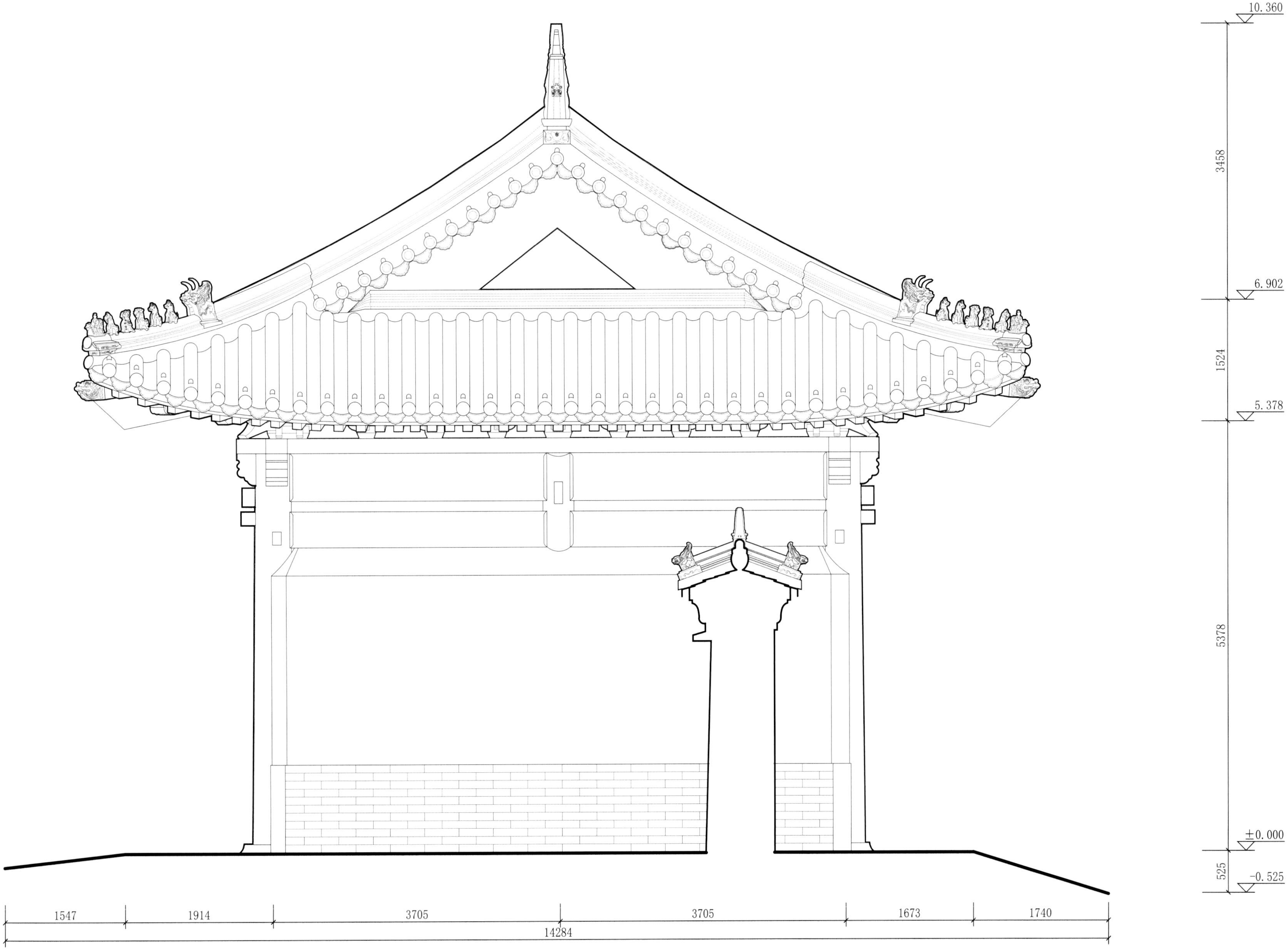

10.360

3458

6.902

1524

5.378

5378

±0.000

525

-0.525

1547 1914 3705 3705 1673 1740

14284

山左里门侧立面图

Side elevation of inside Gate of east of Jingshan Park

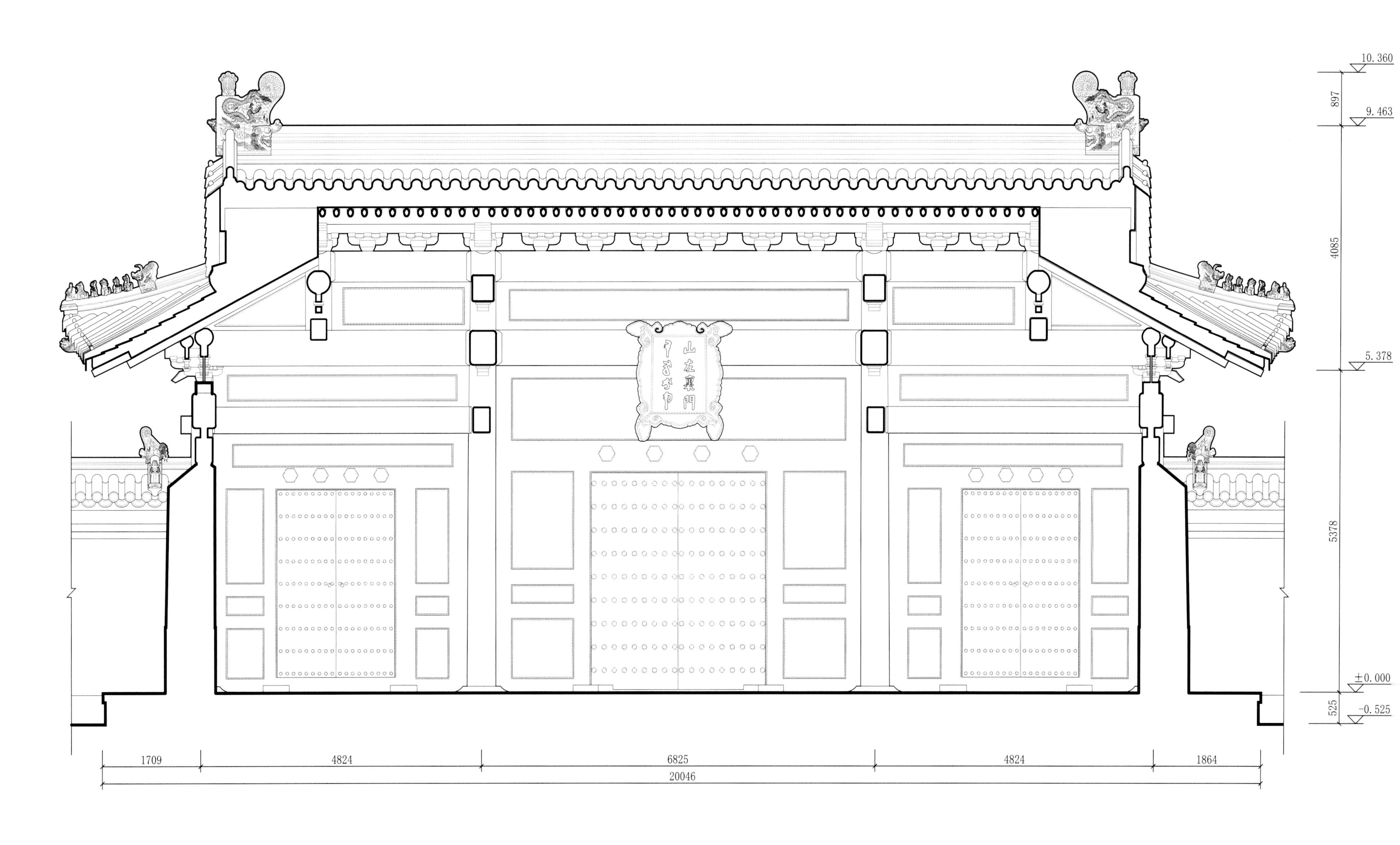

山左里门纵剖面图

Longitudinal section of inside Gate of east of Jingshan Park

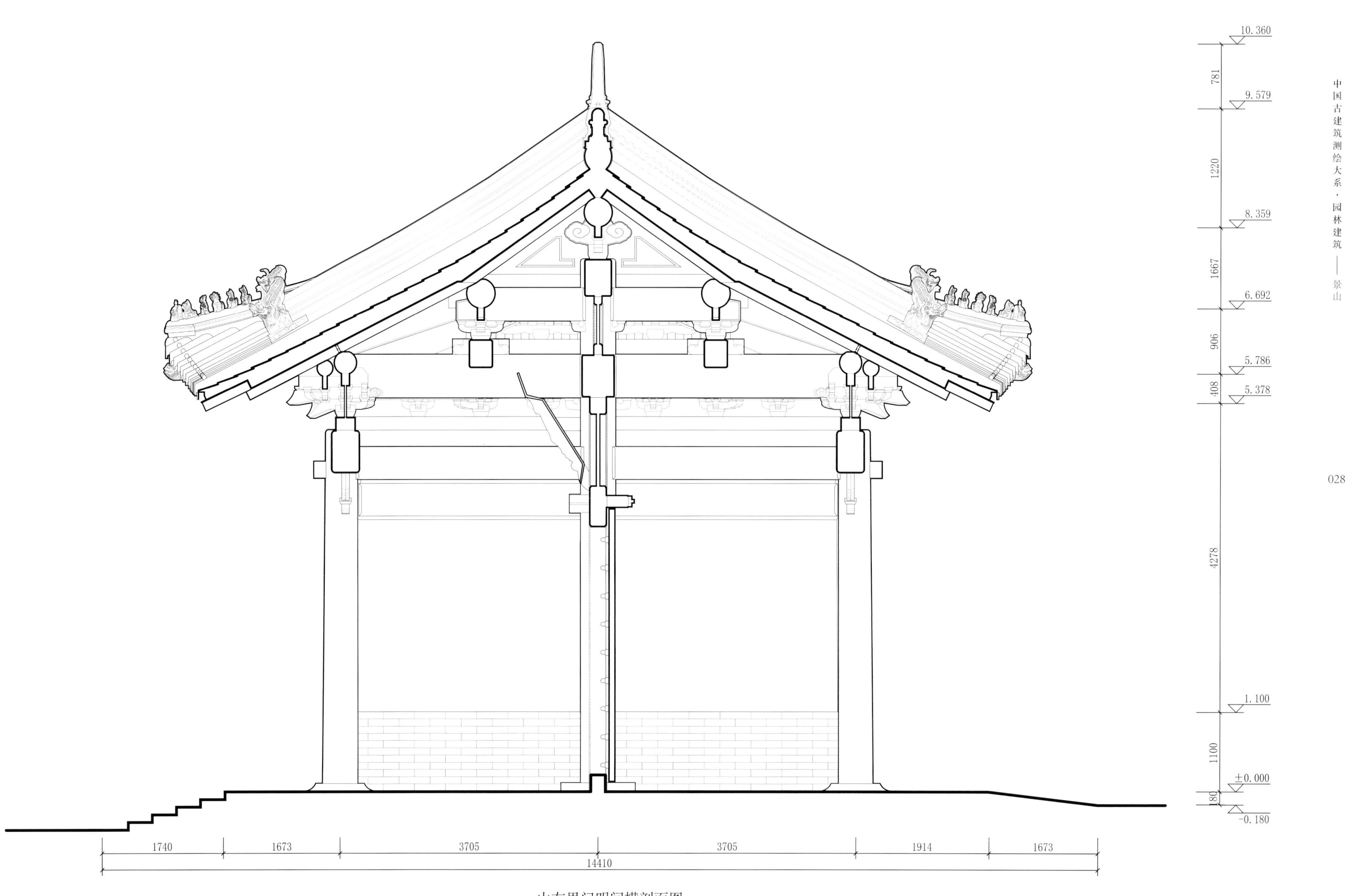

山左里门明间横剖面图

Central bay section of inside Gate of east of Jingshan Park

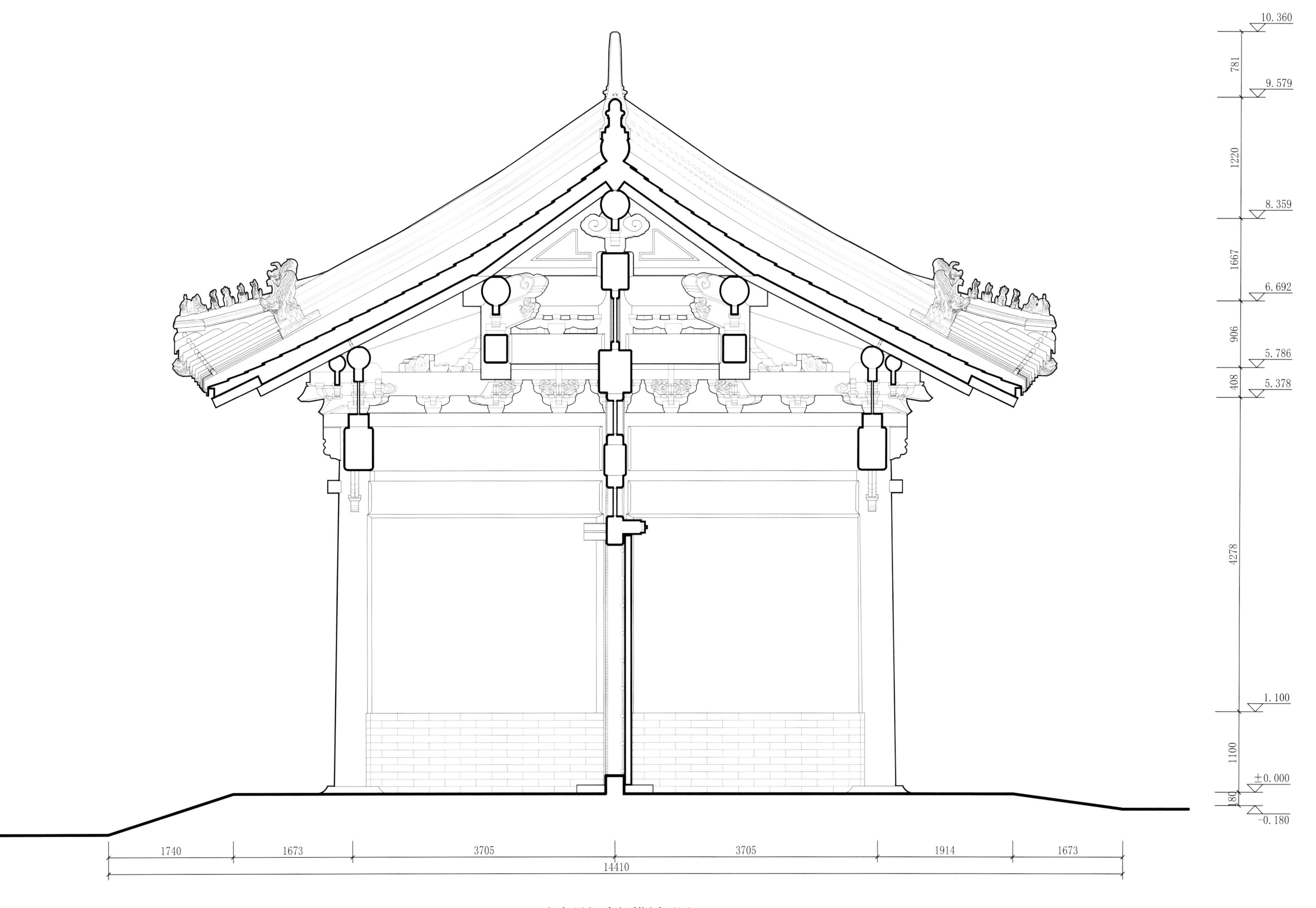

山左里门次间横剖面图

Lateral bay section of inside Gate of east of Jingshan Park

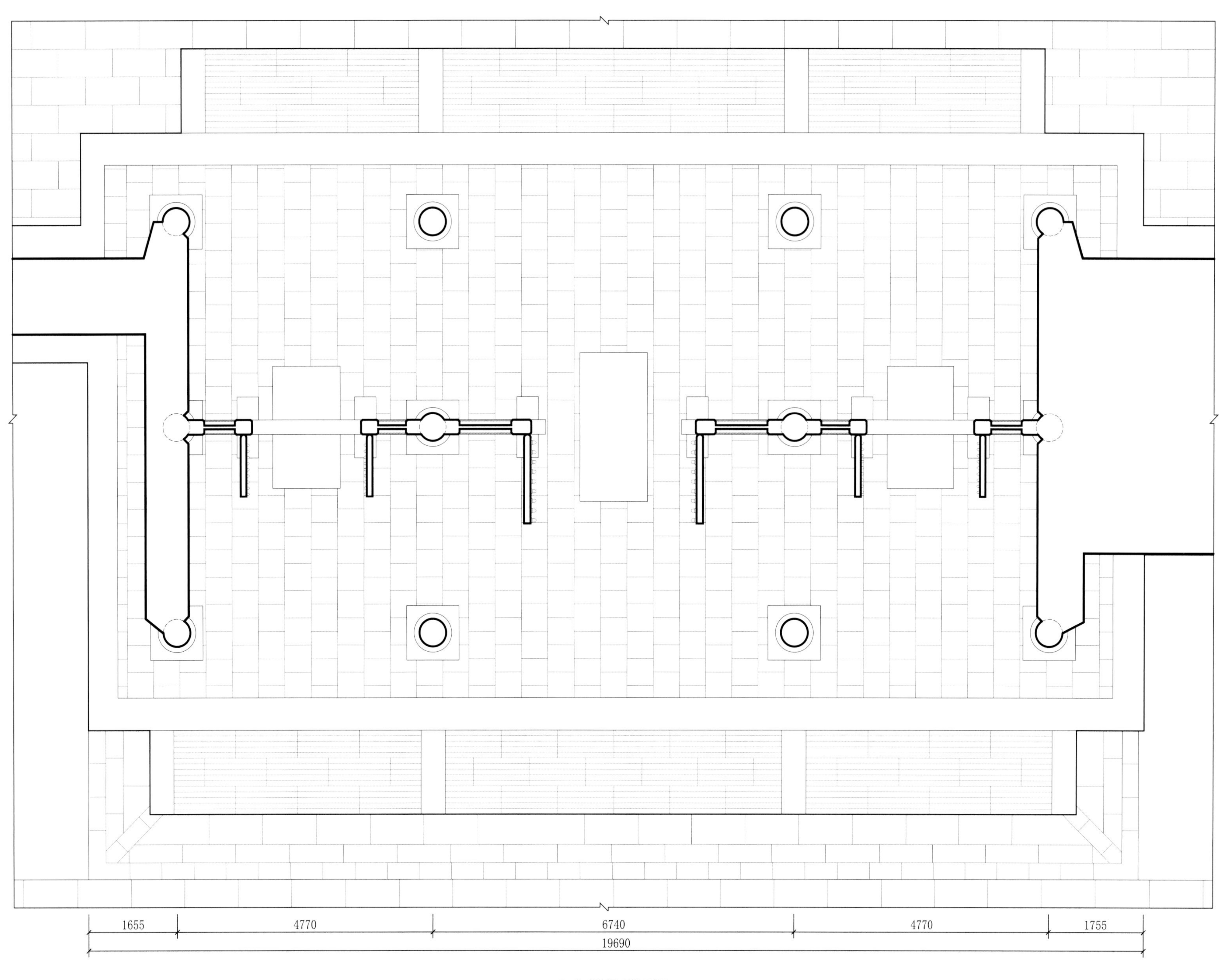

山右里门平面图
Plan of inside Gate of west of Jingshan Park

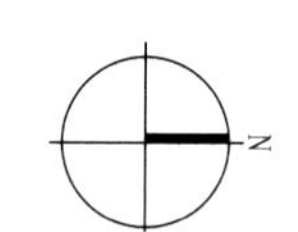

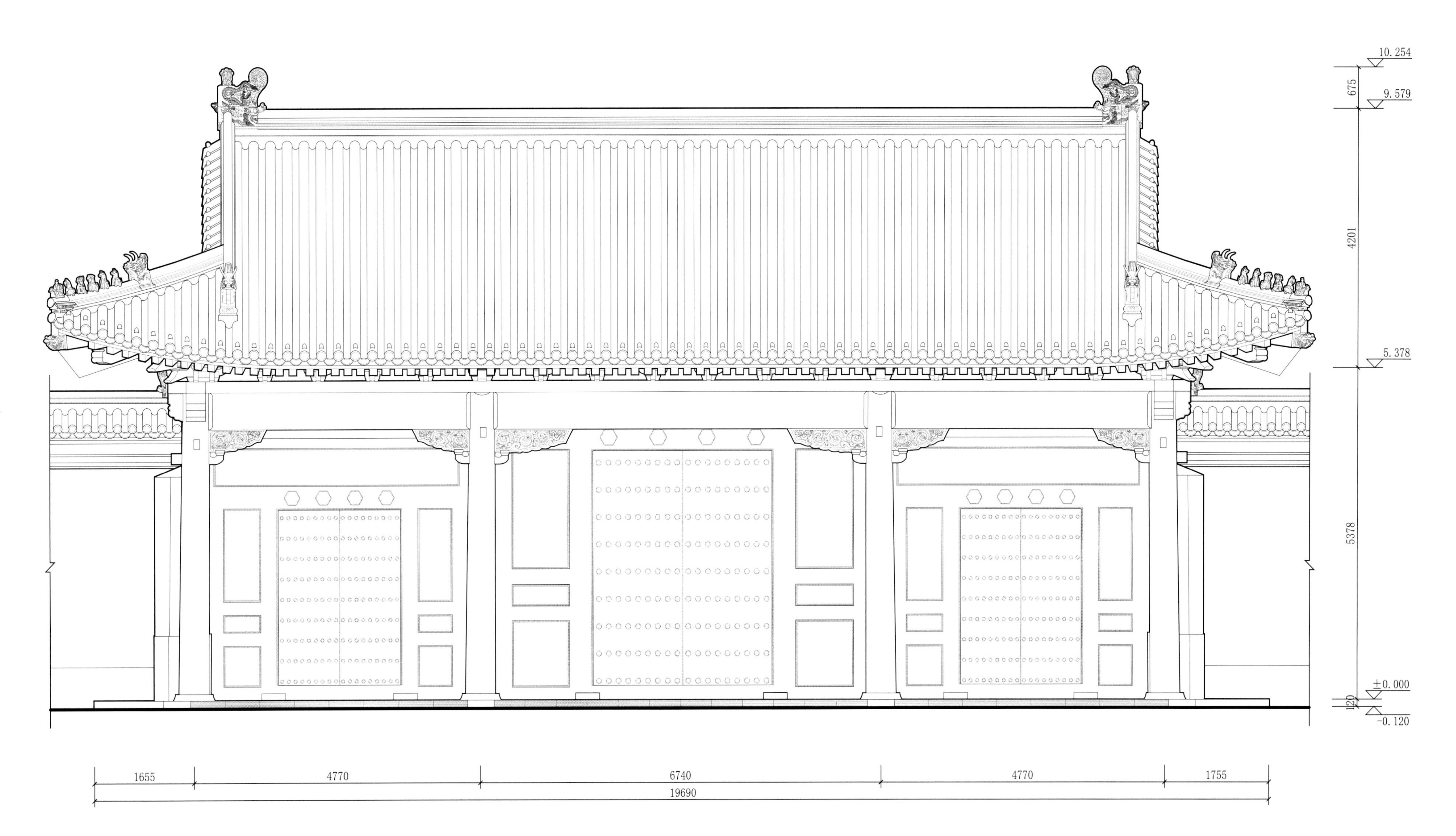

山右里门正立面图

Front elevation of inside Gate of west of Jingshan Park

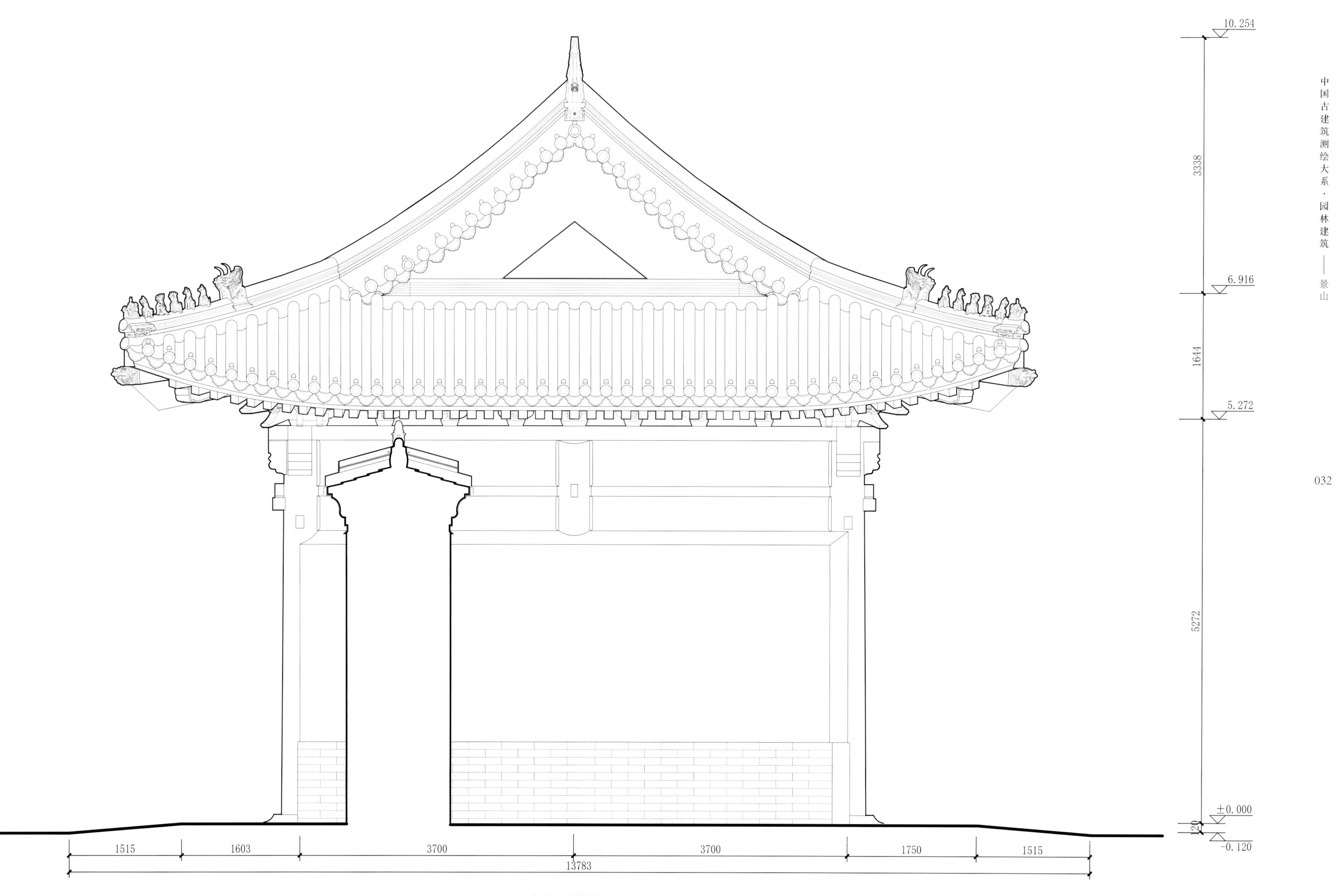

山右里门侧立面图

Side elevation of inside Gate of west of Jingshan Park

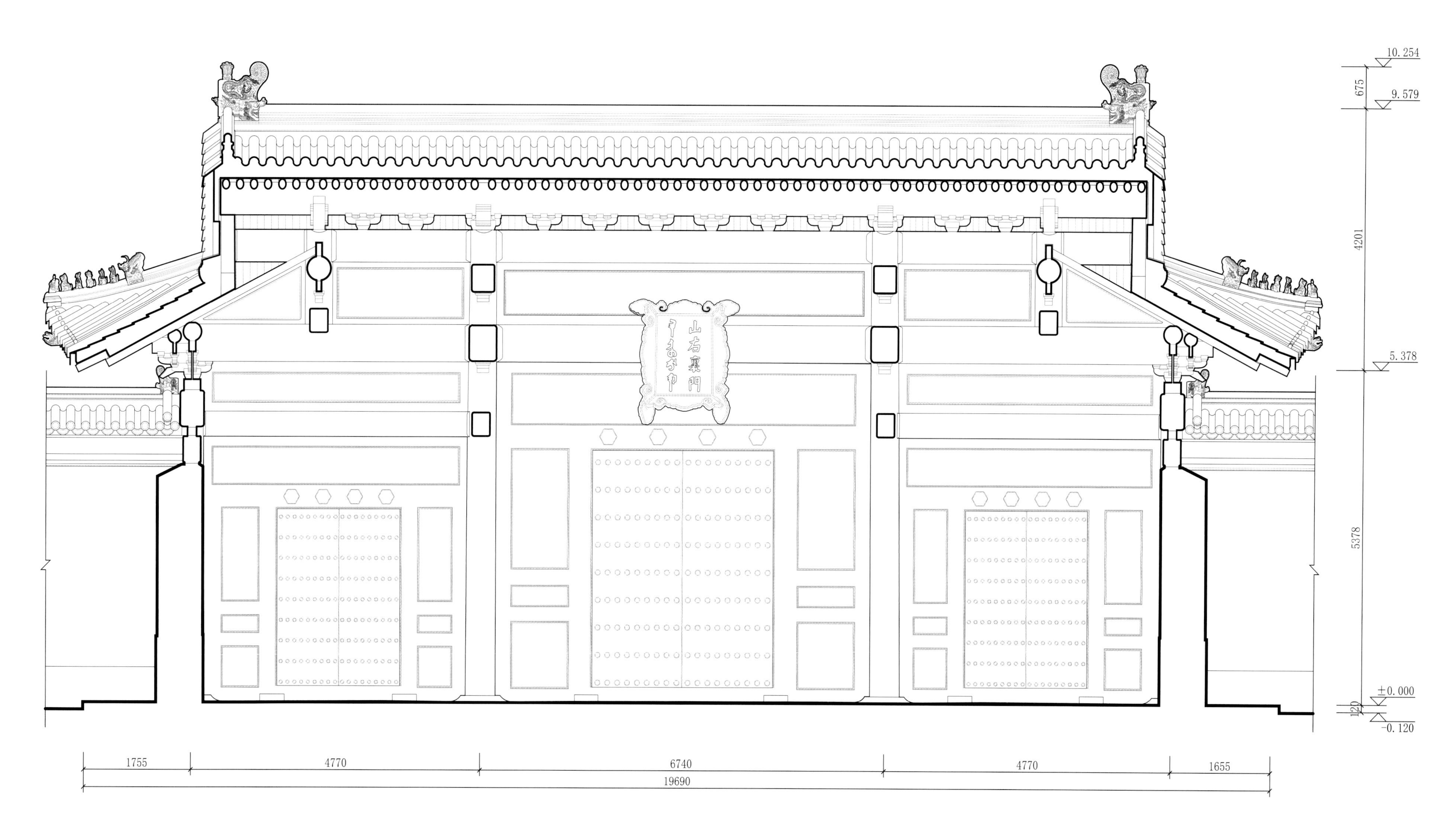

山右里门纵剖面图

Longitudinal section of inside Gate of west of Jingshan Park

10.254
9.579
8.359
6741
5.775
5.378
1.080
±0.000
-0.120

675
1220
1618
866
397
4298
1080
120

1515 1603 3700 3700 1750 1515
13783

山右里门明间横剖面图

Central bay section of inside Gate of west of Jingshan Park

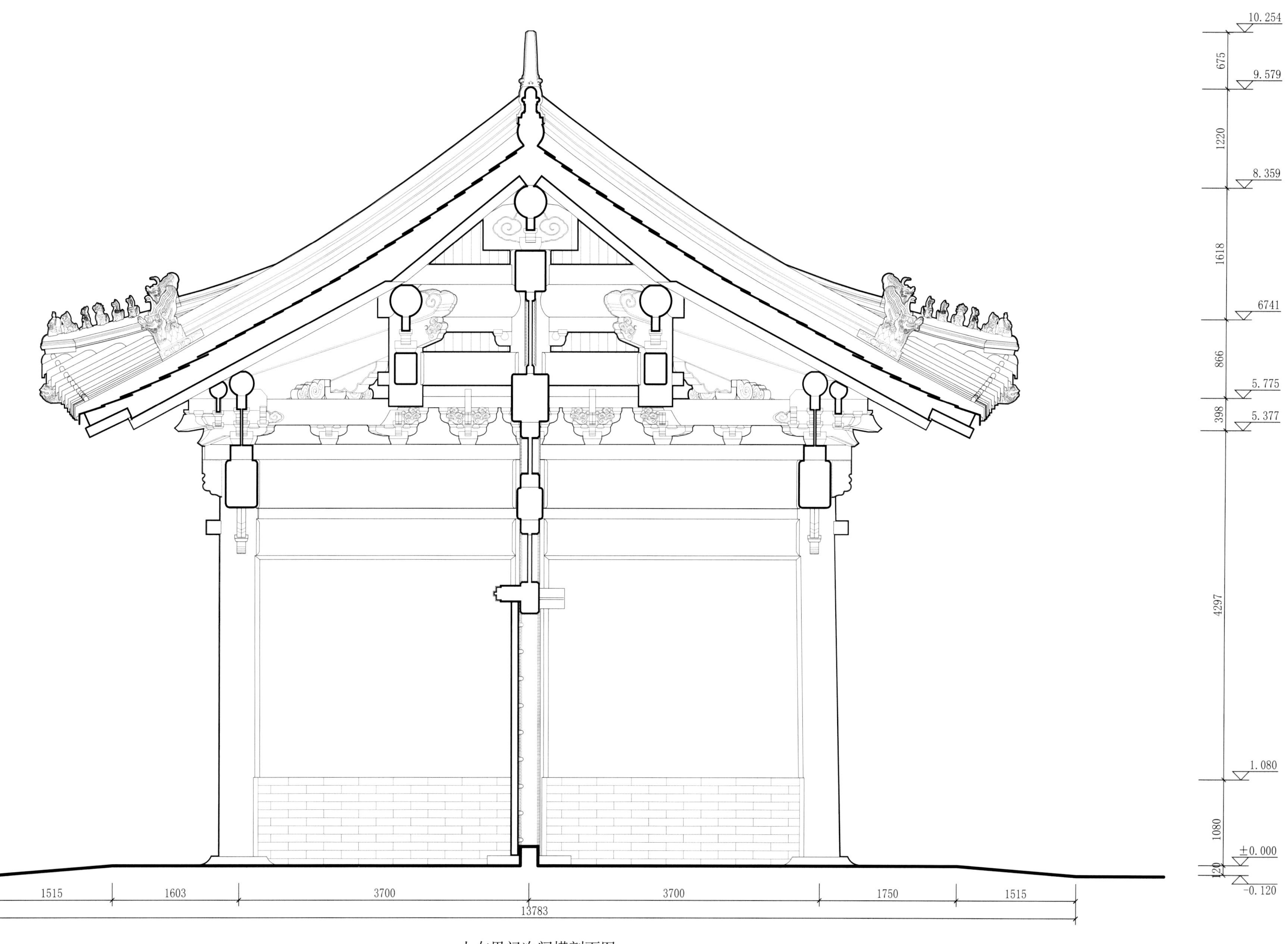

山右里门次间横剖面图

Lateral bay section of inside Gate of west of Jingshan Park

五峰亭
Wufeng Pavilion

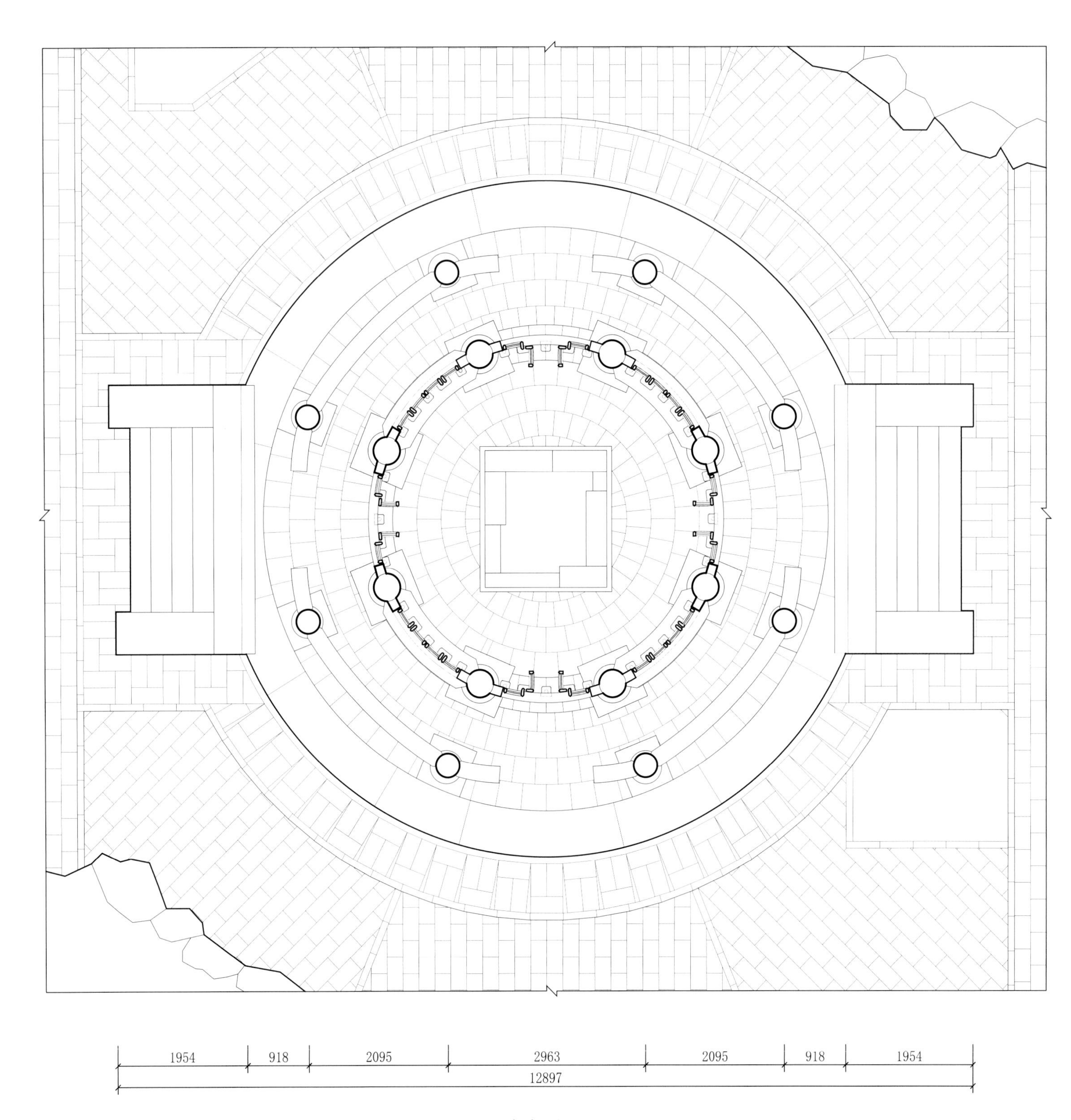

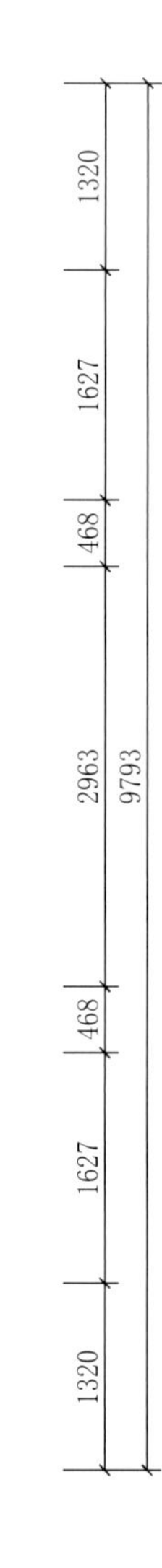

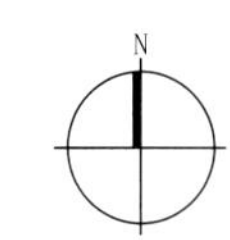

周赏亭平面图
Plan of Zhoushang Pavilion

周赏亭正立面图
Front elevation of Zhoushang Pavilion

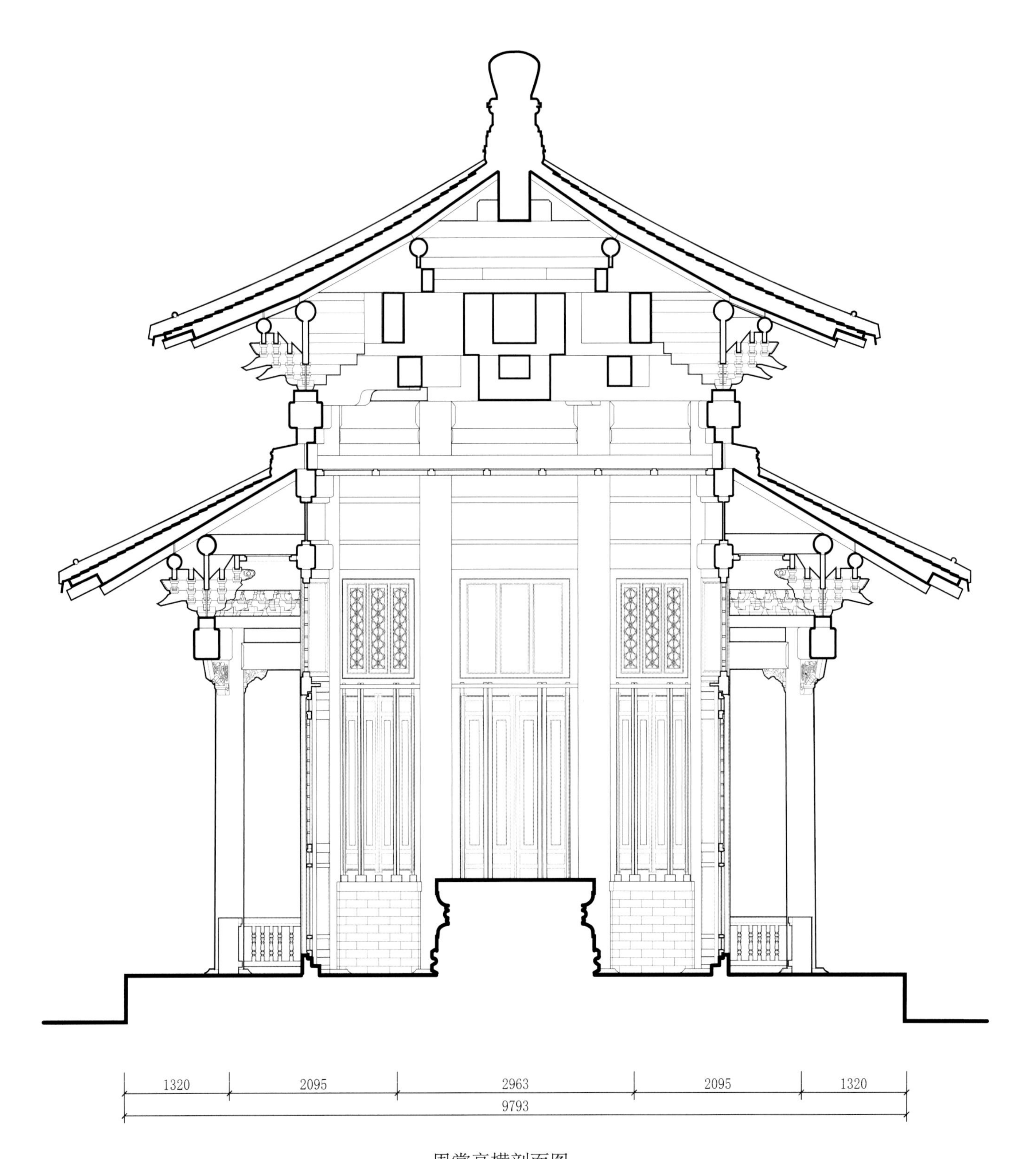

周赏亭横剖面图
Transverse section of Zhoushang Pavilion

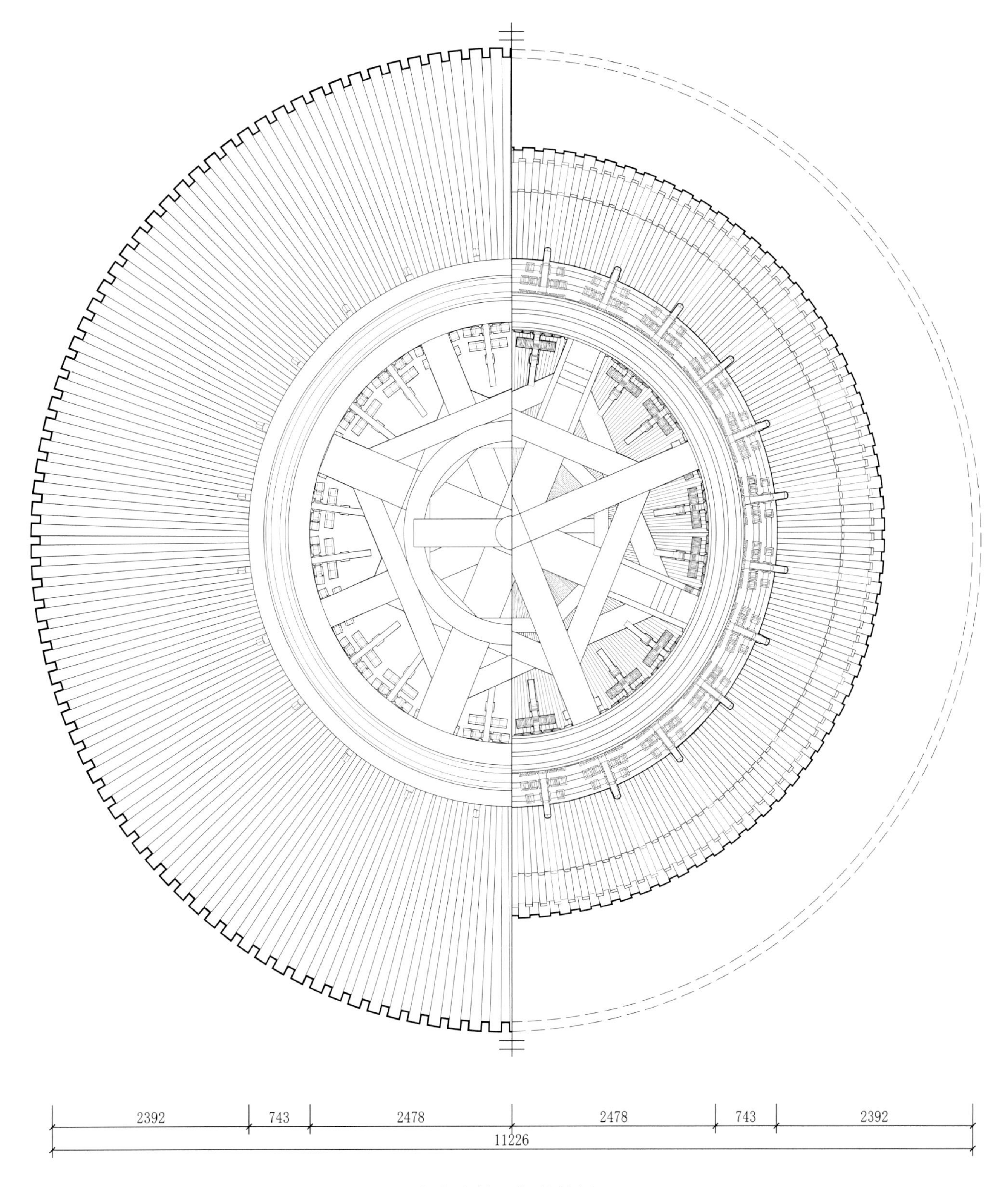

周赏亭上檐梁架仰俯视图
Bottom and vertical view of beams of upper eave of Zhoushang Pavilion

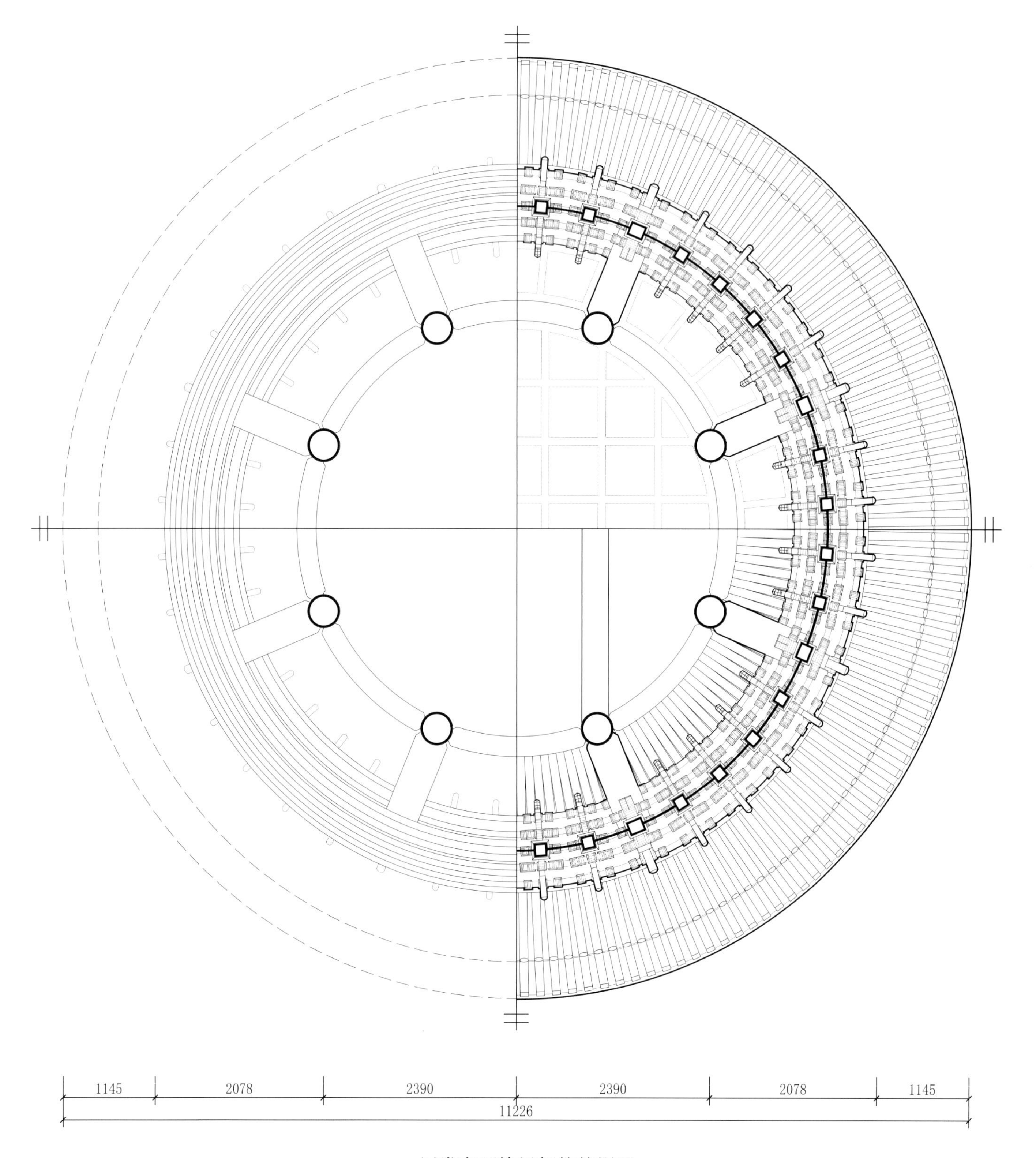

周赏亭下檐梁架仰俯视图
Bottom and vertical view of beams of under eave of Zhoushang Pavilion

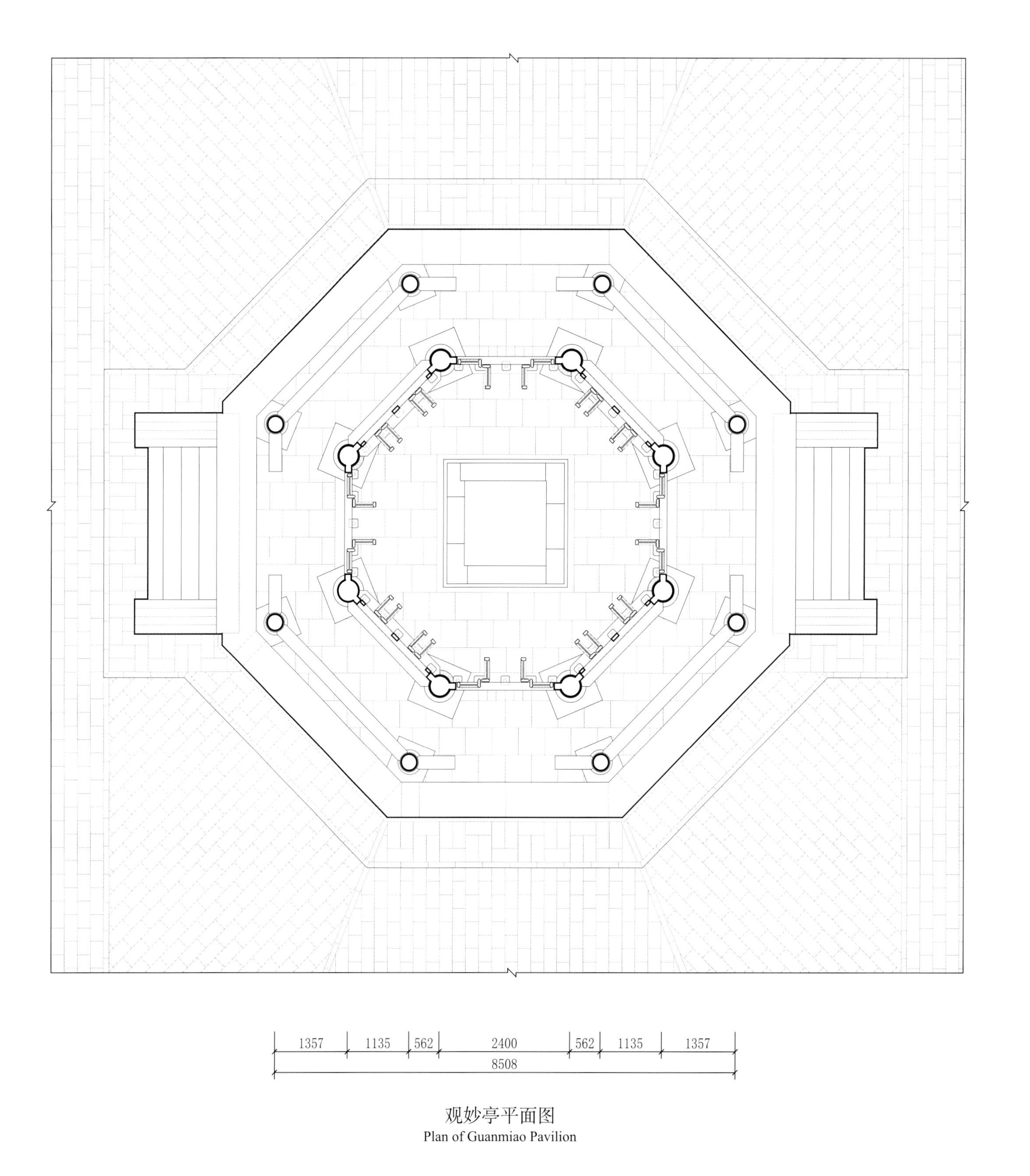

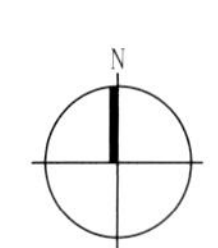

观妙亭平面图
Plan of Guanmiao Pavilion

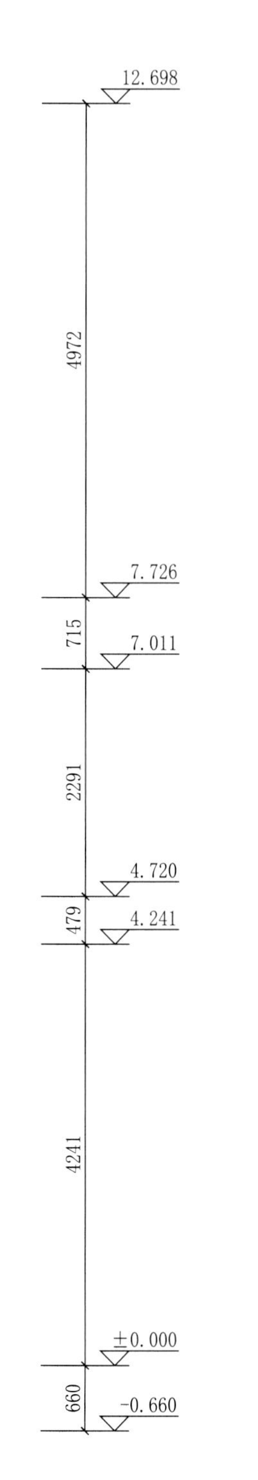

观妙亭正立面图
Front elevation of Guanmiao Pavilion

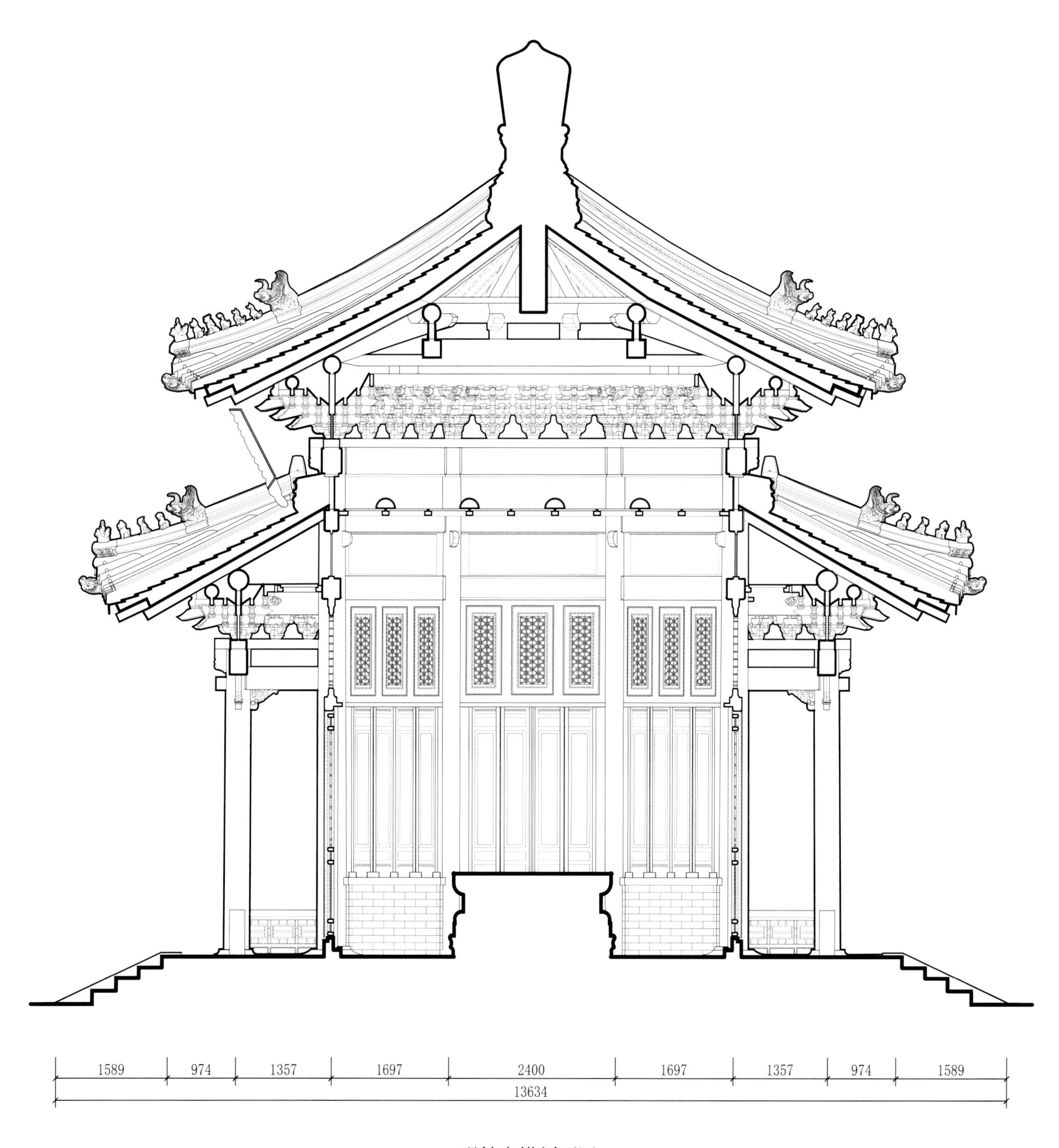

观妙亭横剖面图

Transverse section of Guanmiao Pavilion

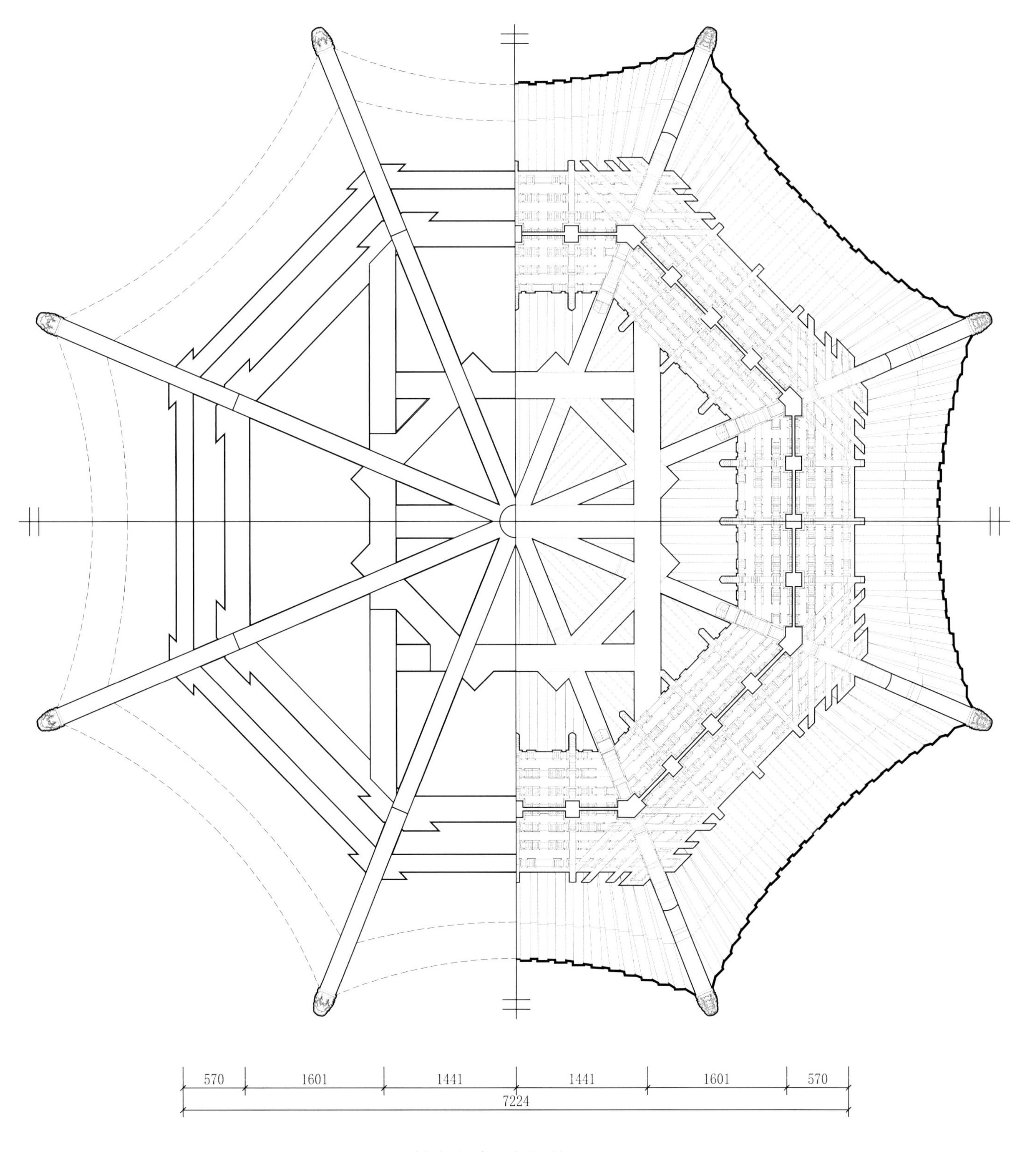

观妙亭上檐梁架仰俯视图
Bottom and vertical view of beams of upper eave of Guanmiao Pavilion

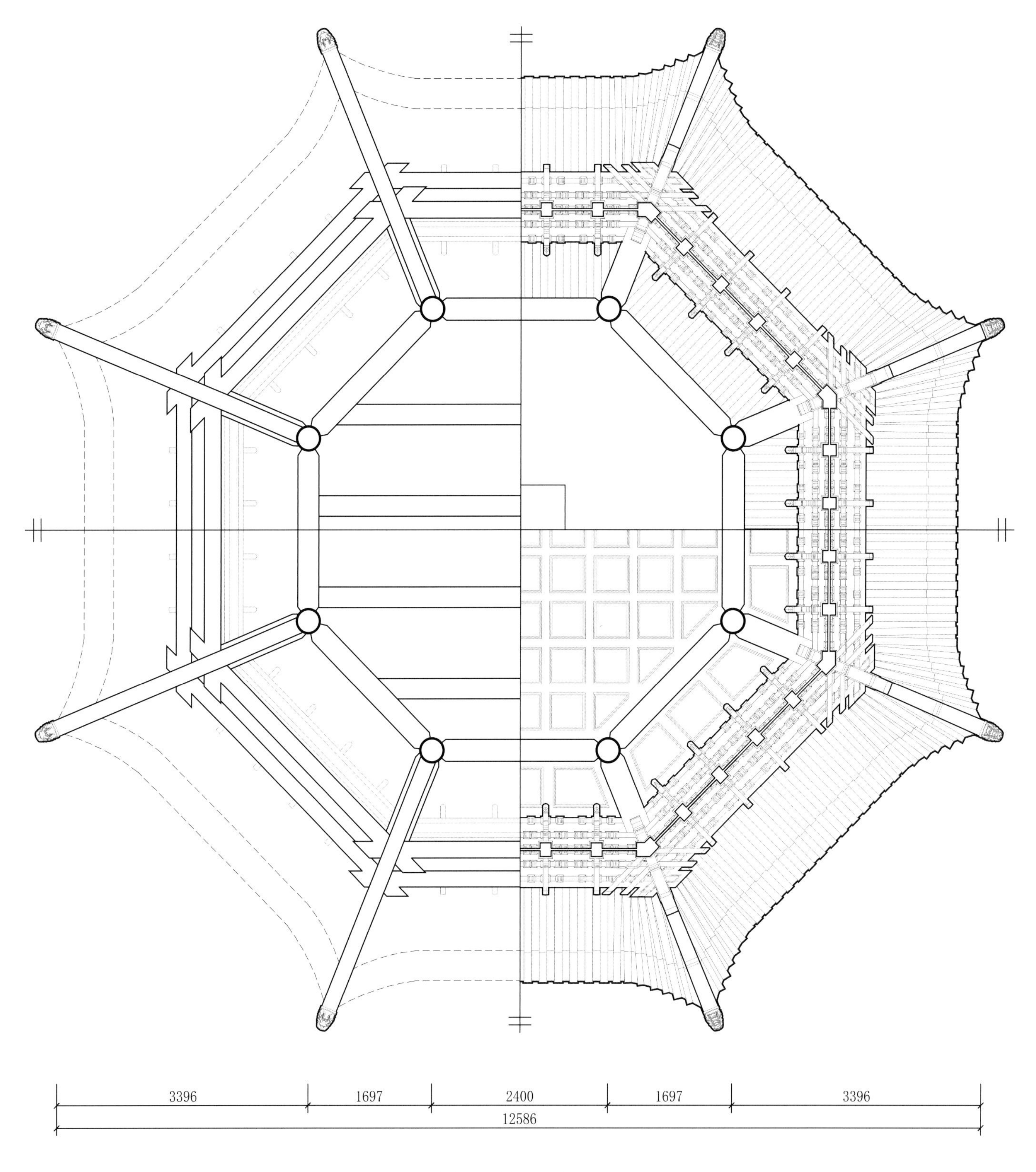

1689
1357
1135
562
2400
11886
562
1135
1357
1689

观妙亭下檐梁架仰俯视图

Bottom and vertical view of beams of under eave of Guanmiao Pavilion

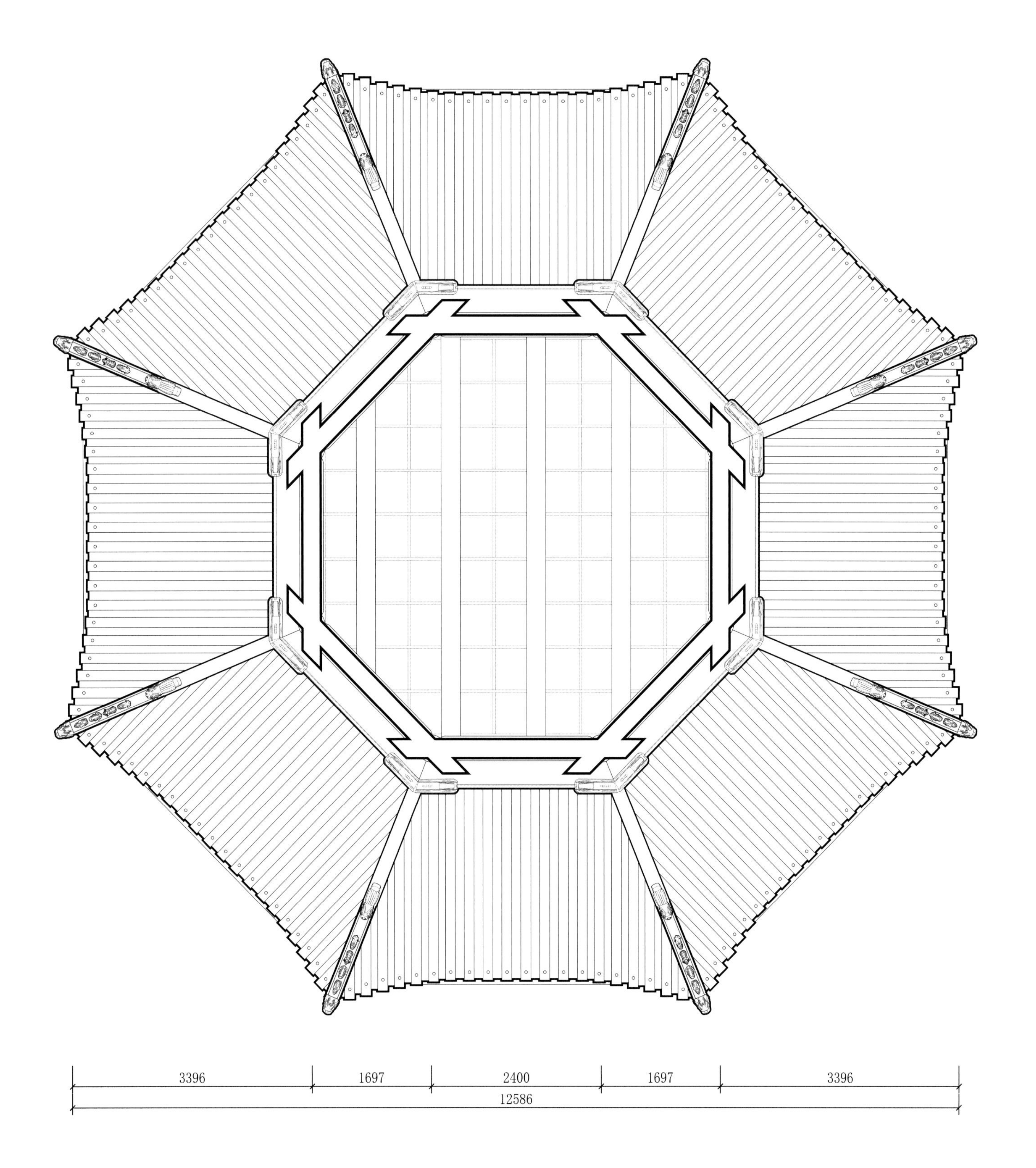

观妙亭屋顶仰视图
Roof Bottom view of Guanmiao Pavilion

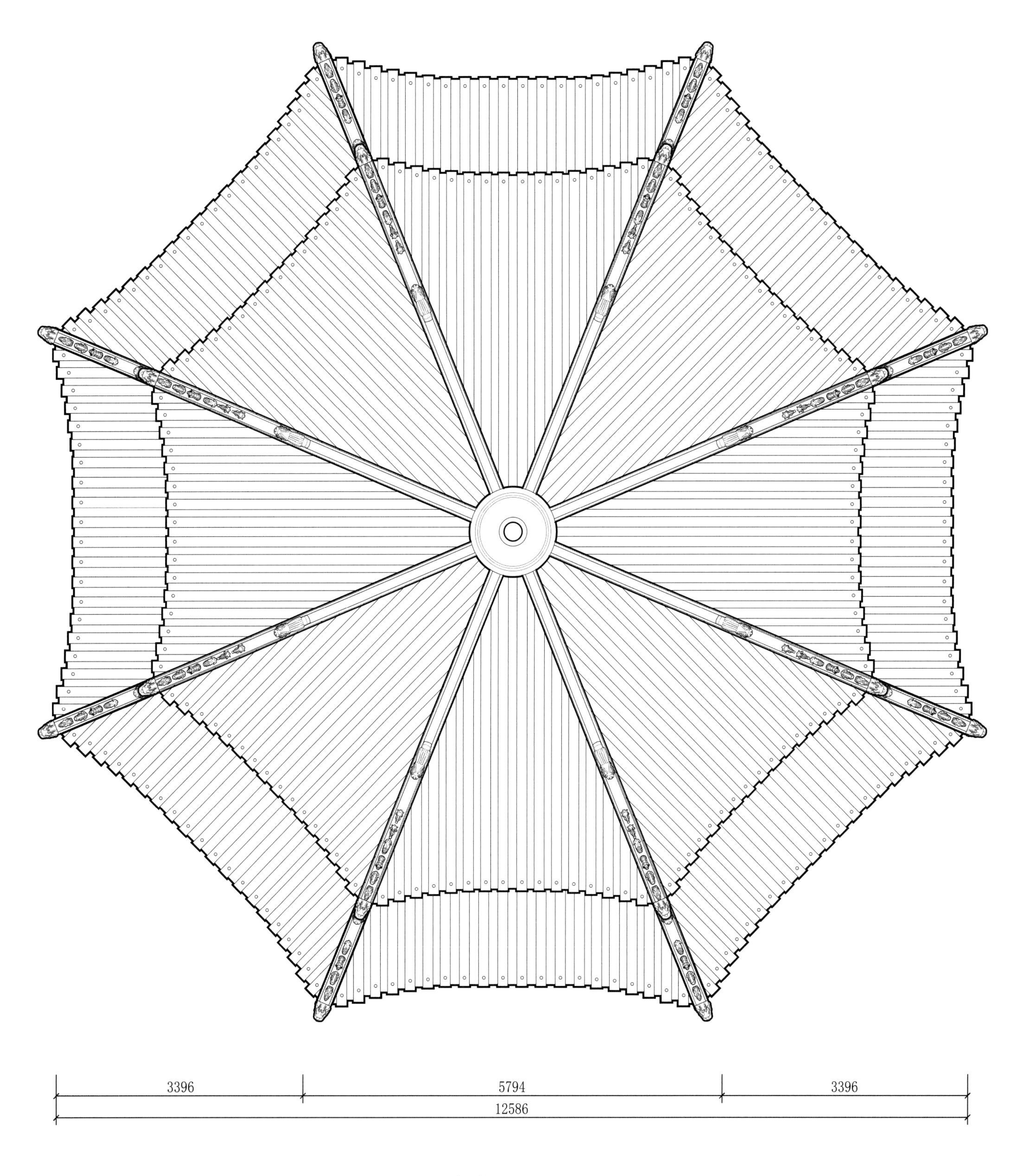

观妙亭屋顶平面图
Roof Plan of Guanmiao Pavilion

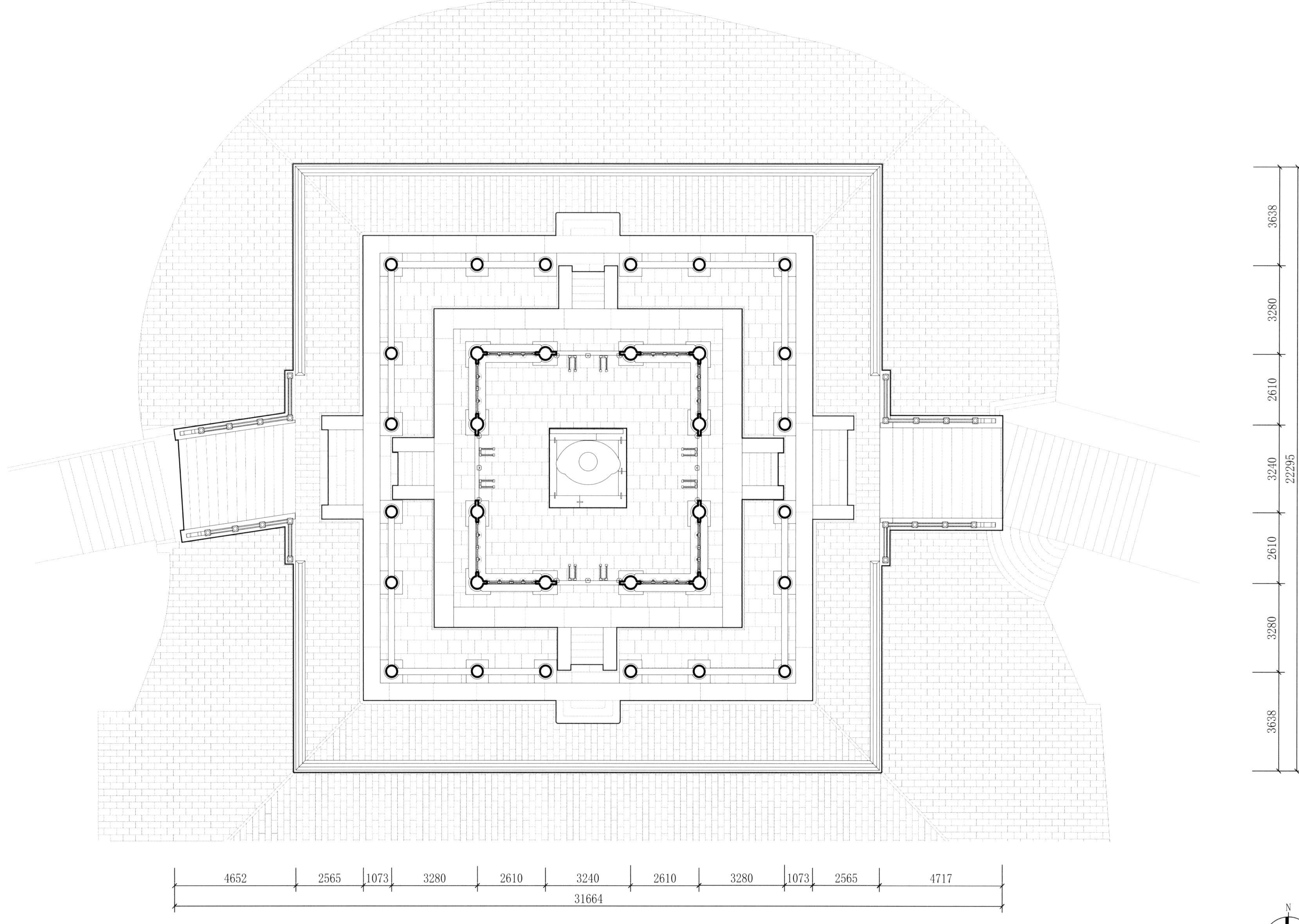

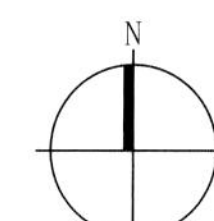

万春亭平面图
Plan of Everlasting Spring Pavilion

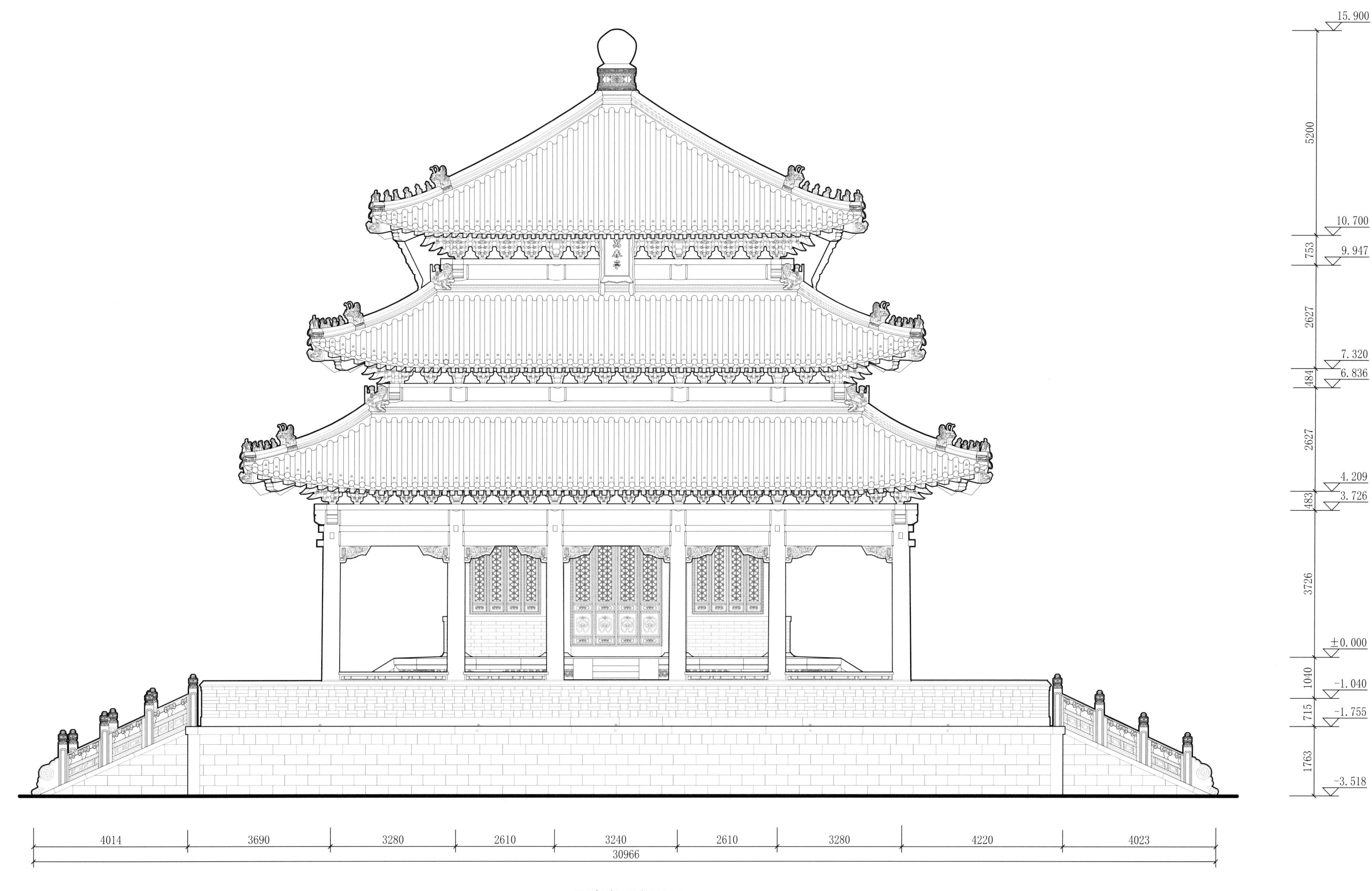

万春亭正立面图

Front elevation of Everlasting Spring Pavilion

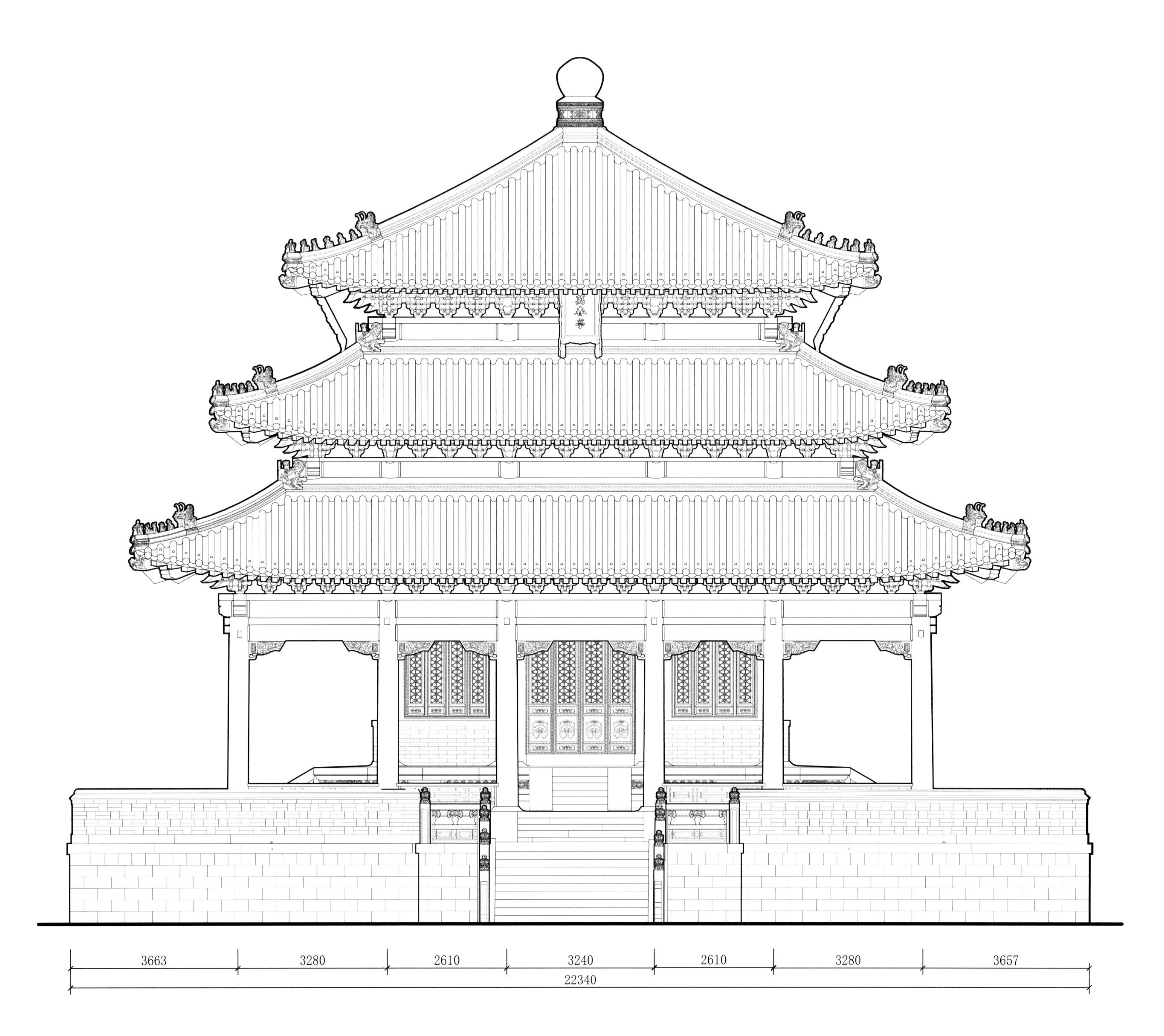

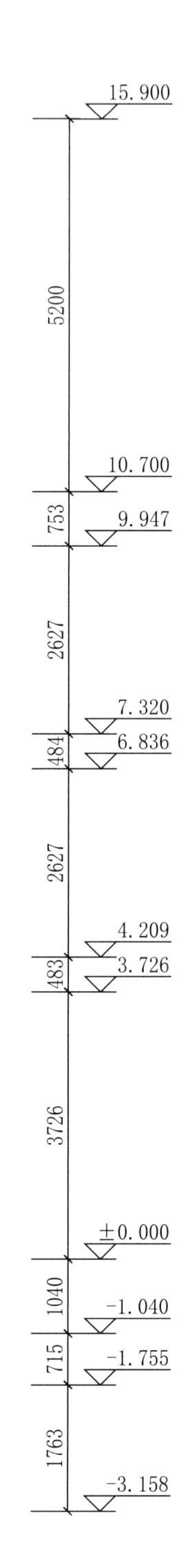

万春亭侧立面图

Side elevation of Everlasting Spring Pavilion

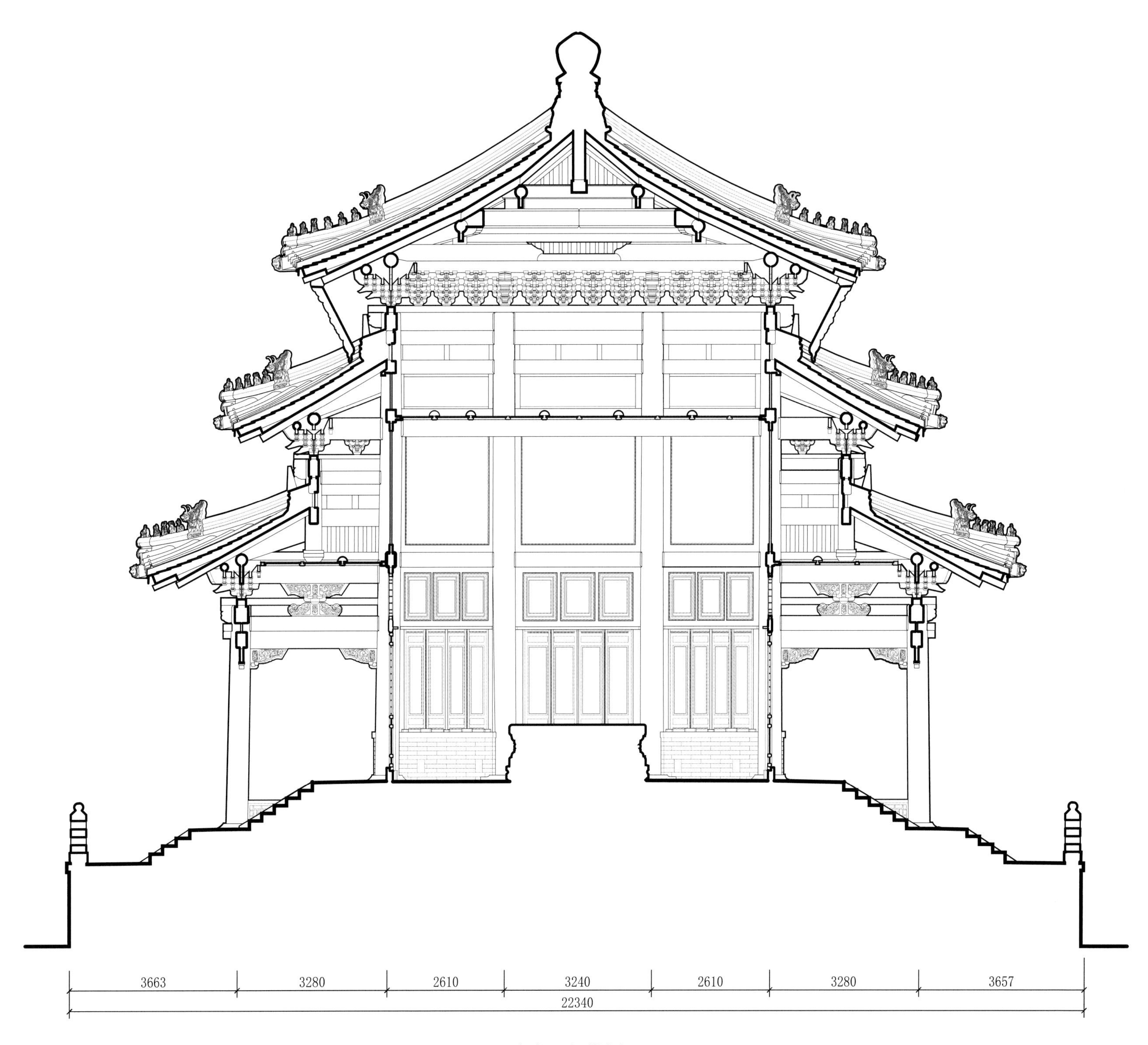

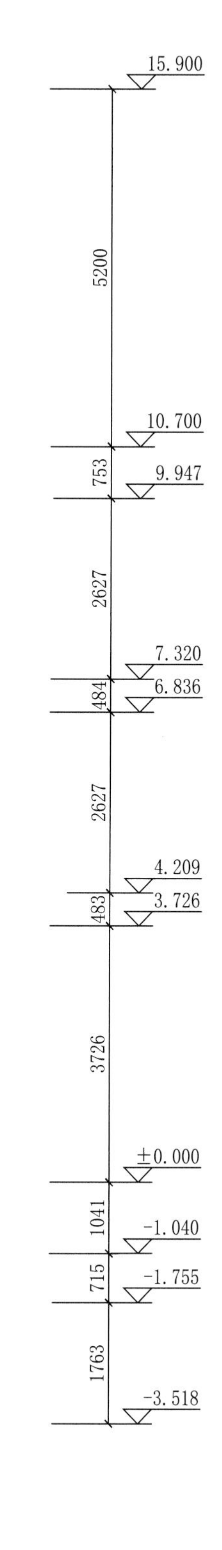

万春亭明间横剖面图

Central bay section of Everlasting Spring Pavilion

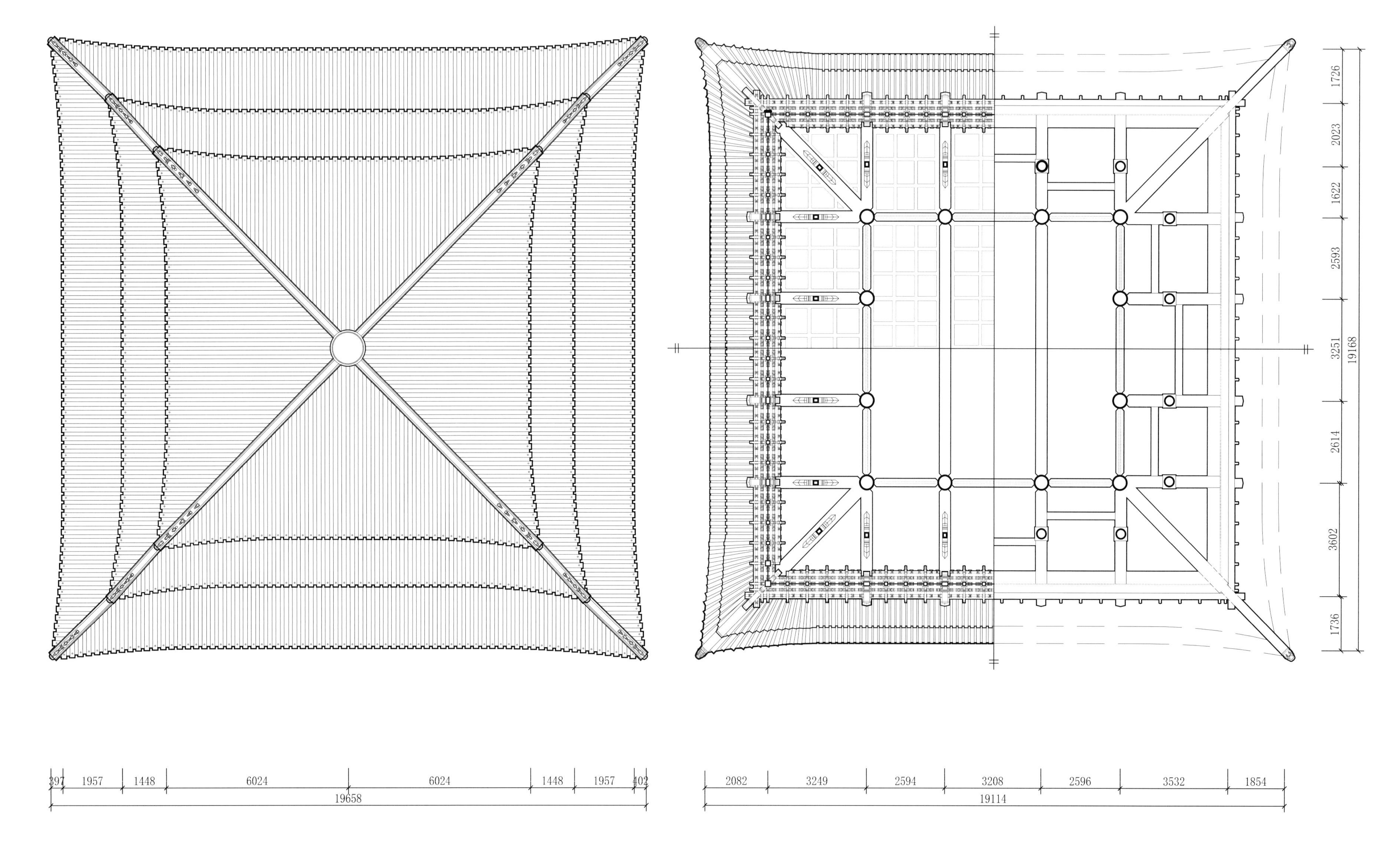

万春亭屋顶平面图
Roof plan of Everlasting Spring Pavilion

万春亭下檐梁架仰俯视图
Bottom and vertical view of beams of under eave of Everlasting Spring Pavilion

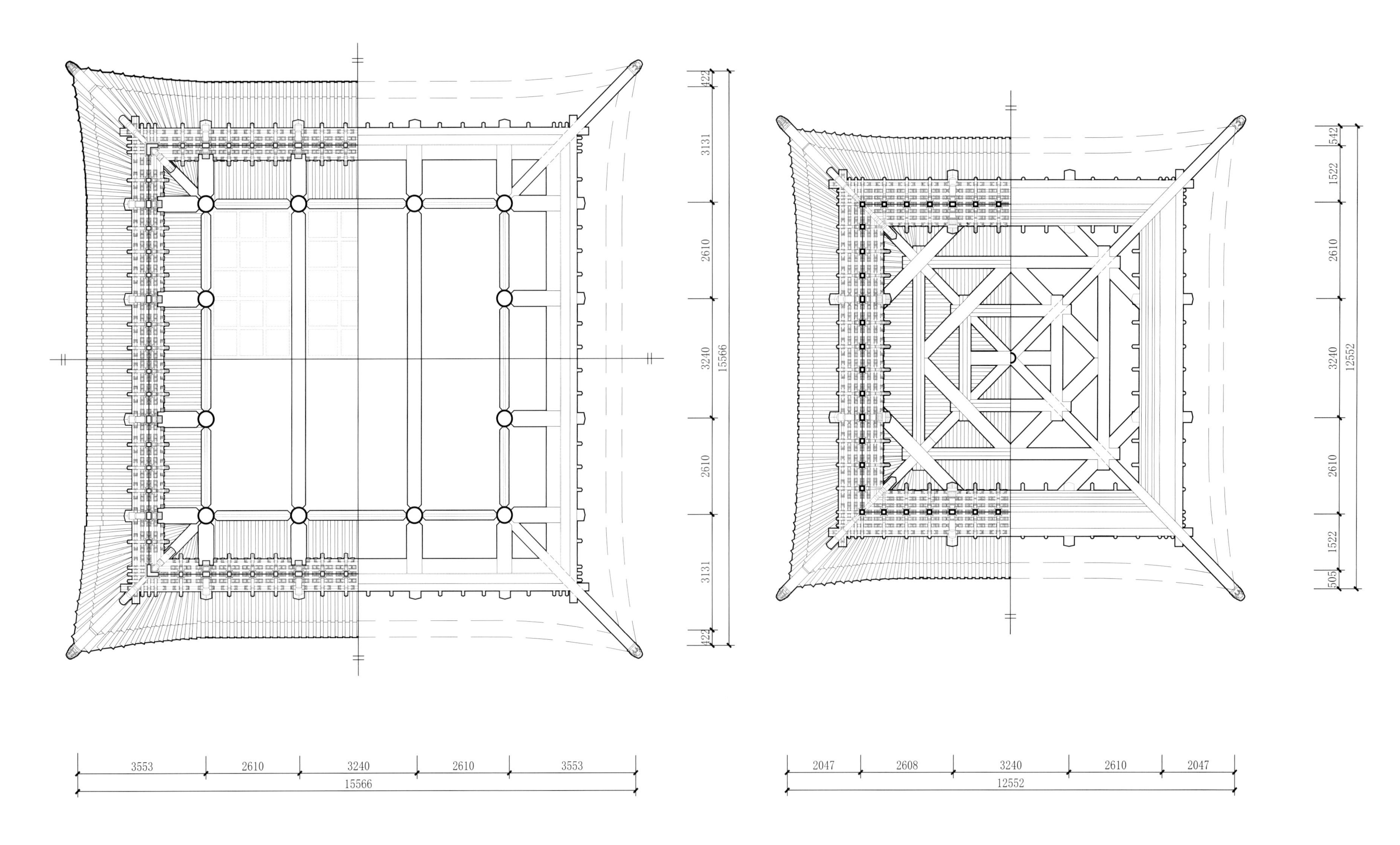

万春亭一重檐梁架仰俯视图

Bottom and vertical view of beams of double eave roof of Everlasting Spring Pavilion

万春亭上檐梁架仰俯视图

Bottom and vertical view of beams of upper eave of Everlasting Spring Pavilion

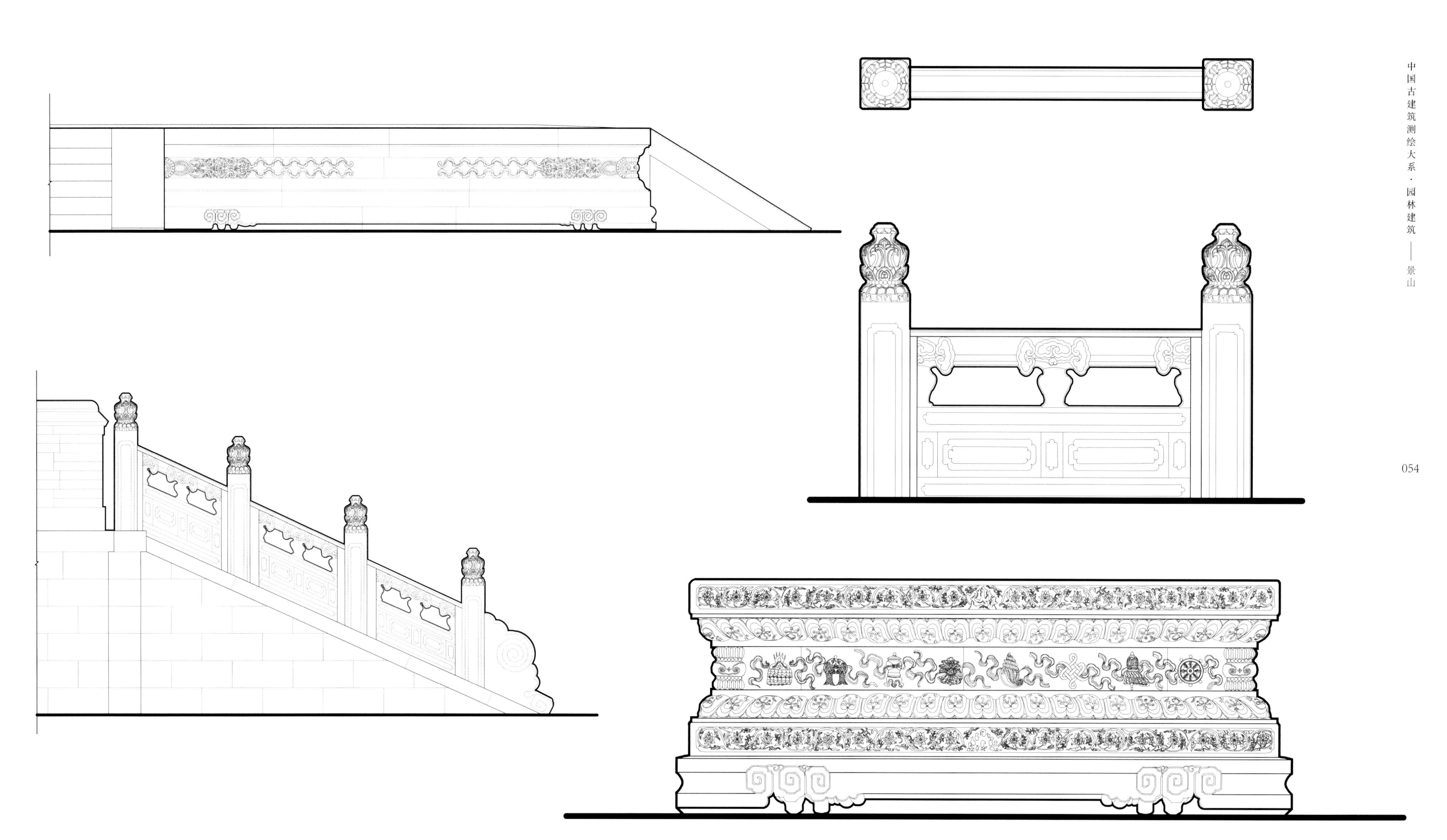

万春亭须弥座台基望柱栏板大样图
Detial drawing of stone Xumizuo stylobate and baluster and frieze panel of Everlasting Spring Pavilion

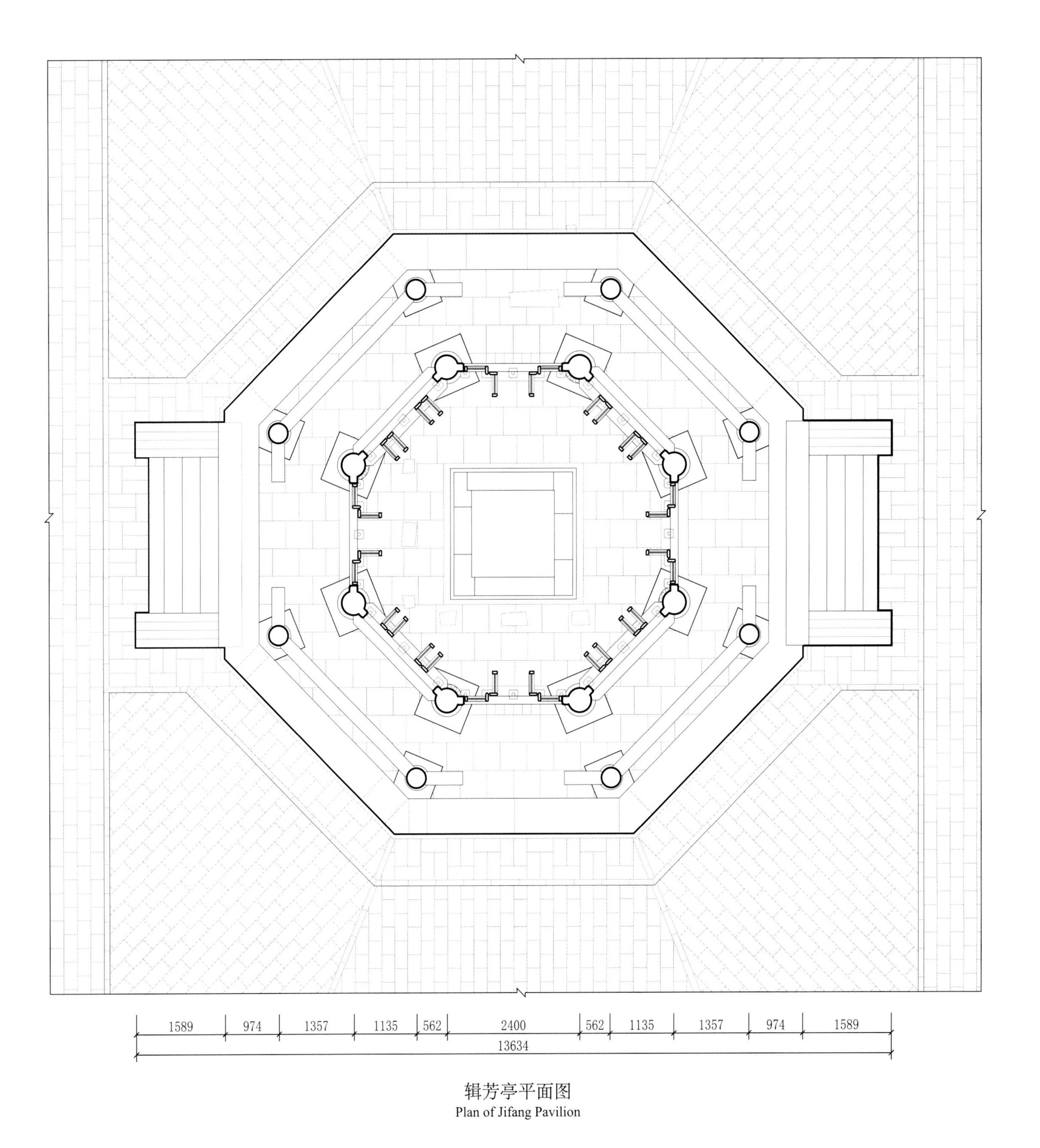

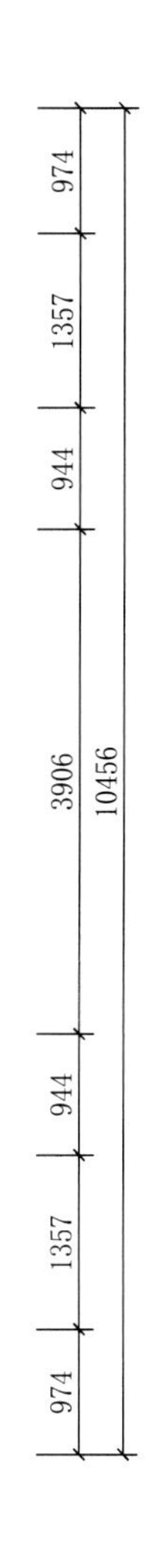

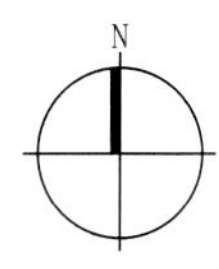

辑芳亭平面图

Plan of Jifang Pavilion

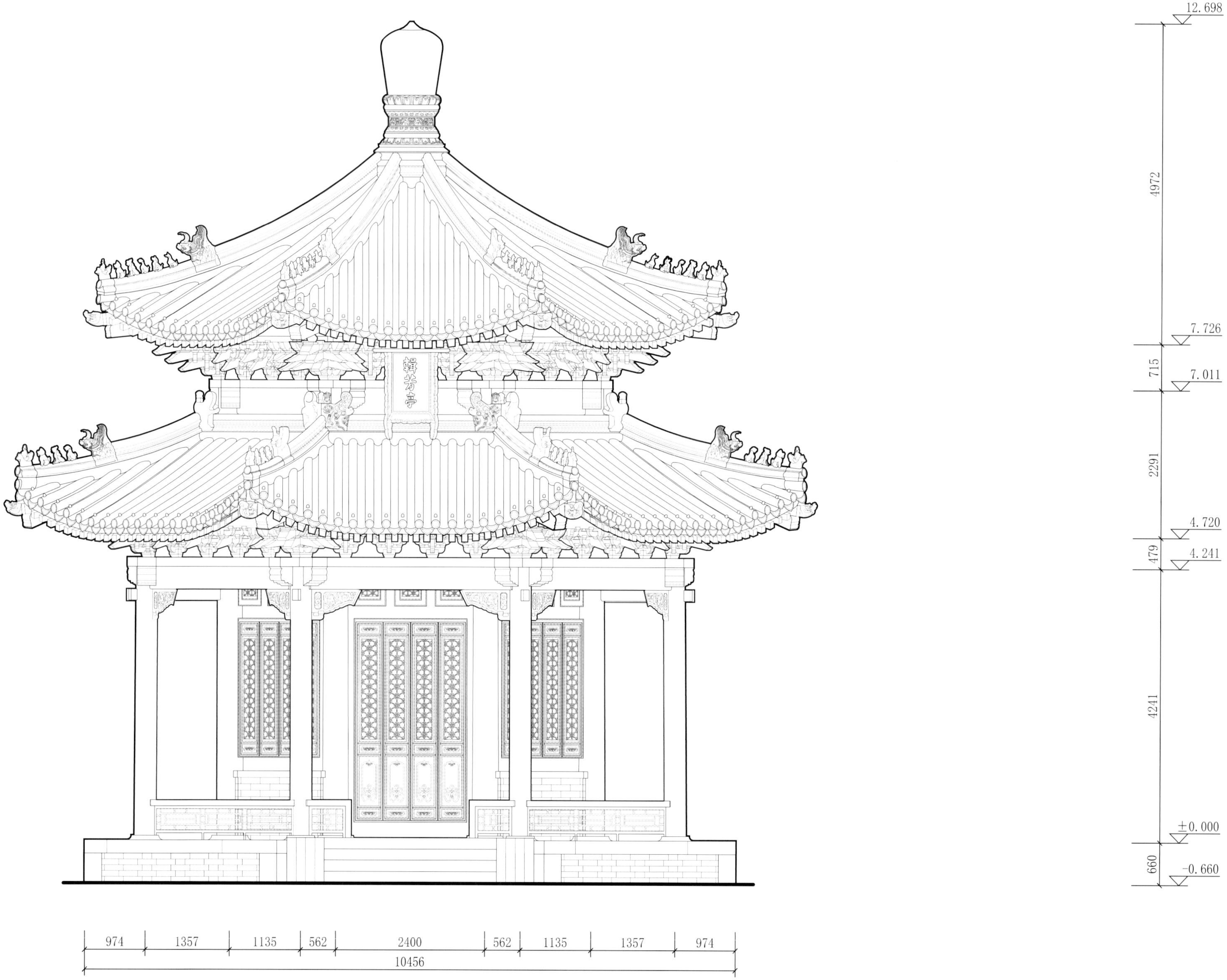

辑芳亭正立面图

Front elevation of Jifang Pavilion

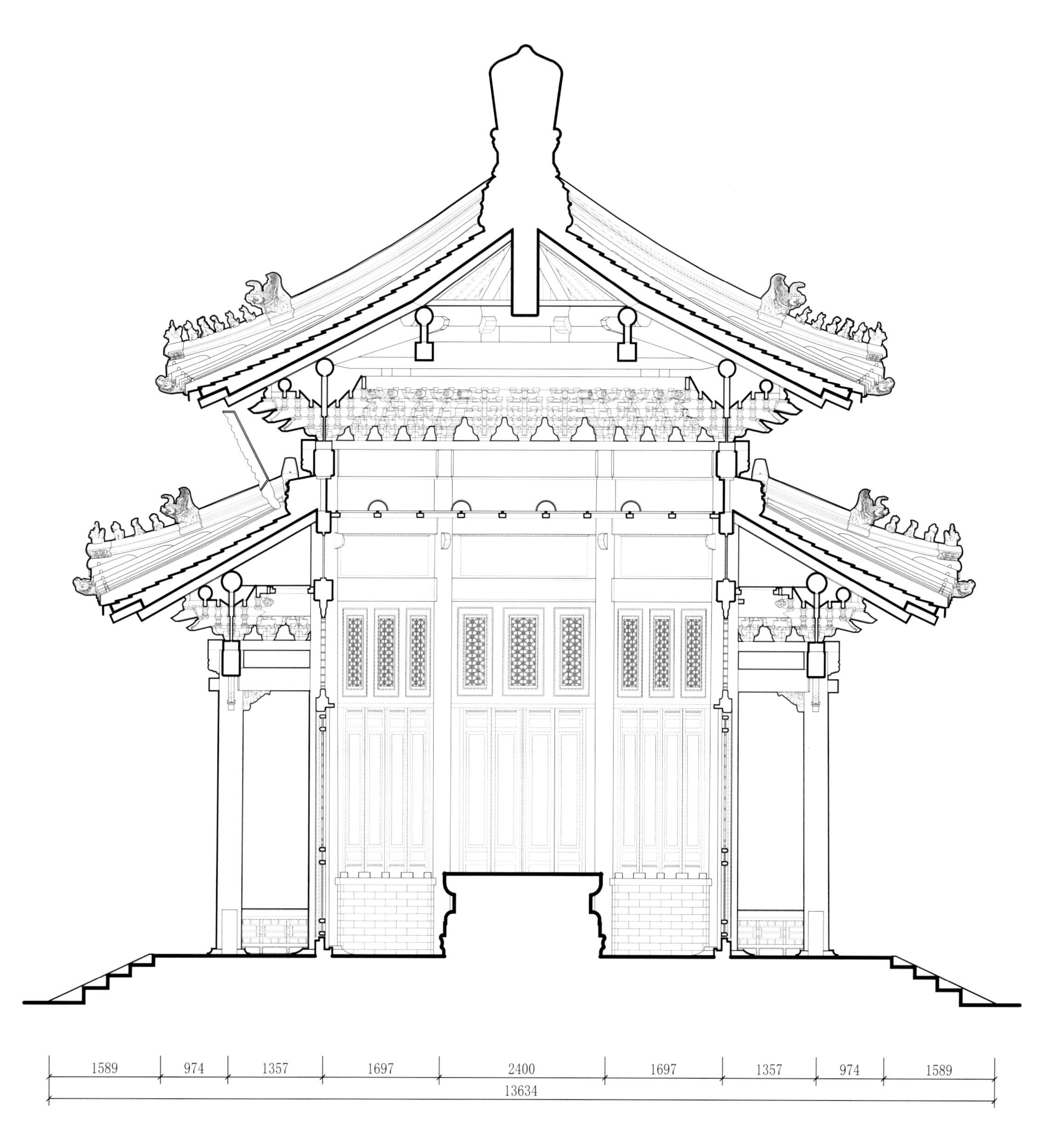

辑芳亭横剖面图

Central bay section of Jifang Pavilion

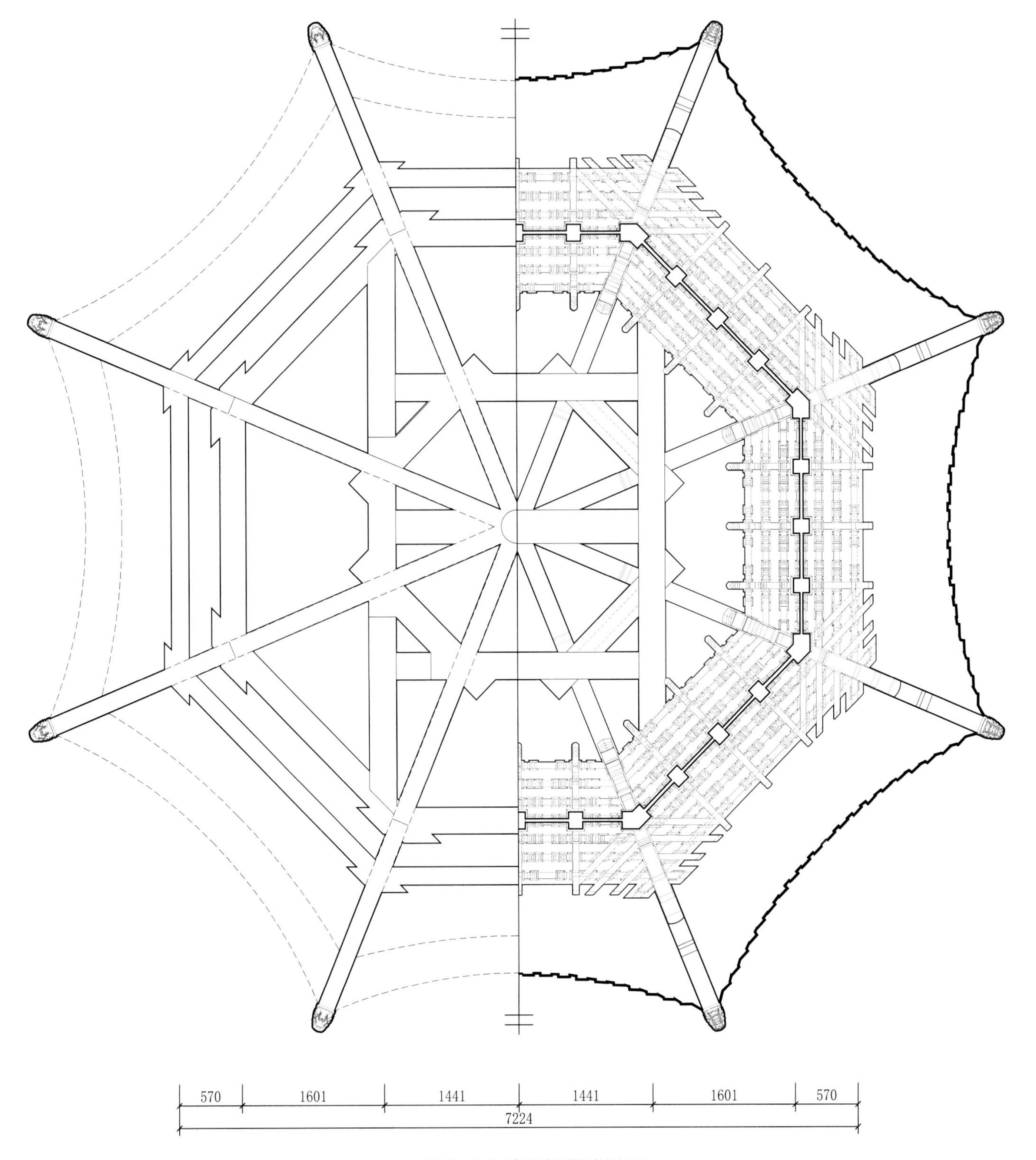

辑芳亭上檐梁架仰俯视图
Bottom and vertical view of beams of upper eave of Jifang Pavilion

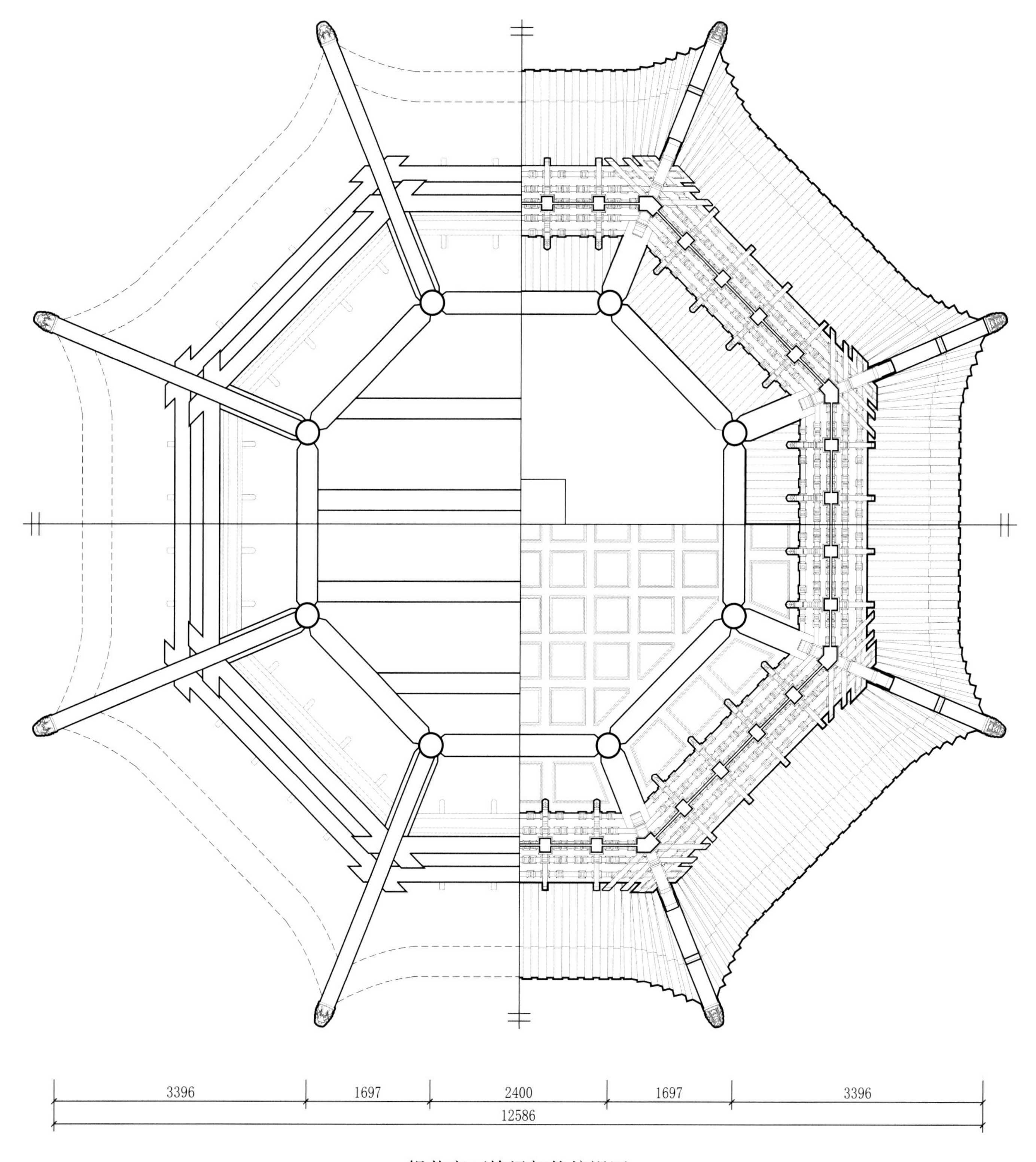

辑芳亭下檐梁架仰俯视图

Bottom and vertical view of beams of under eave of Jifang Pavilion

辑芳亭屋顶平面图
Roof Plan of Jifang Pavilion

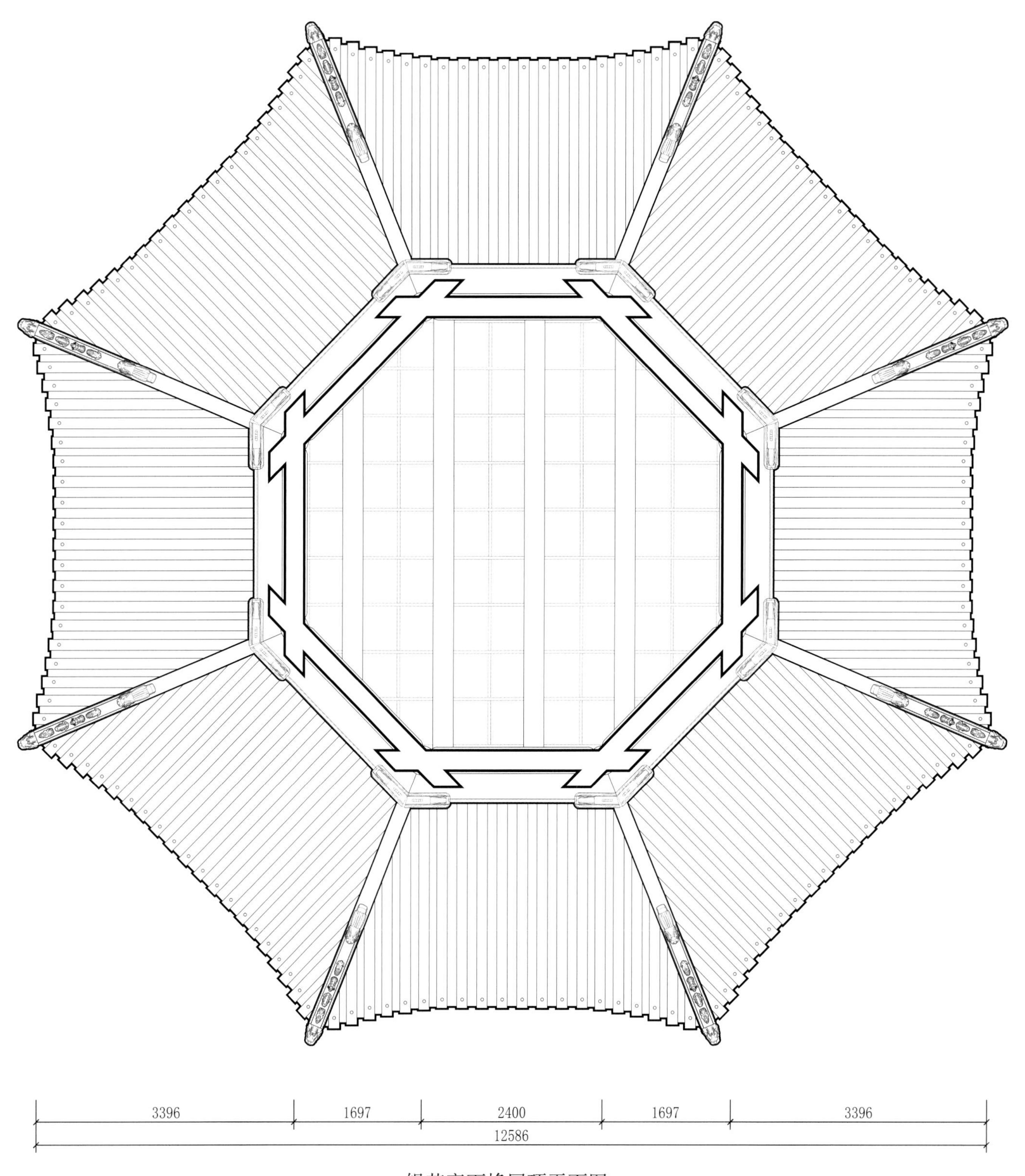

辑芳亭下檐屋顶平面图
Roof Plan of under eave of Jifang Pavilion

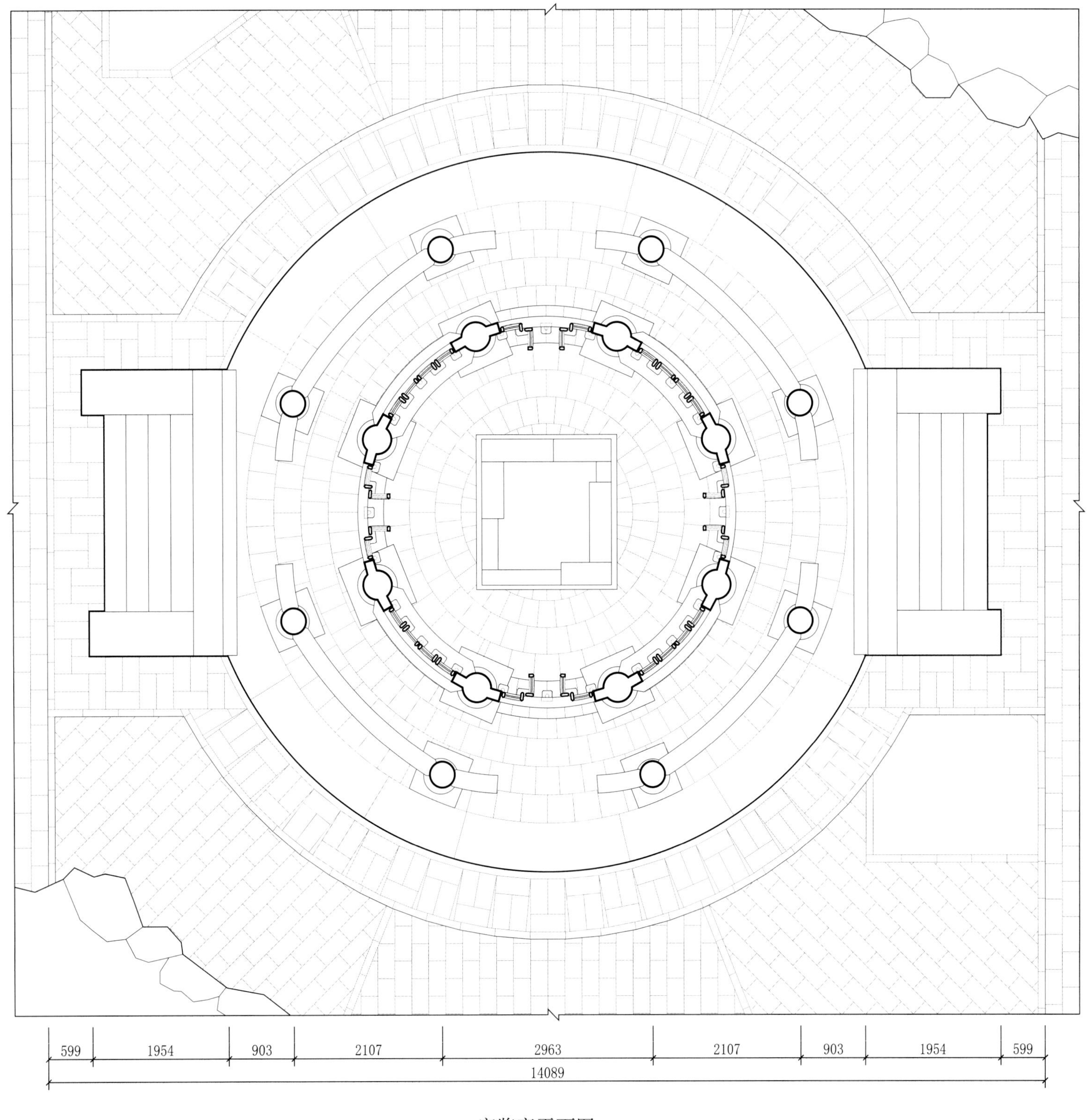

1320
1627
468
2963
9793
468
1627
1320

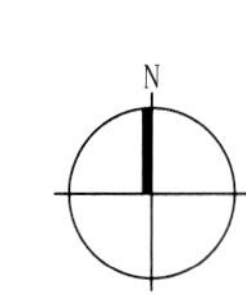

富览亭平面图
Plan of Fulan Pavilion

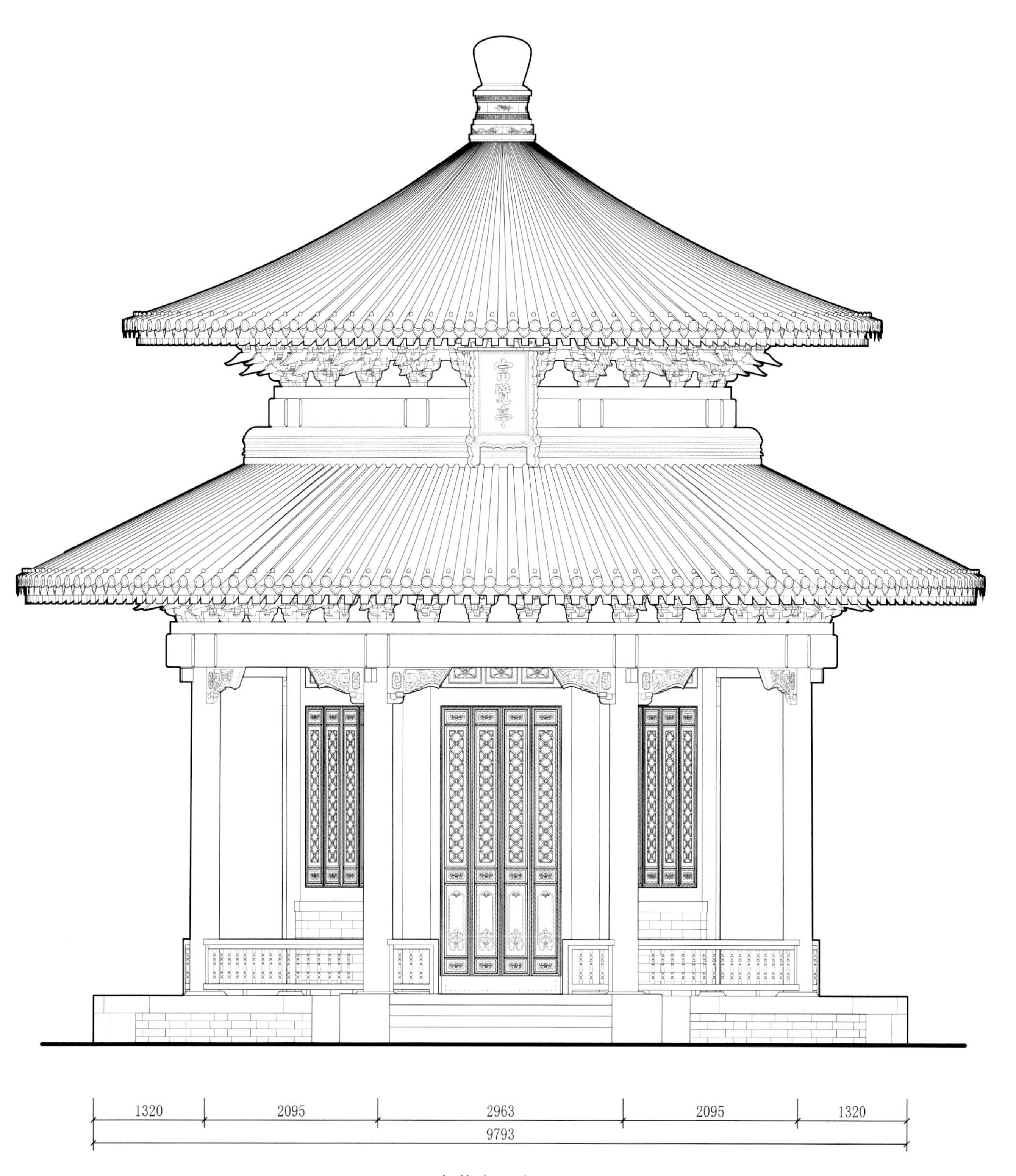

富览亭正立面图

Front elevation of Fulan Pavilion

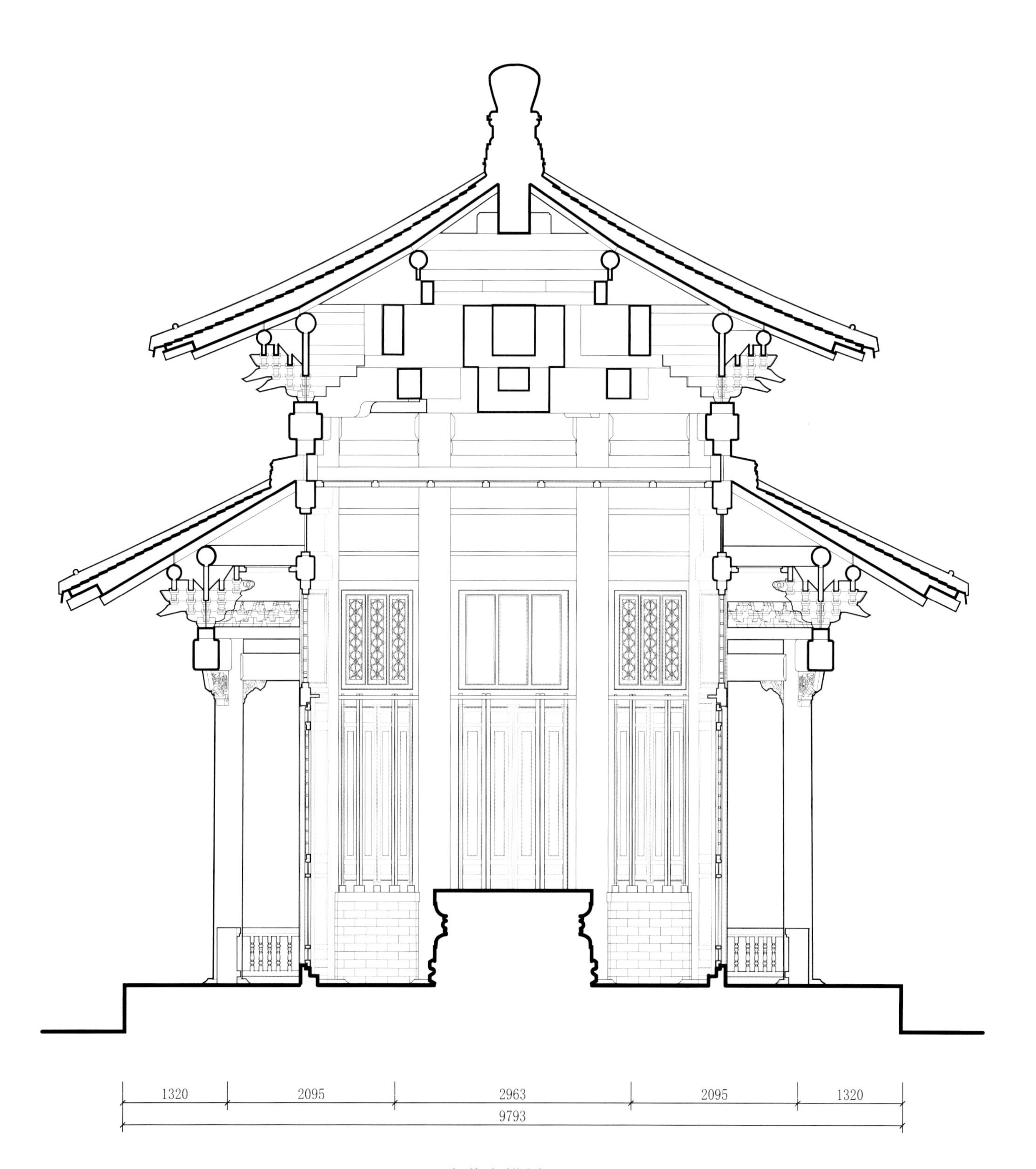

富览亭横剖面图
Central bay section of Fulan Pavilion

11.154
2505
8.649
1090
7.559
650
6.909
1860
5.049
526
4.523
3418
1.105
1105
±0.000
561
-0.561

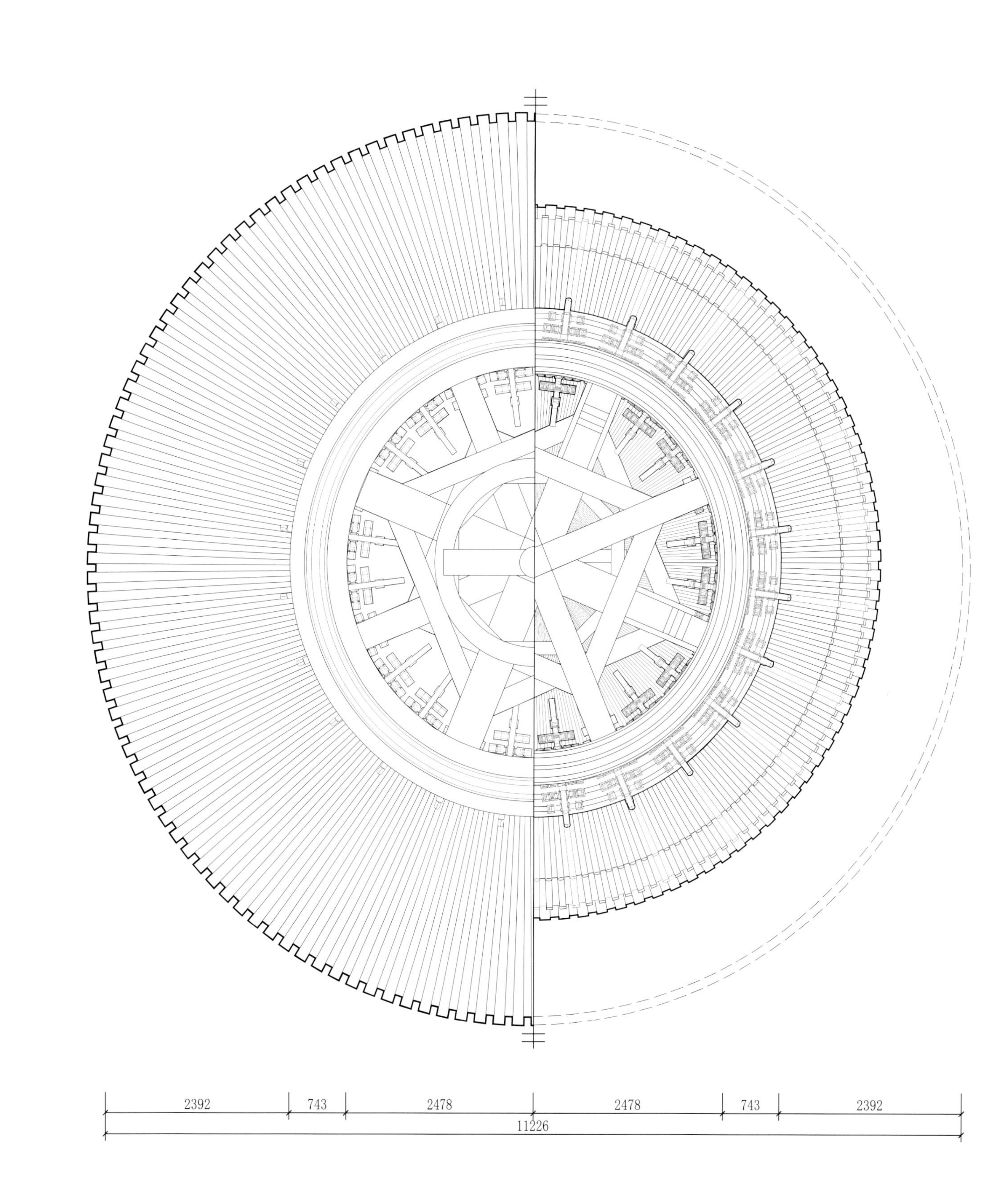

富览亭上檐梁架仰俯视图
Bottom and vertical view of beams of upper eave of Fulan Pavilion

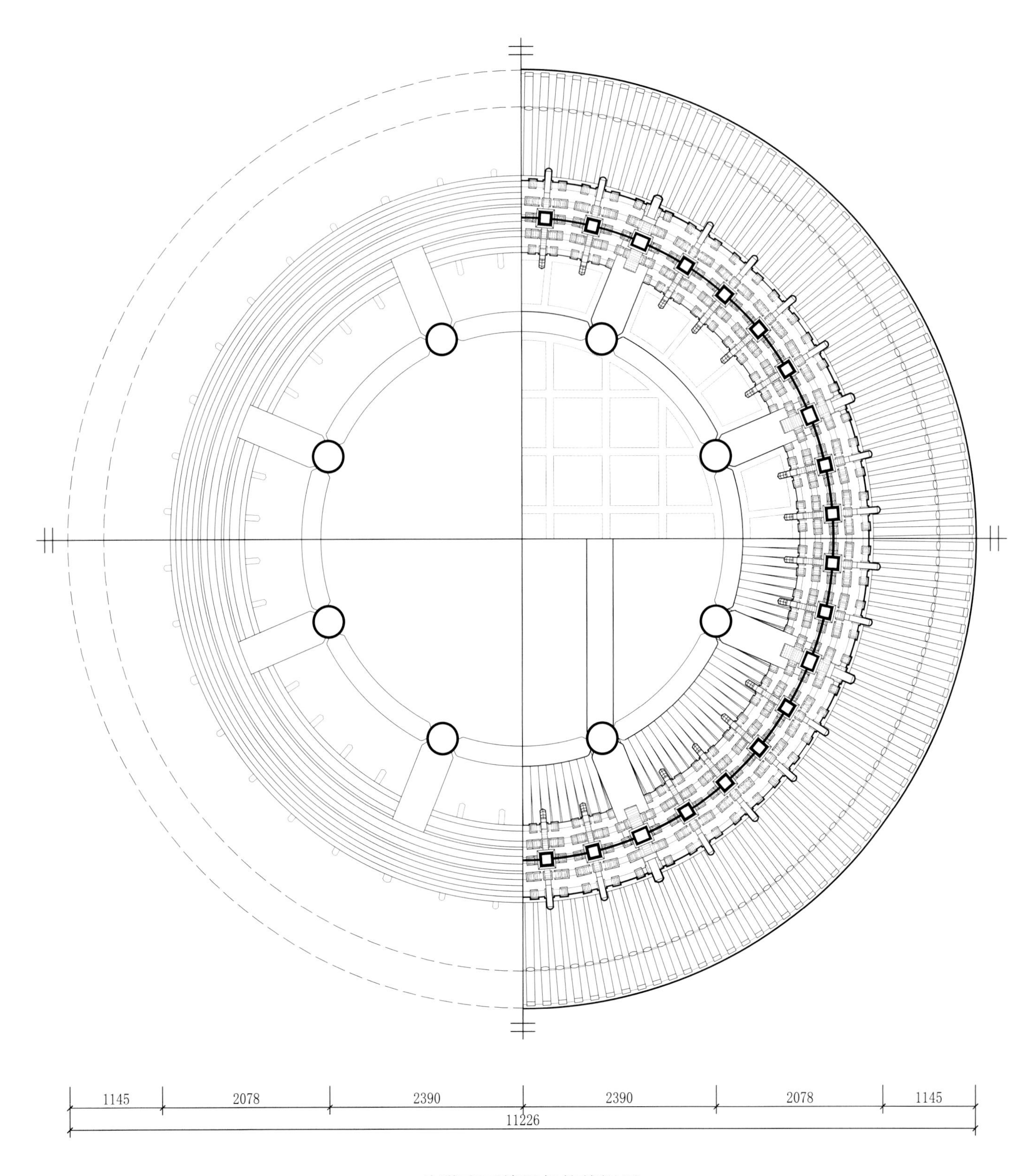

1145
2078
2390
2390
2078
1145
11226

富览亭下檐梁架仰俯视图
Bottom and vertical view of beams of under eave of Fulan Pavilion

楼阁
Building

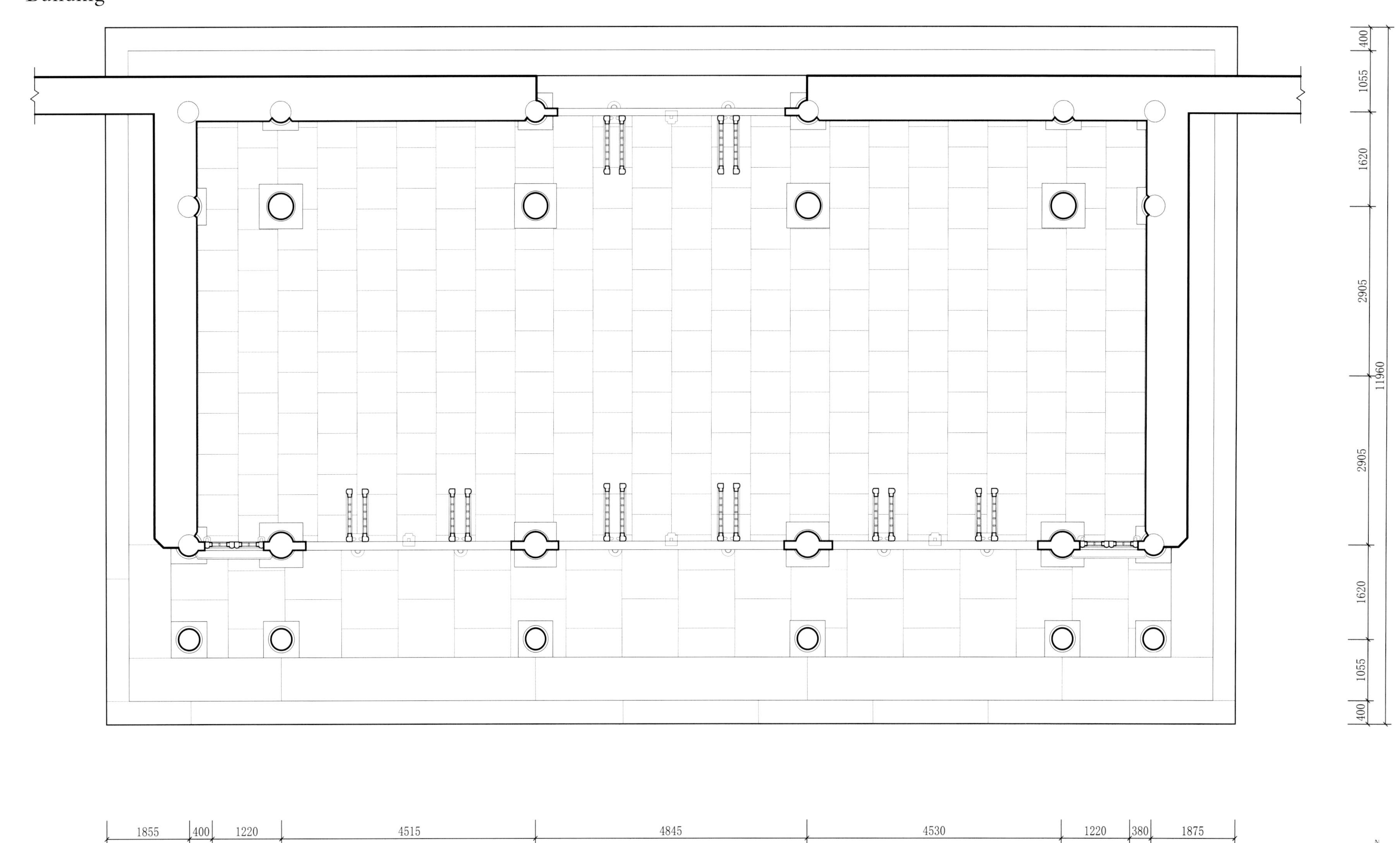

绮望楼一层平面图
Ground floor of the Qiwang Building

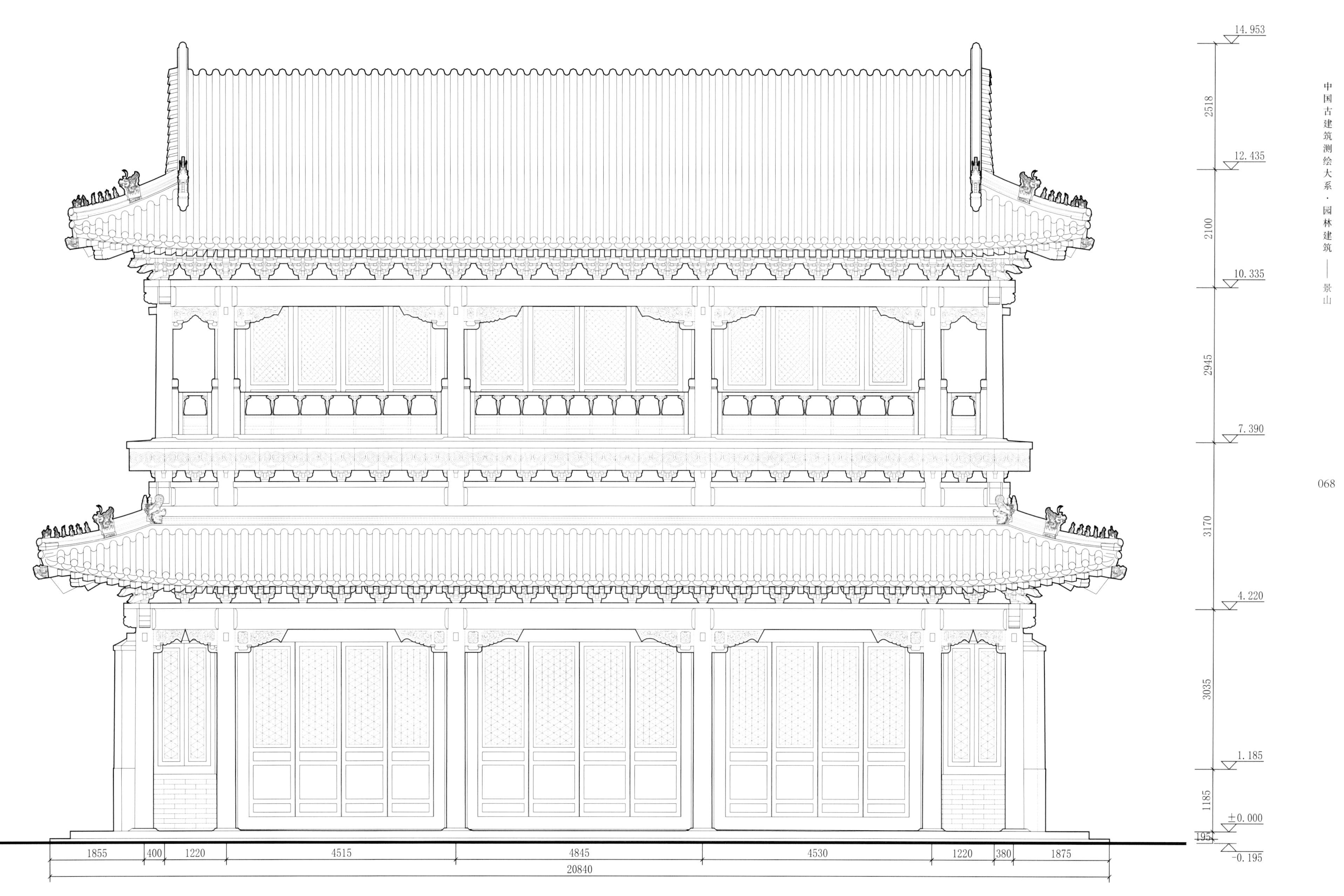

绮望楼南立面图

South Elevation of the Qiwang Building

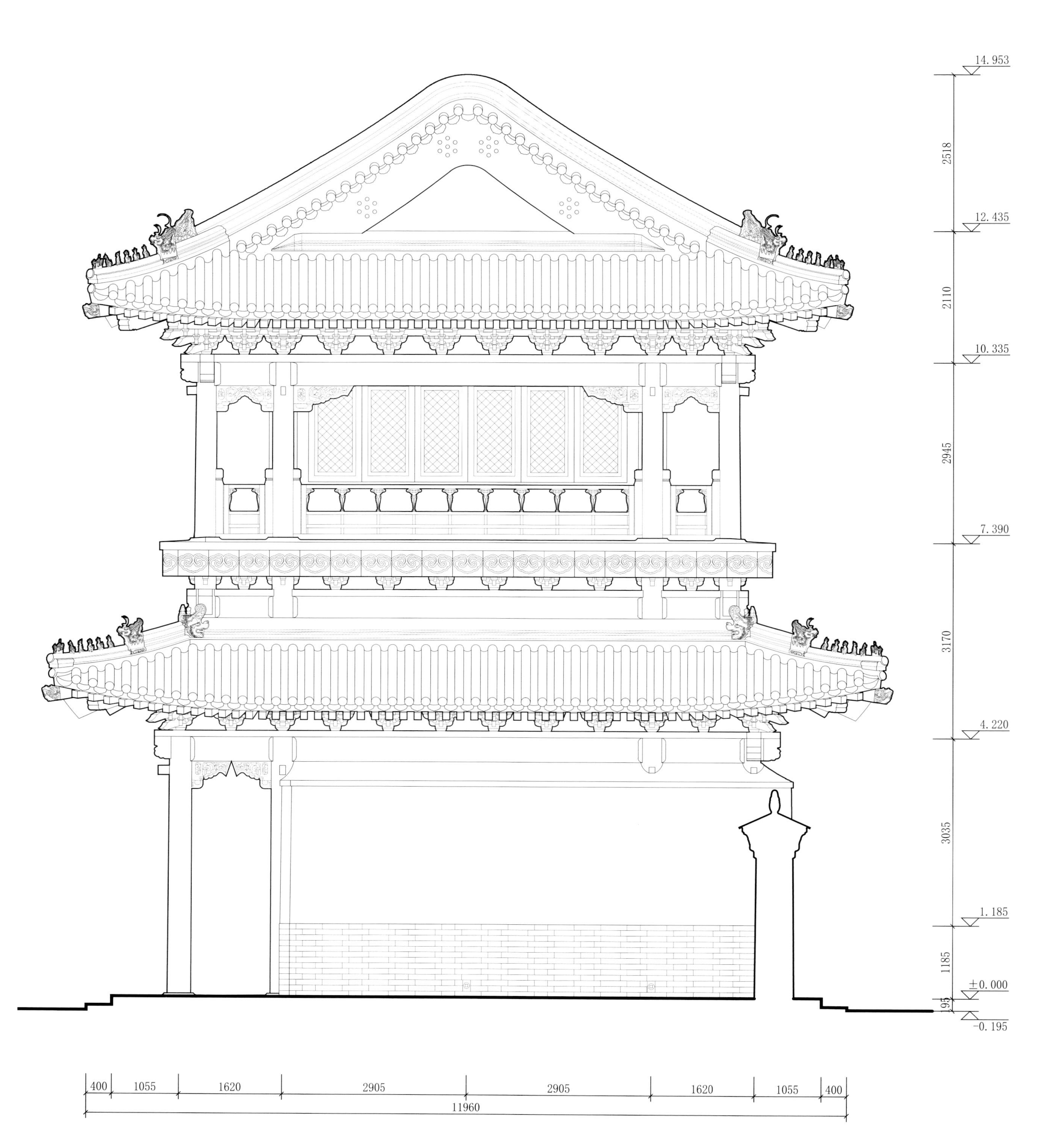

绮望楼东立面图
East Elevation of the Qiwang Building

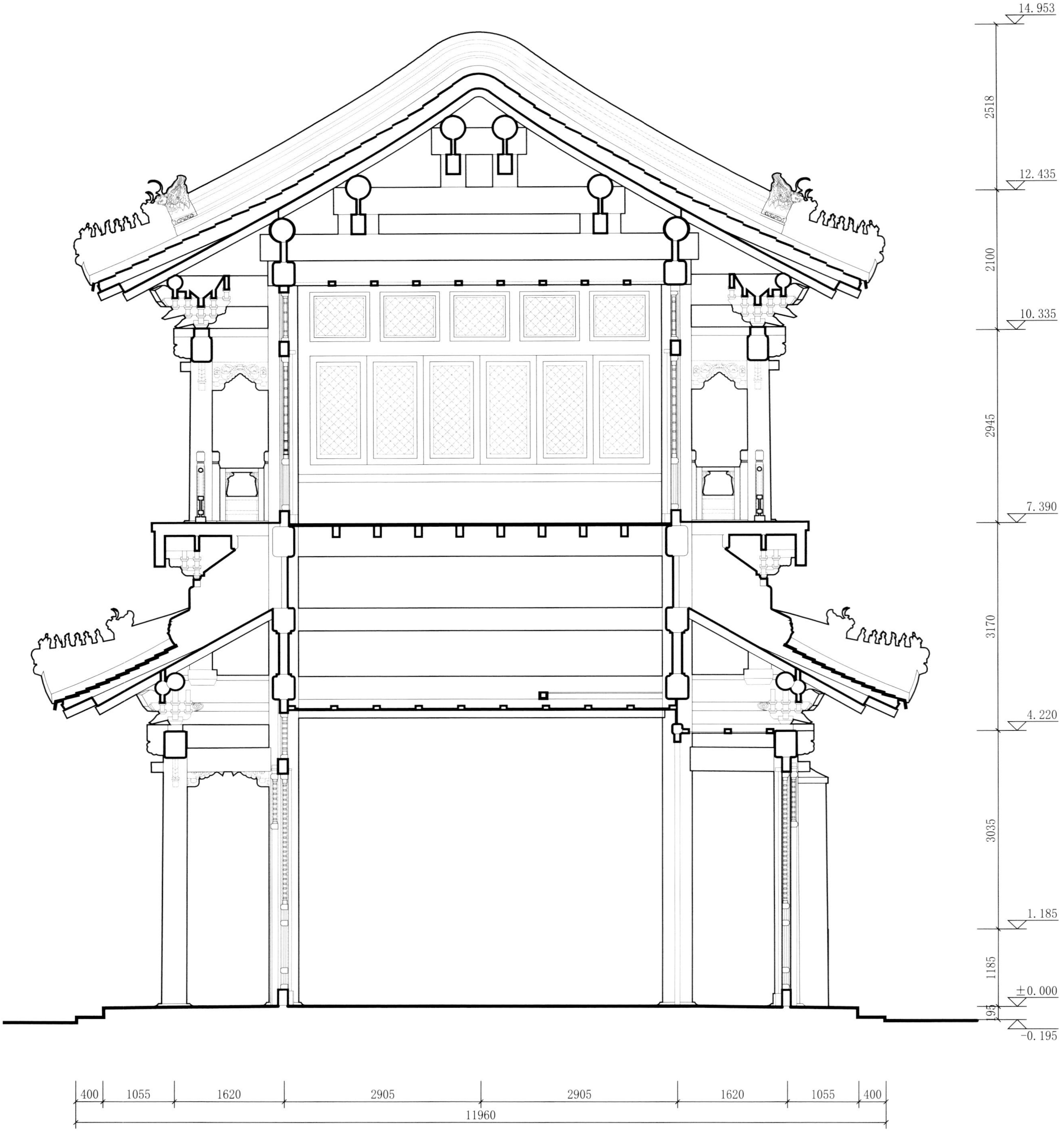

绮望楼明间横剖面图

Central bay section of the Qiwang Building

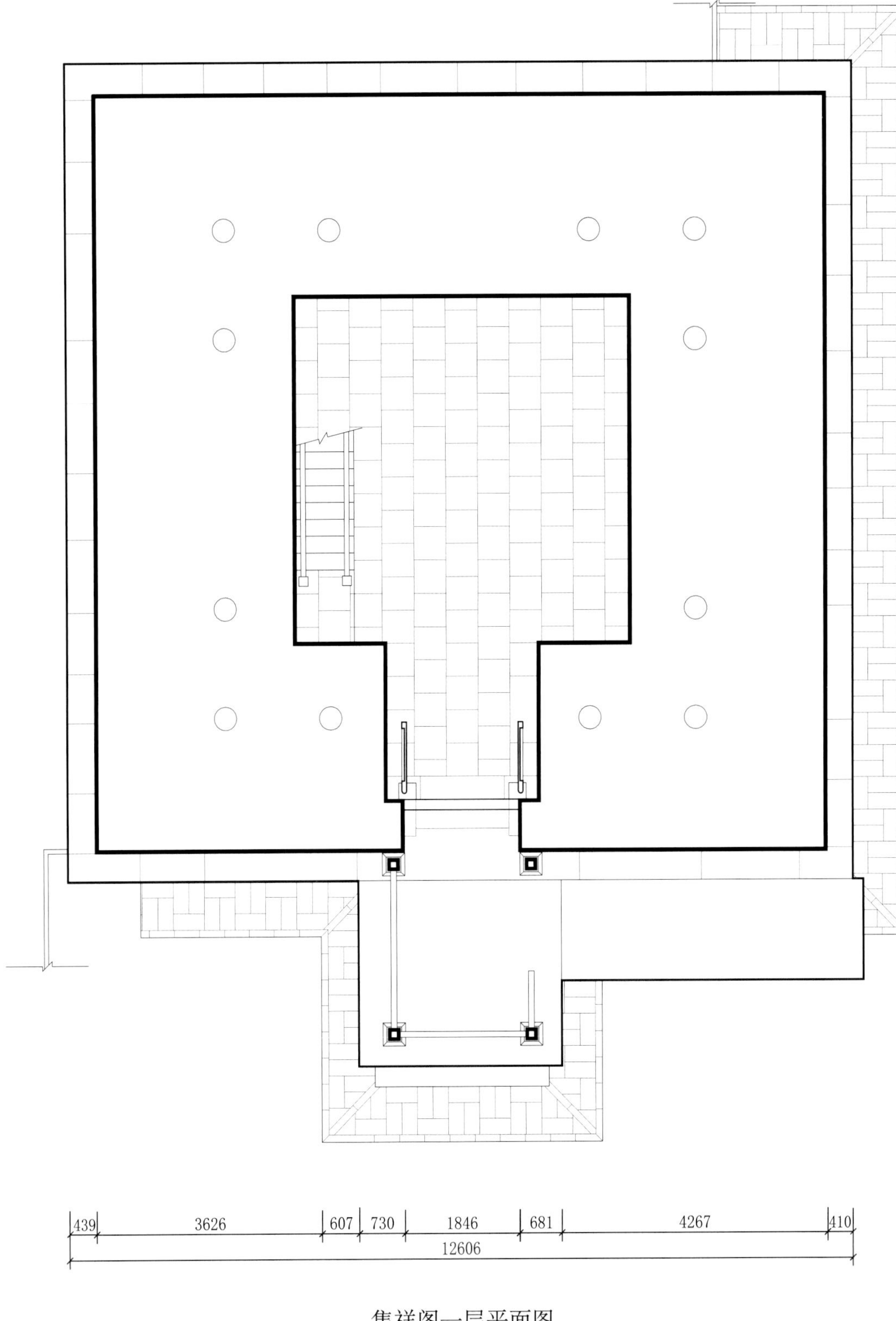

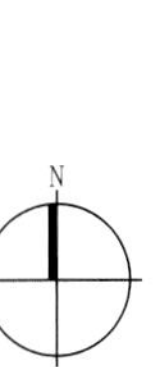

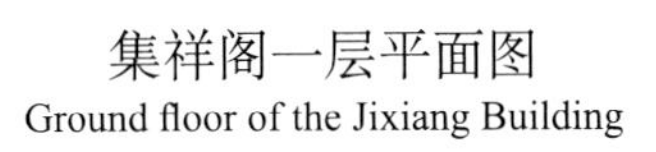

集祥阁一层平面图
Ground floor of the Jixiang Building

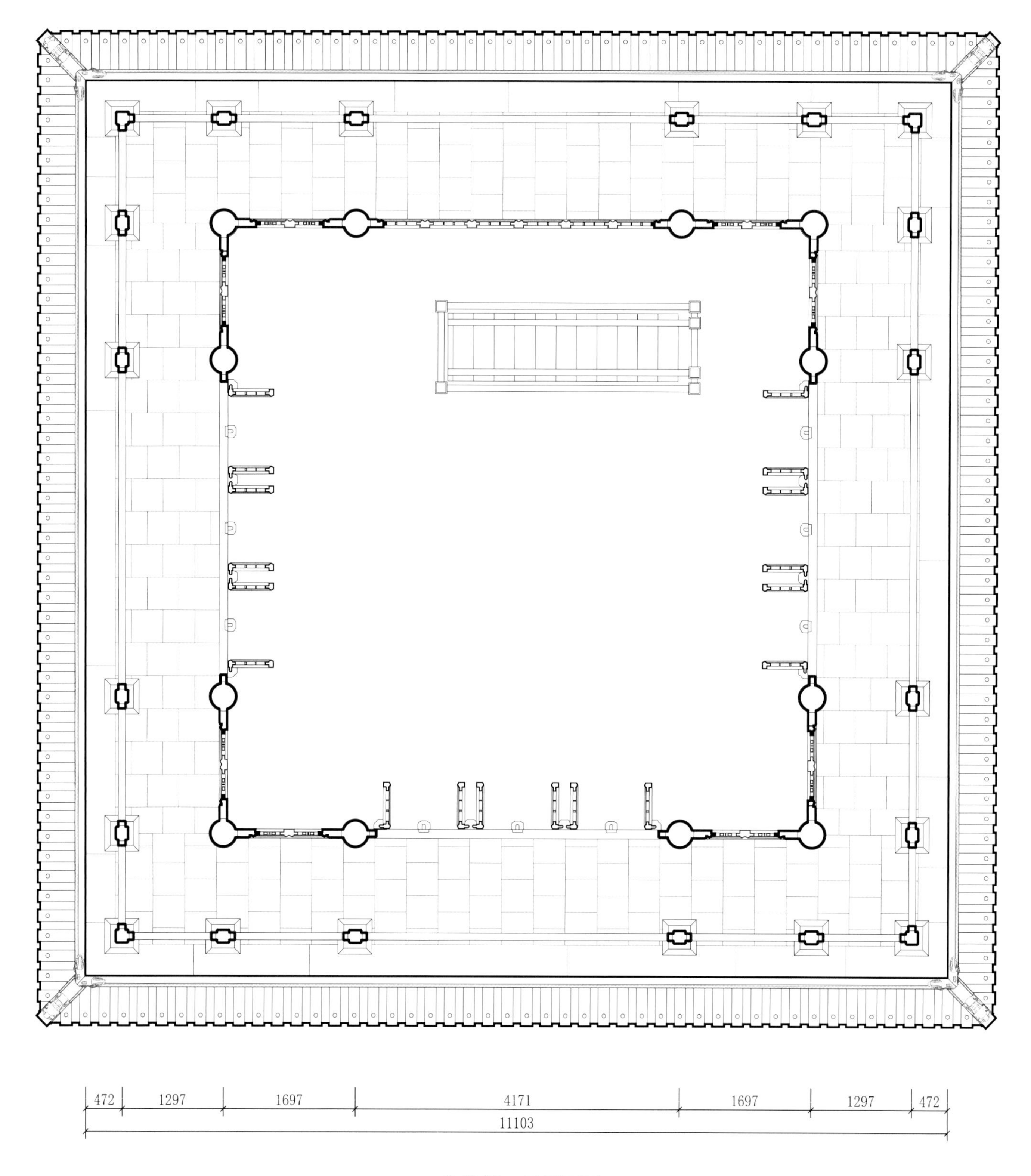

集祥阁二层平面图
Second floor of the Jixiang Building

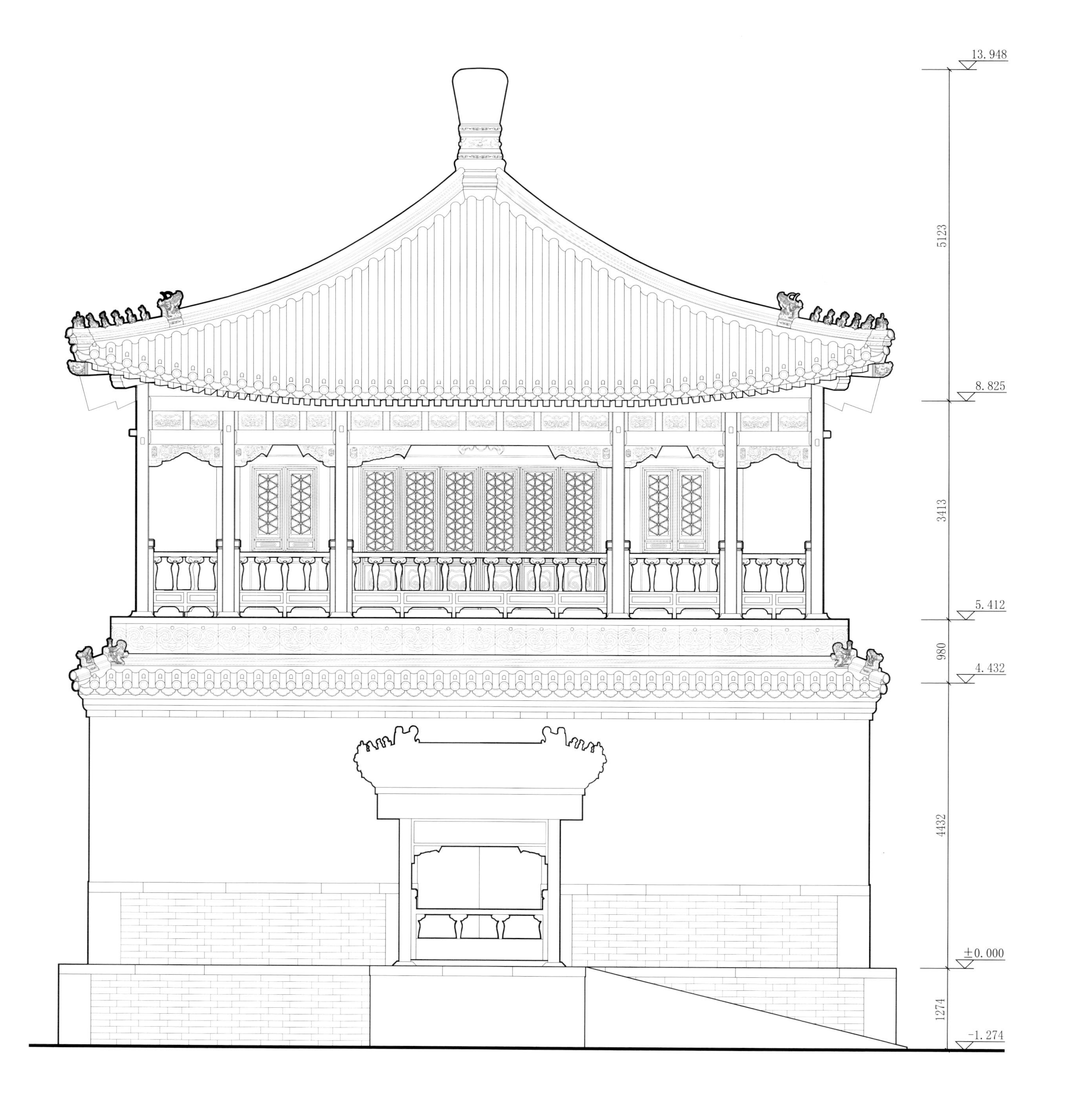

集祥阁南立面图
South Elevation of the Jixiang Building

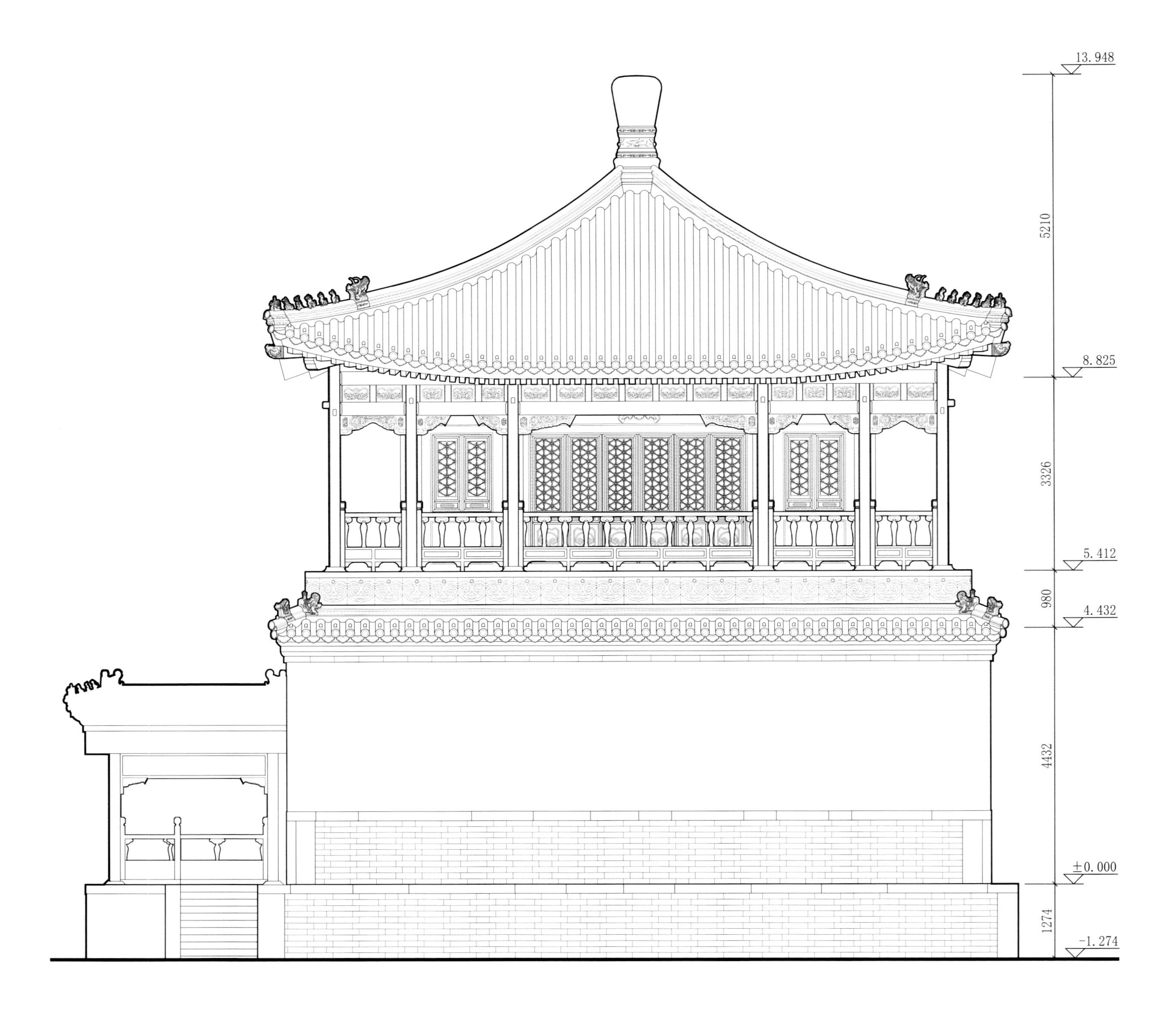

集祥阁东立面图
East Elevation of the Jixiang Building

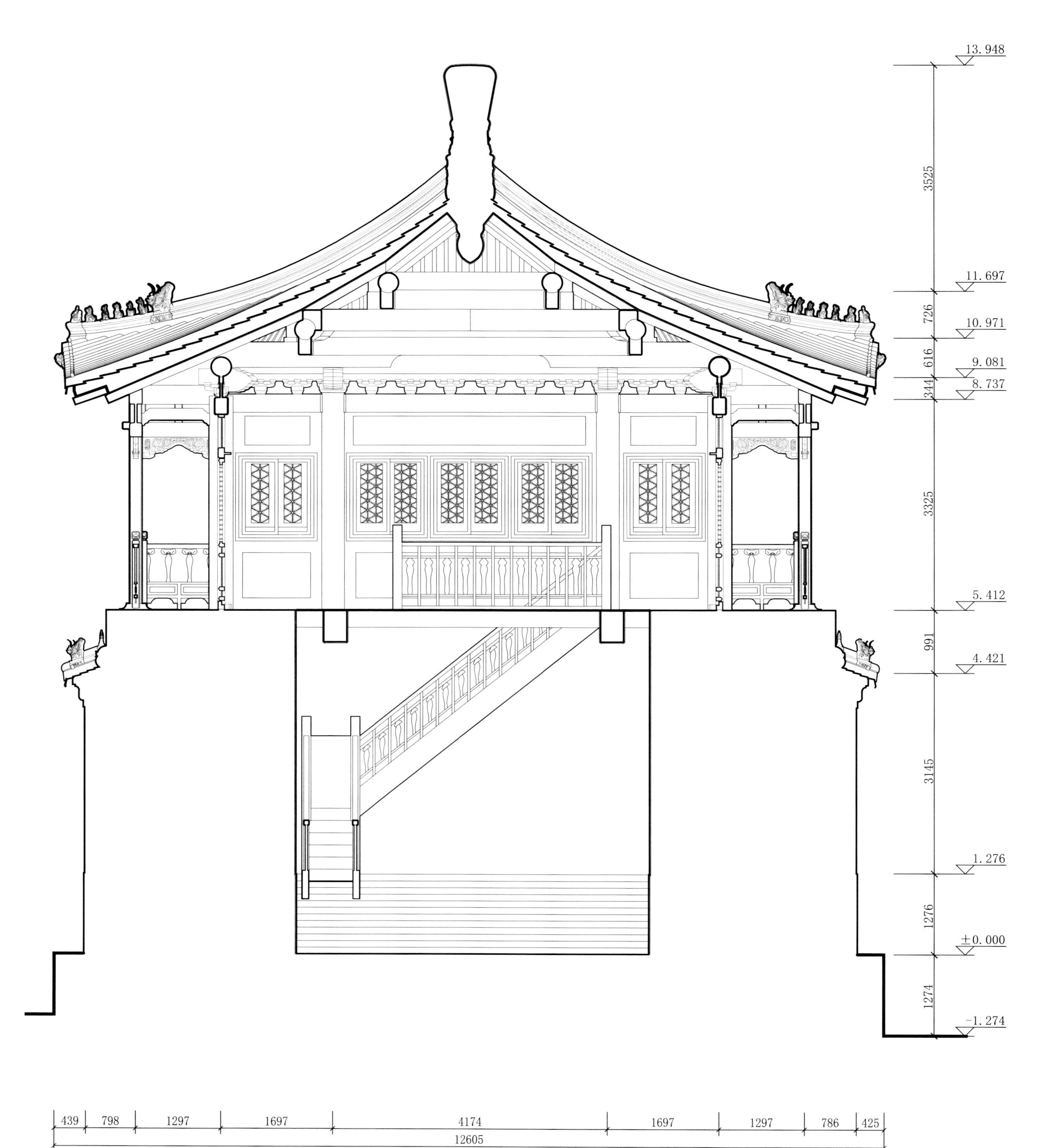

集祥阁剖面图（一）
Section I of the Jixiang Building

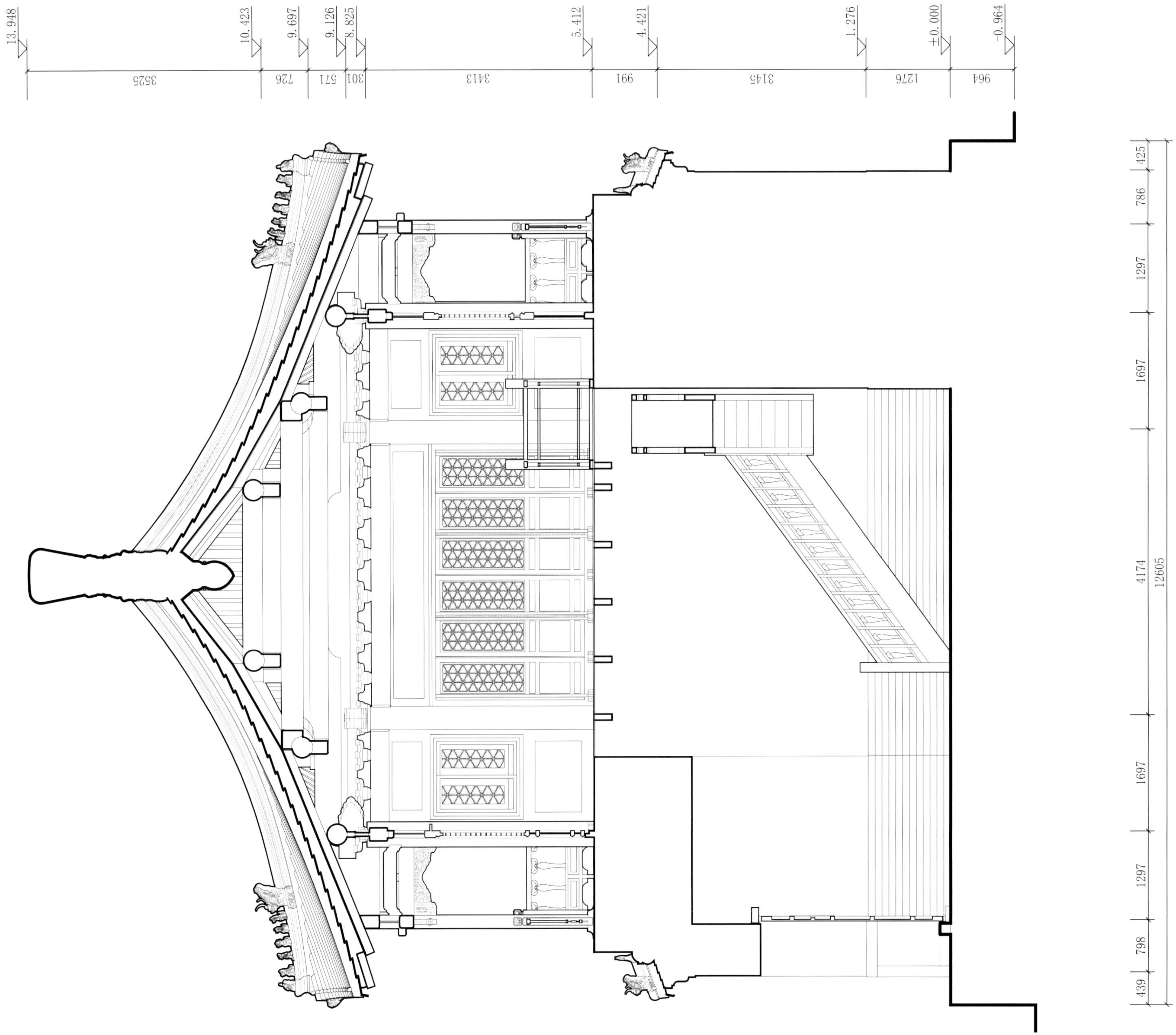

集祥阁剖面图（二）
Section II of the Jixiang Building

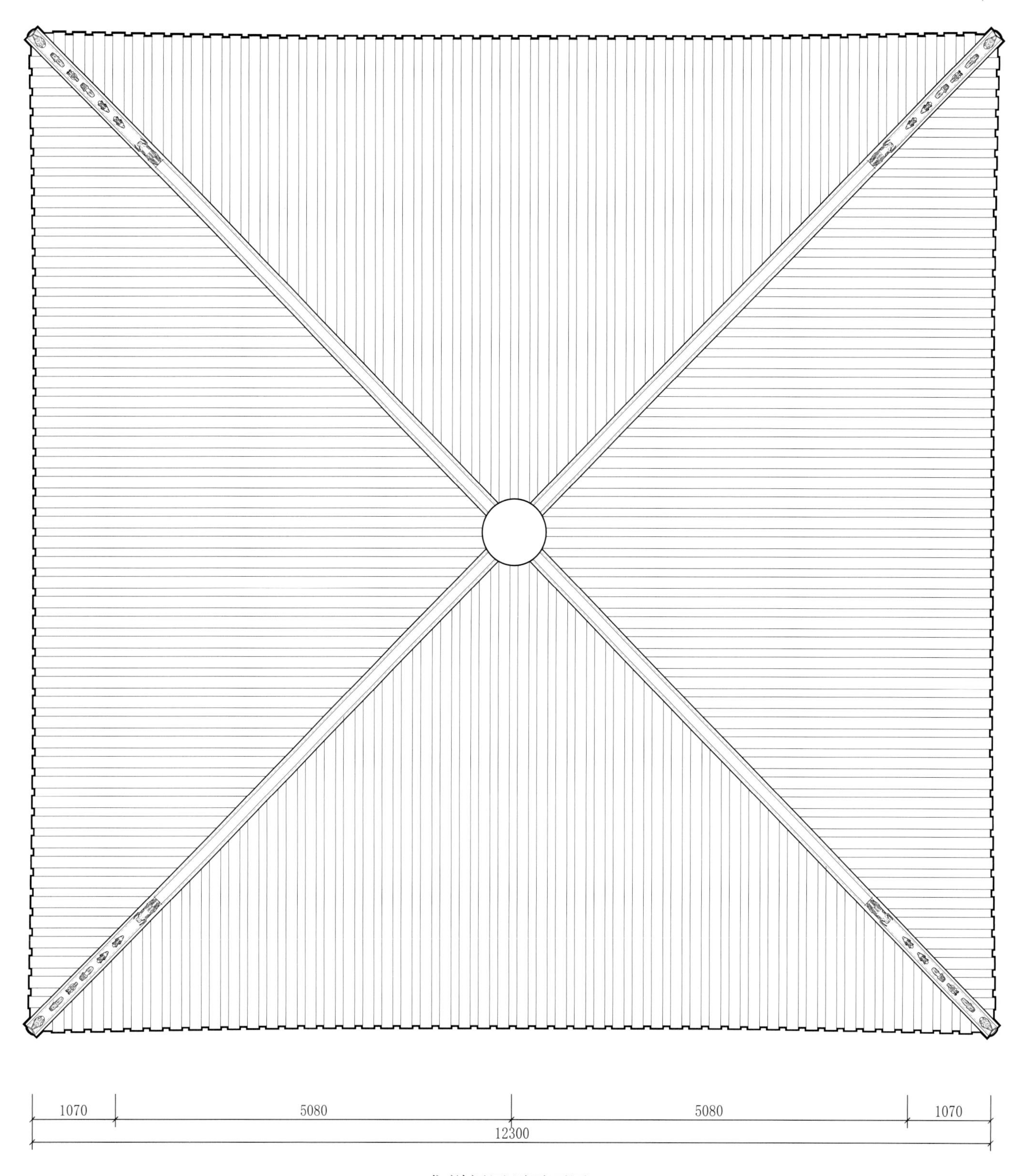

集祥阁屋顶平面图
Roof plan of the Jixiang Building

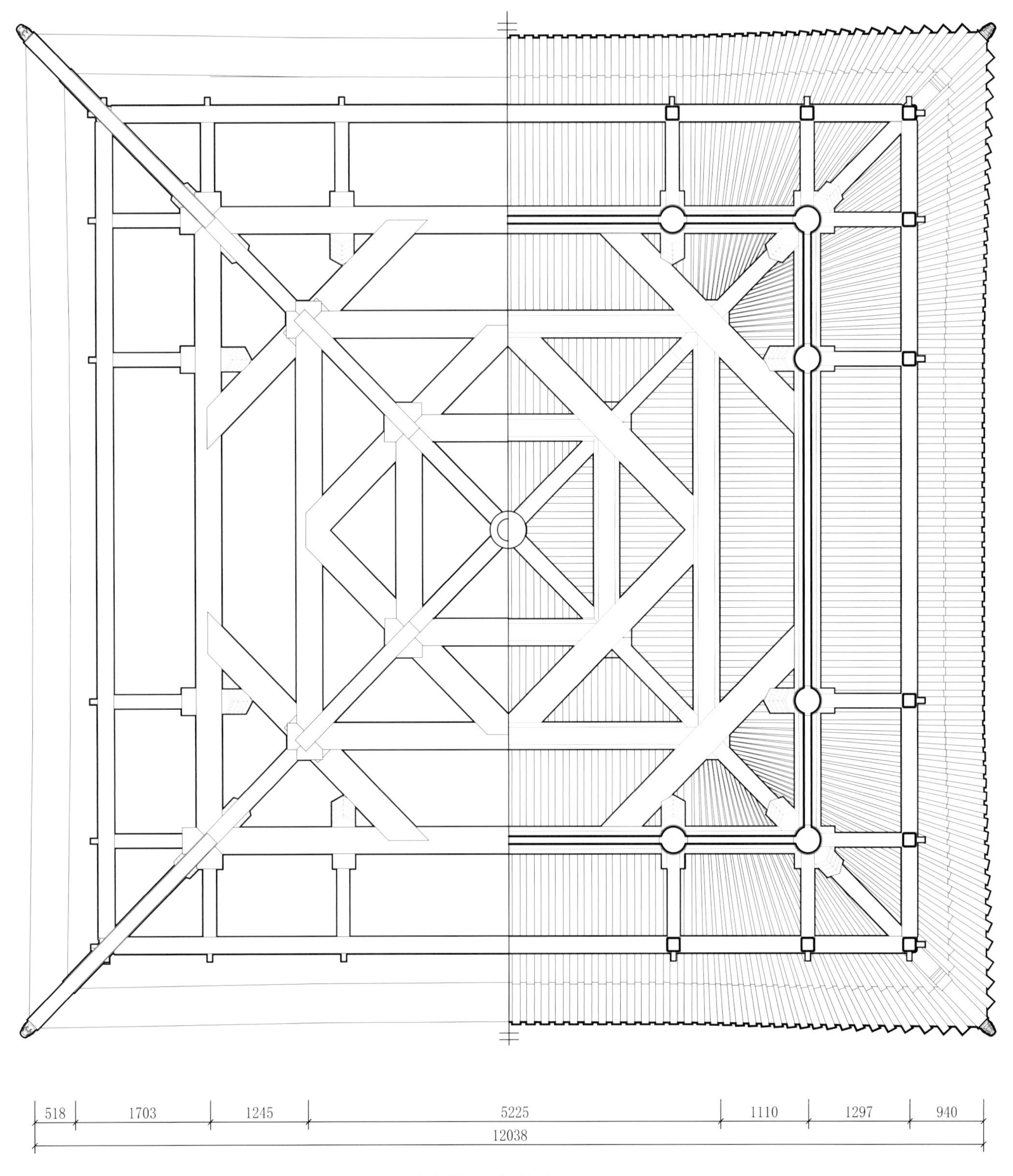

集祥阁梁架仰俯视图
Bottom and vertical view of beams of the Jixiang Building

寿皇殿建筑群

The Groups of Shouhuang Hall

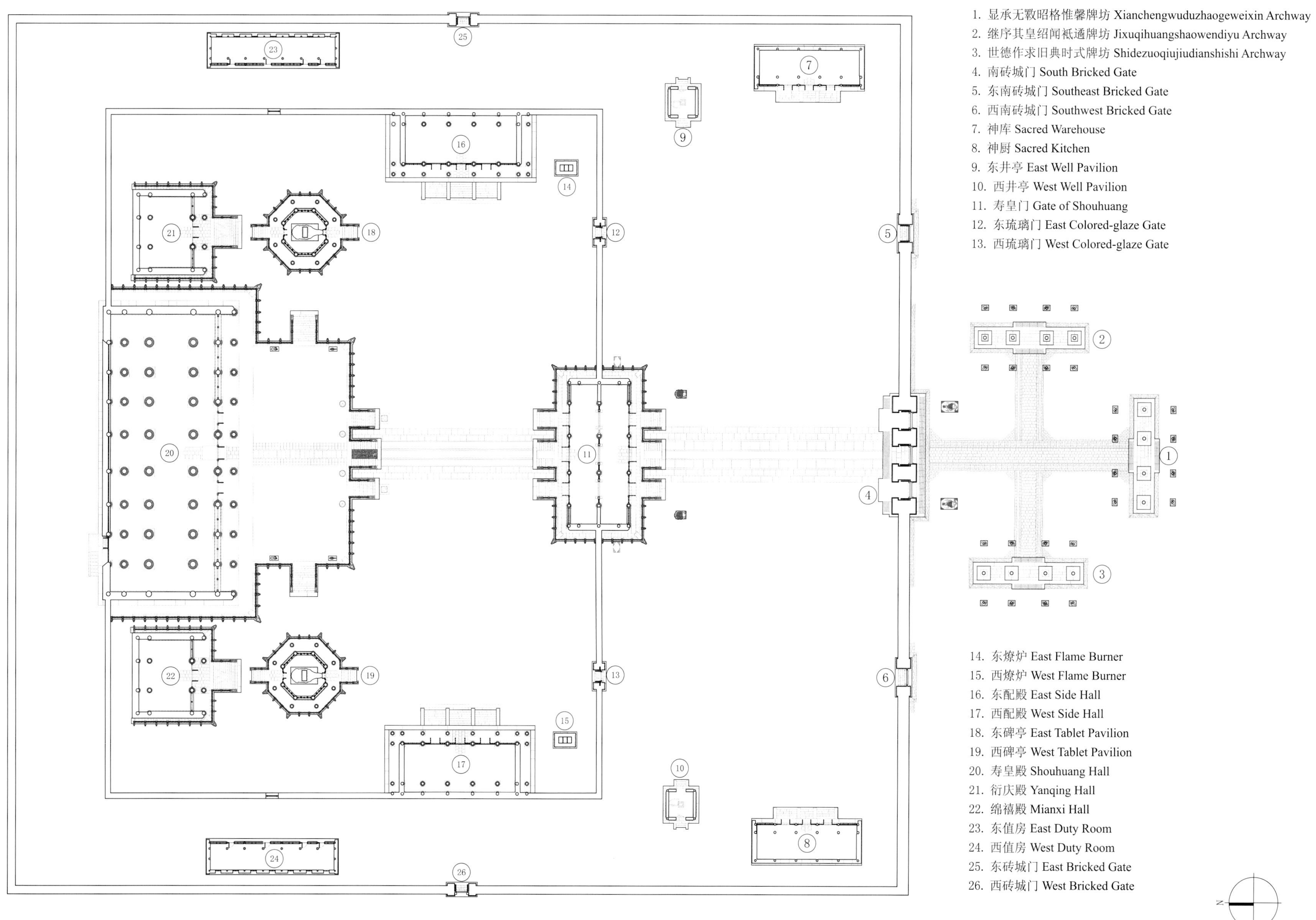

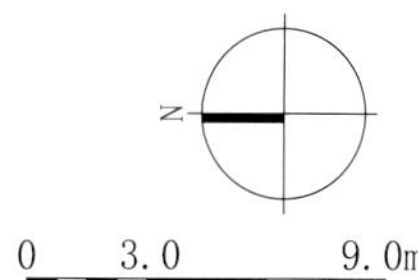

寿皇殿建筑群平面图
Plan of the Groups of Shouhuang Hall

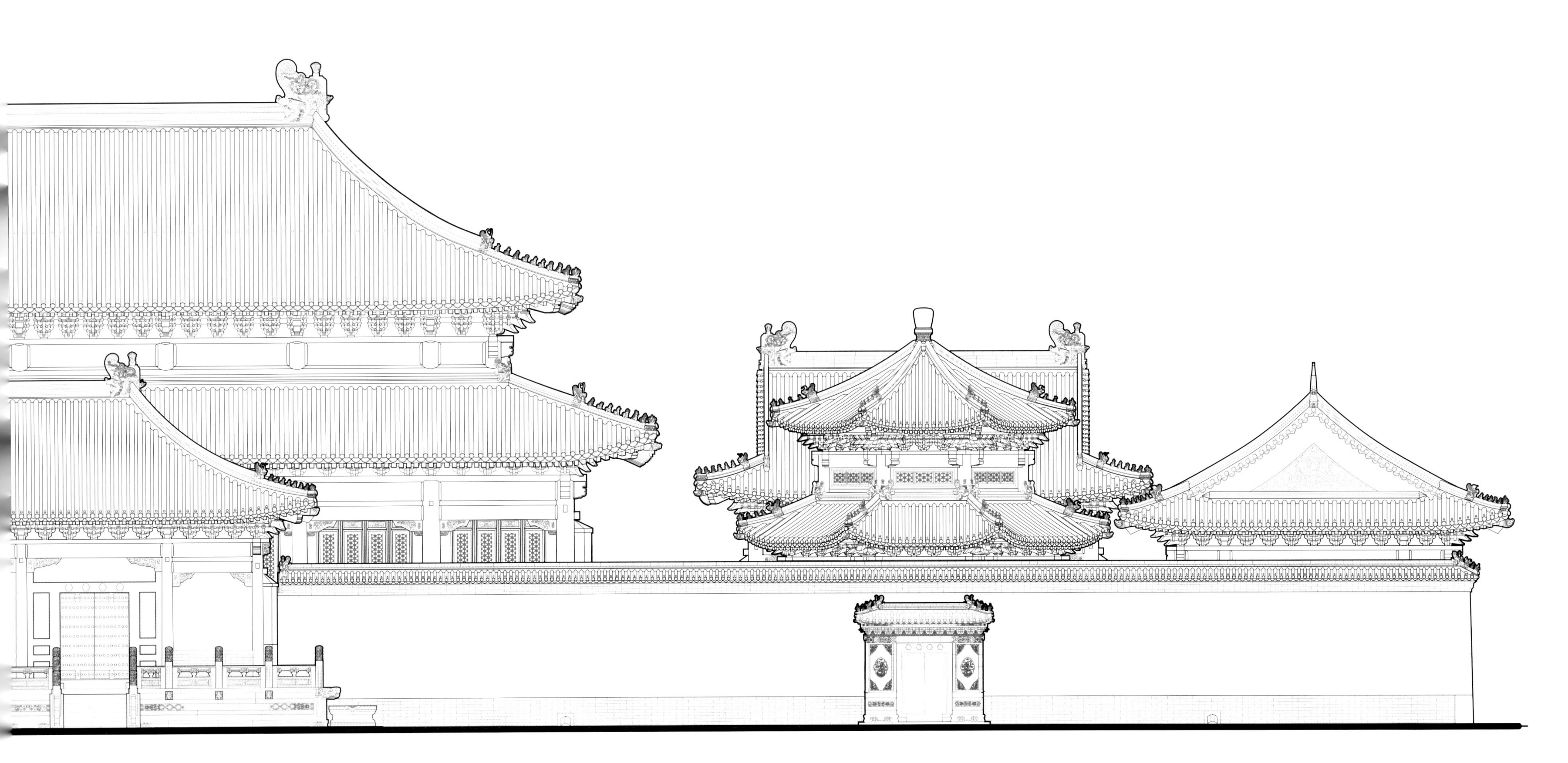

0 1.5 4.5m

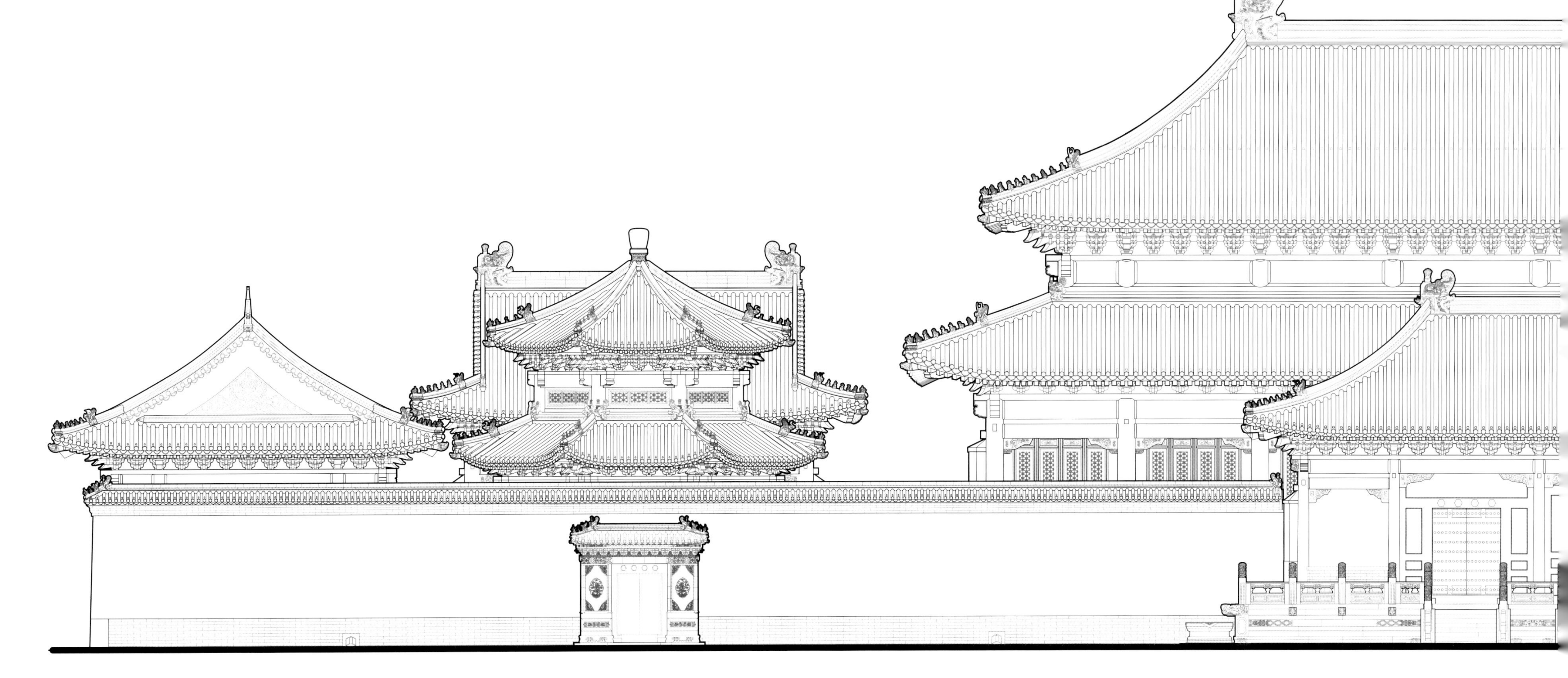

寿皇殿建筑群内院南立面图
South Elevation of the groups of Shouhuang Hall

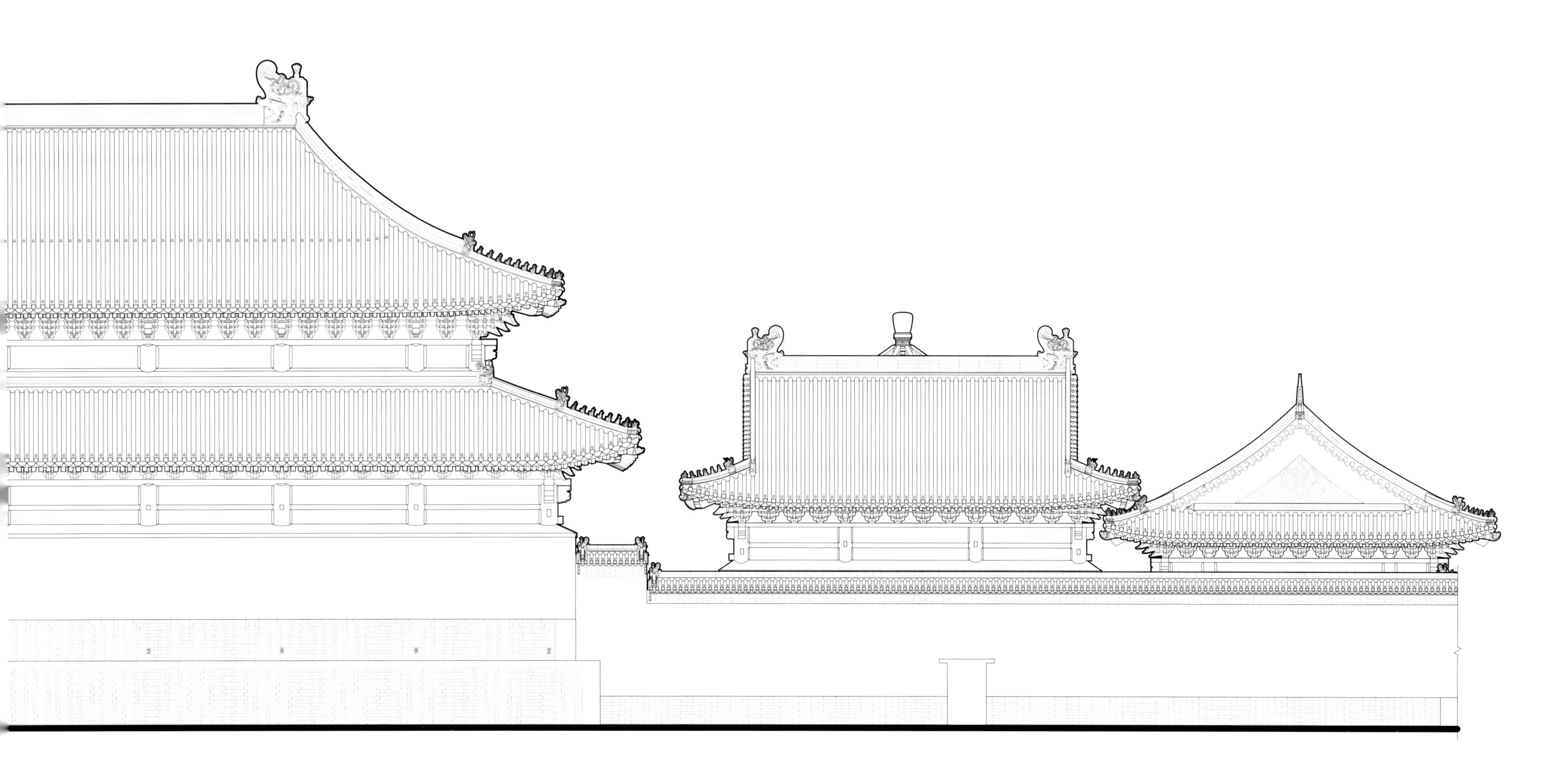

0
1.5
4.5m

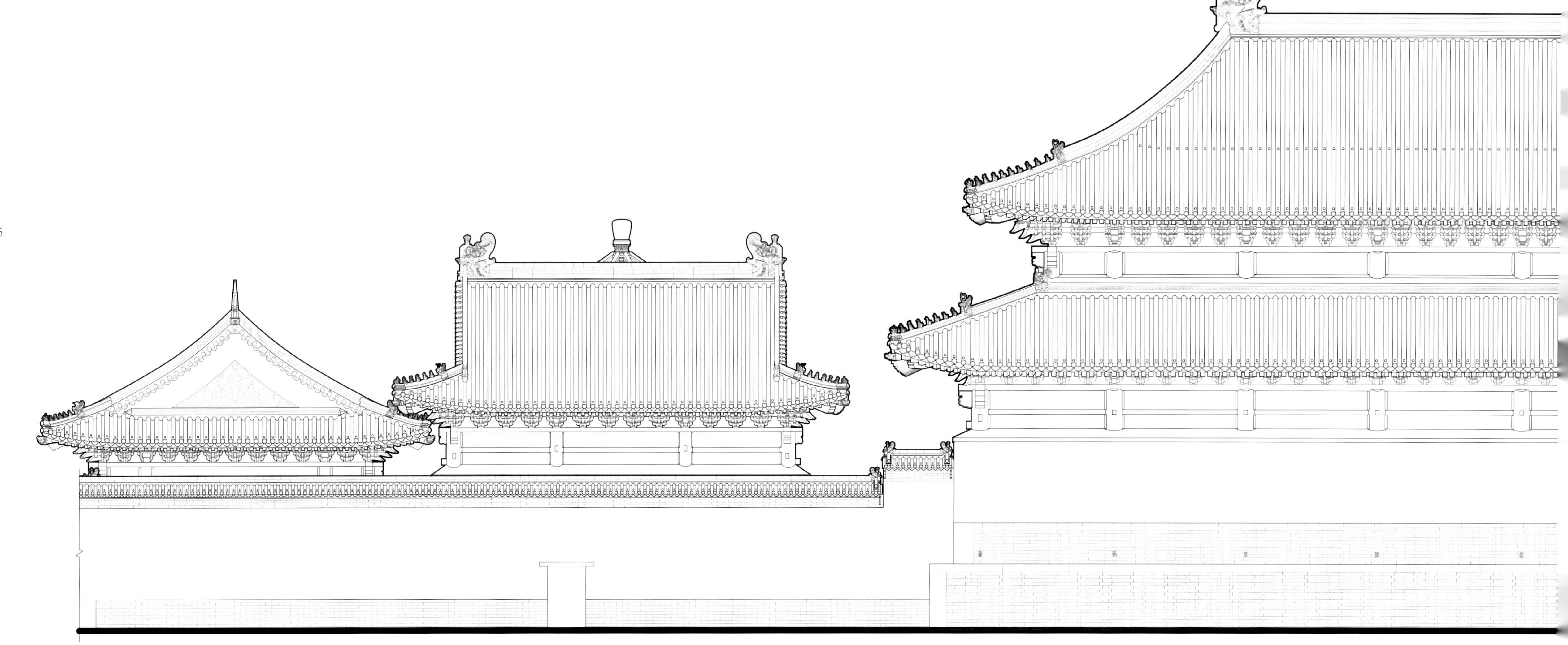

寿皇殿建筑群内院北立面图
North Elevation of the groups of Shouhuang Hall

0 1.2 3.6m

寿皇殿建筑群内院东立面图

East Elevation of the groups of Shouhuang Hall

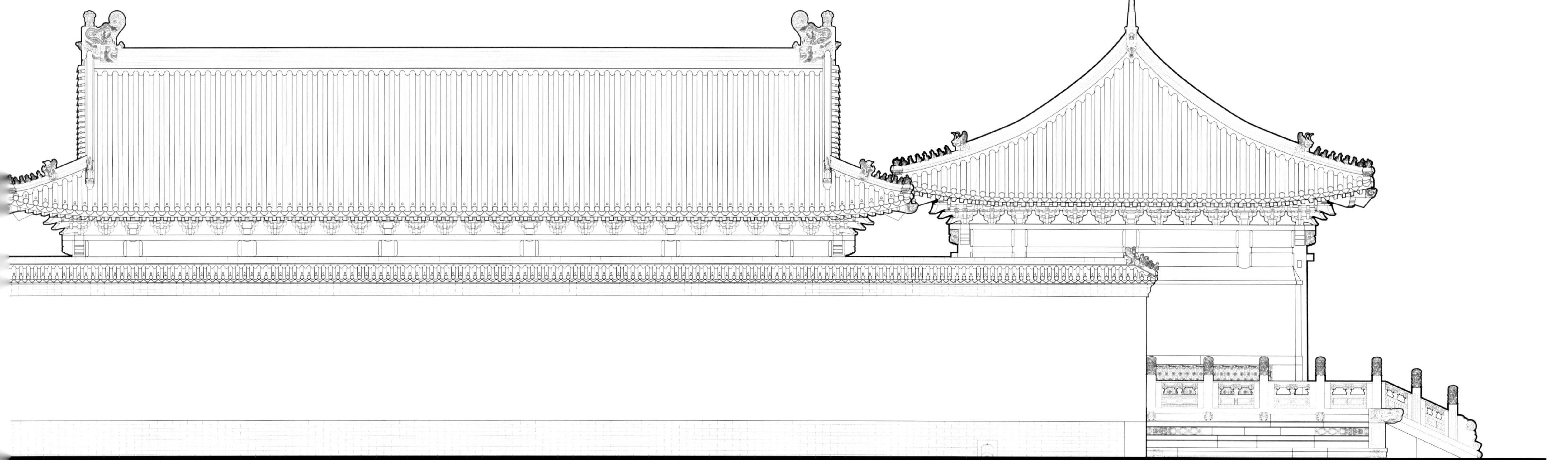
0
1.2
3.6m

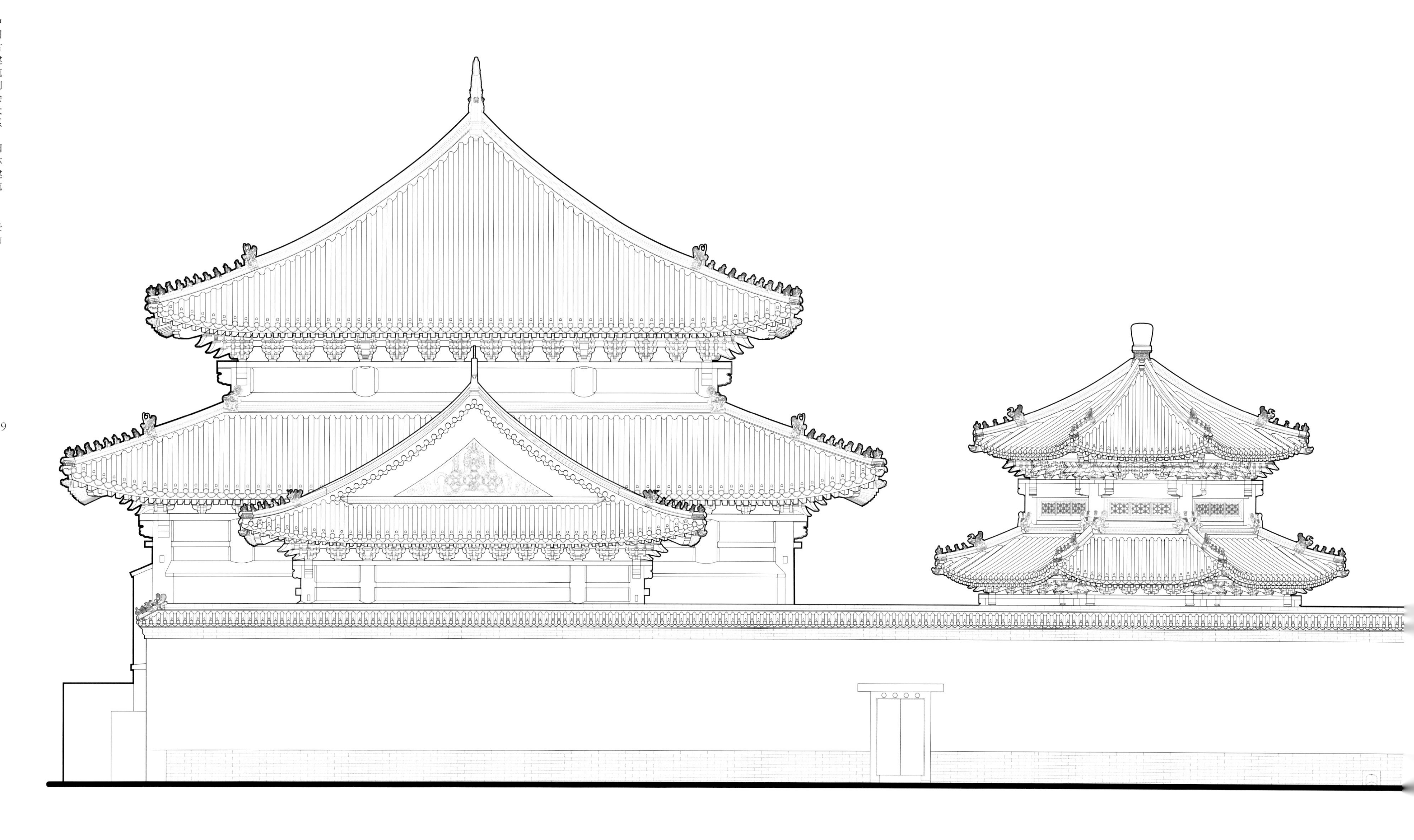

寿皇殿建筑群内院西立面图

West Elevation of the groups of Shouhuang Hall

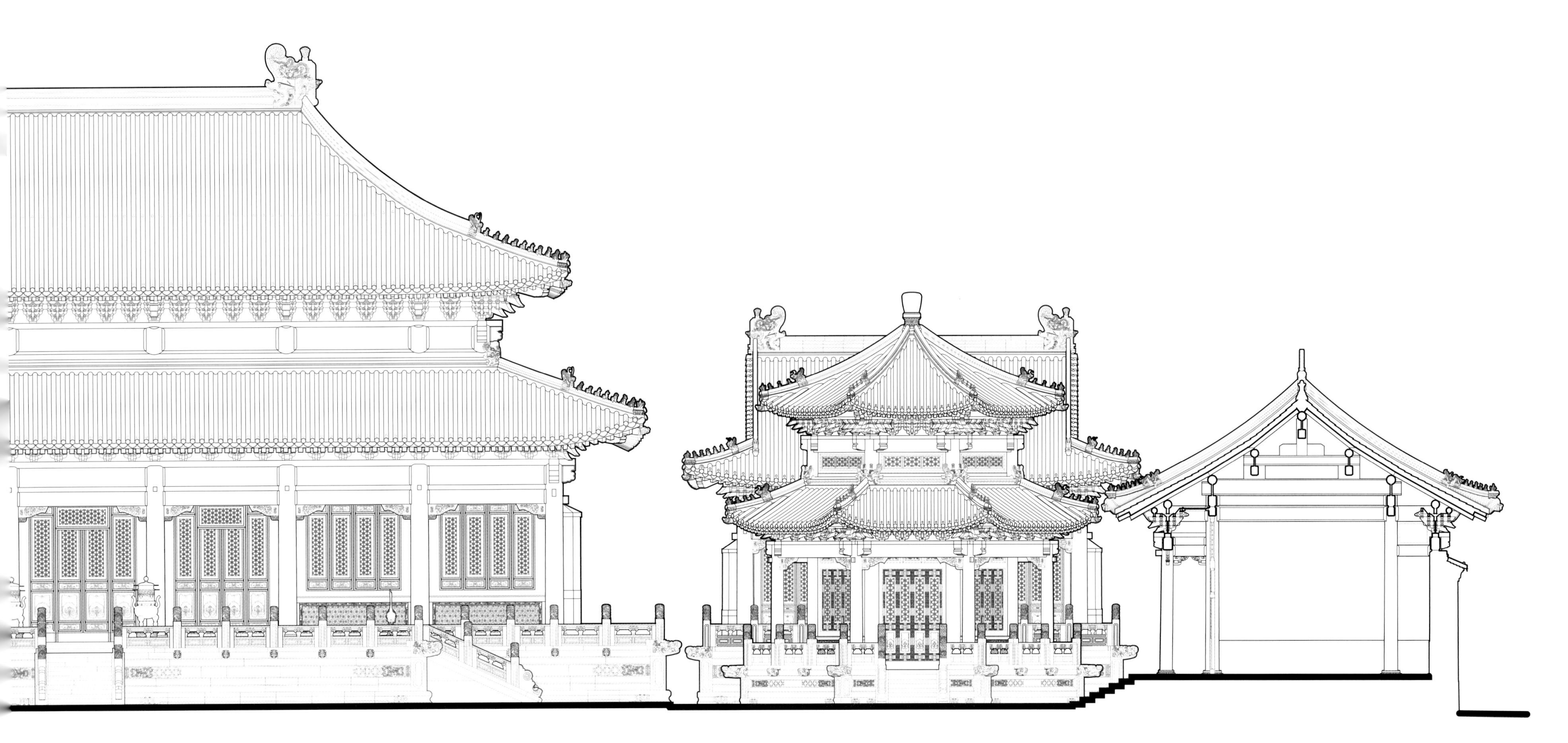
0 1.5 4.5m

寿皇殿建筑群总横剖面图

Transverse section of the groups of Shouhuang Hall

0
2.5
7.5m

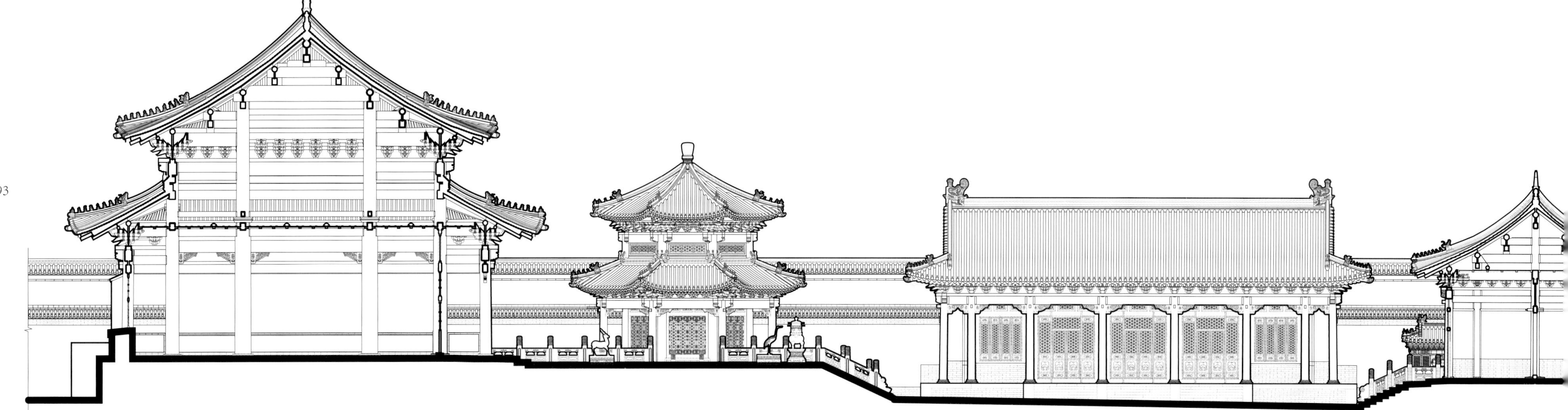

寿皇殿建筑群总纵剖面图

Longitudinal section of the groups of Shouhuang Hall

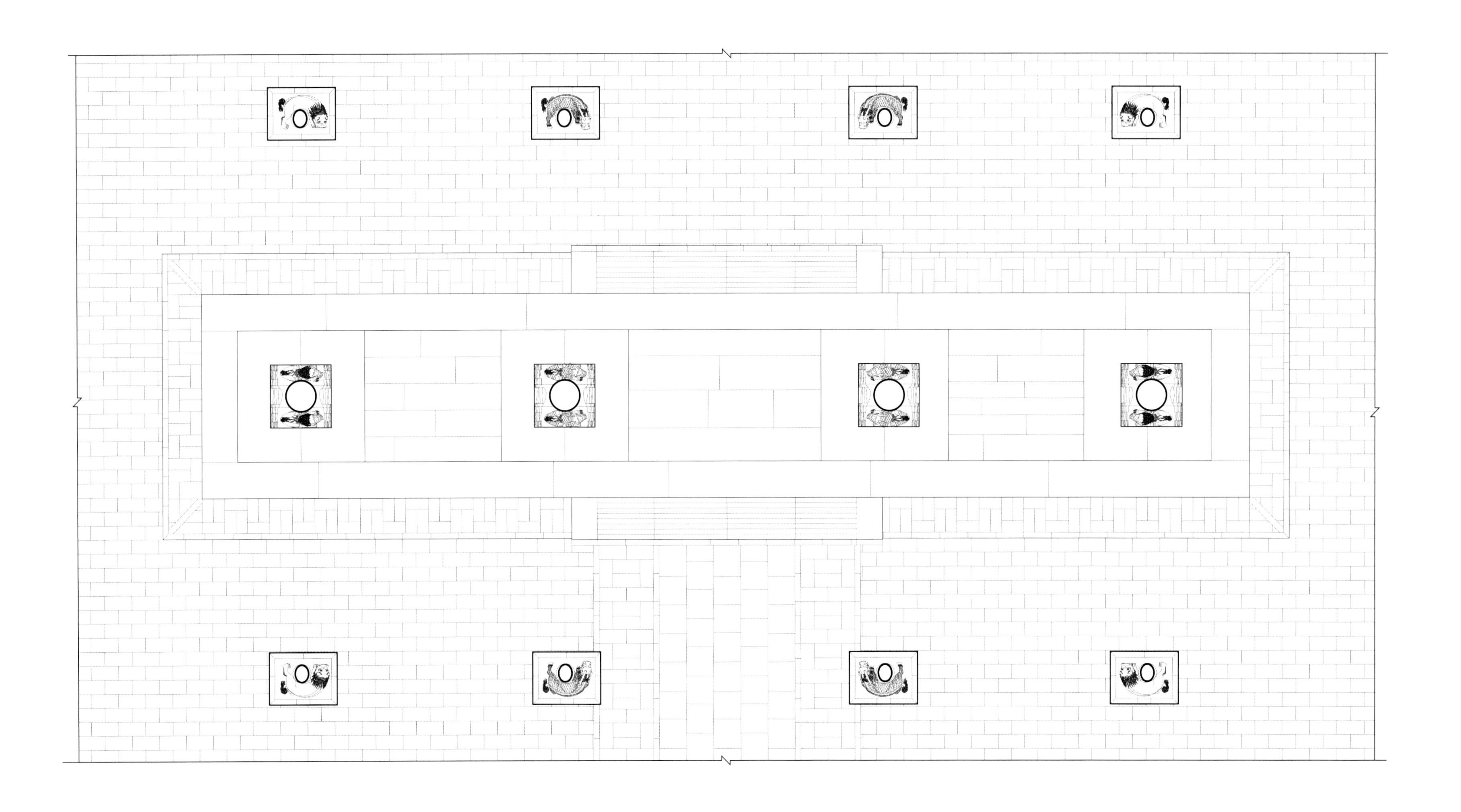

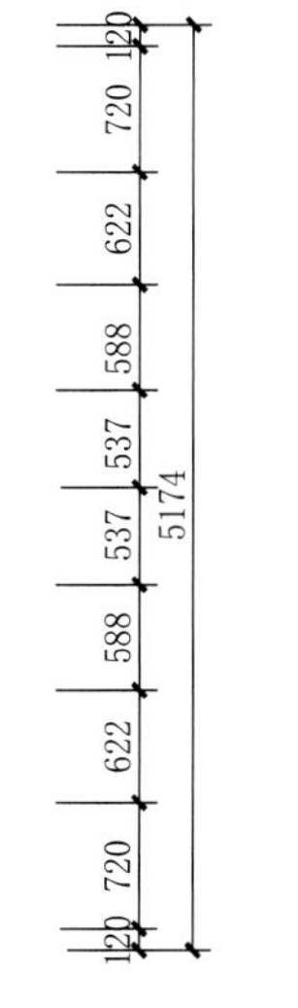

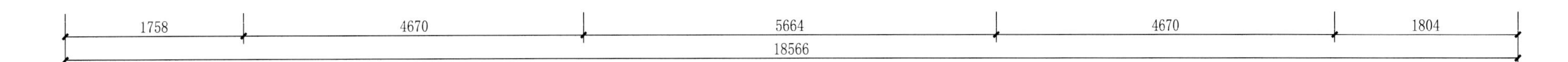

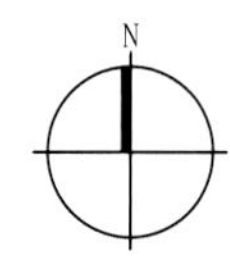

显承无斁昭格惟馨牌坊平面图

Plan of Xianchengwuduzhaogeweixin Archway

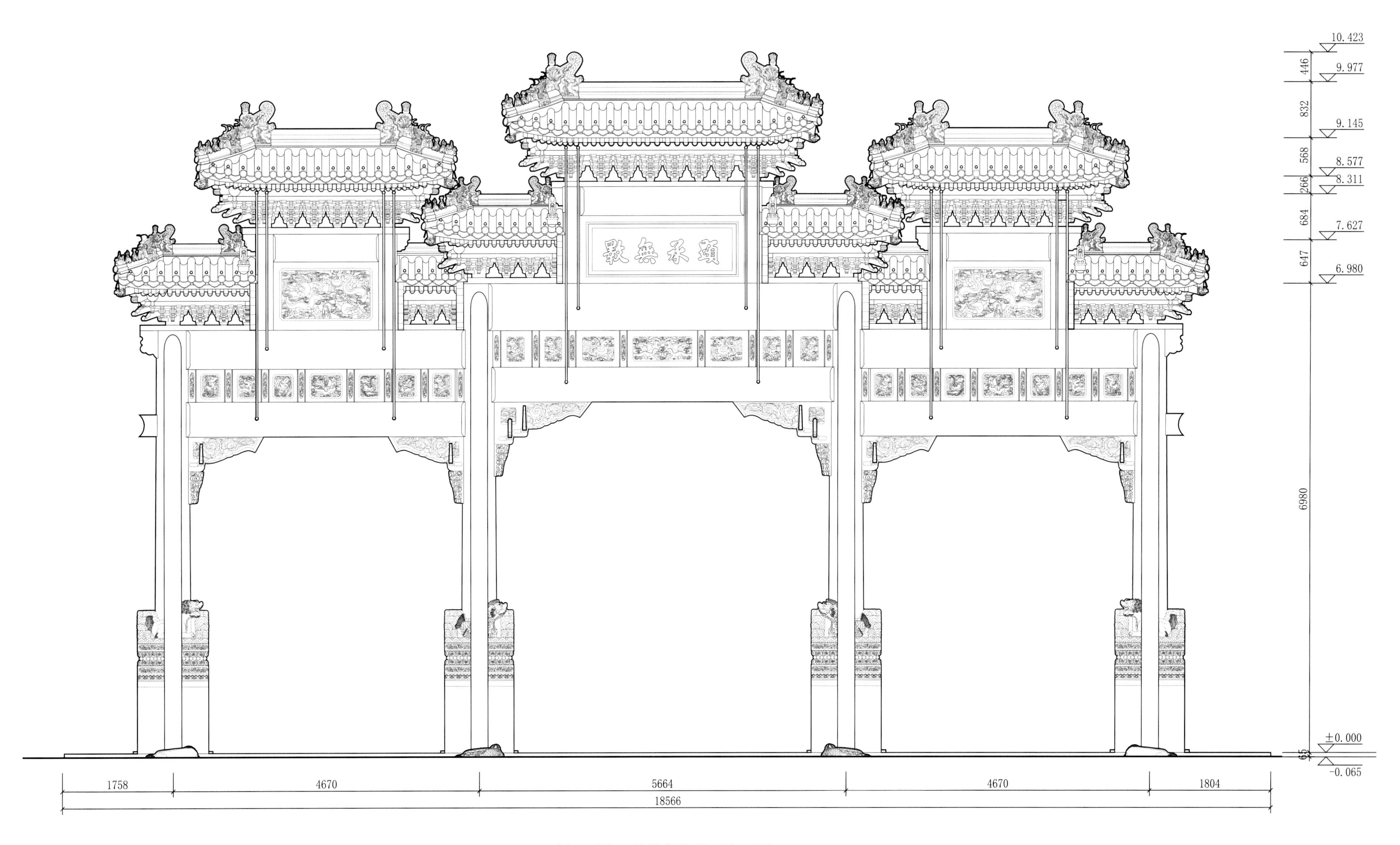

显承无斁昭格惟馨牌坊正立面图

Front elevation of Xianchengwuduzhaogeweixin Archway

显承无斁昭格惟馨牌坊背立面图
Rear elevation of Xianchengwuduzhaogeweixin Archway

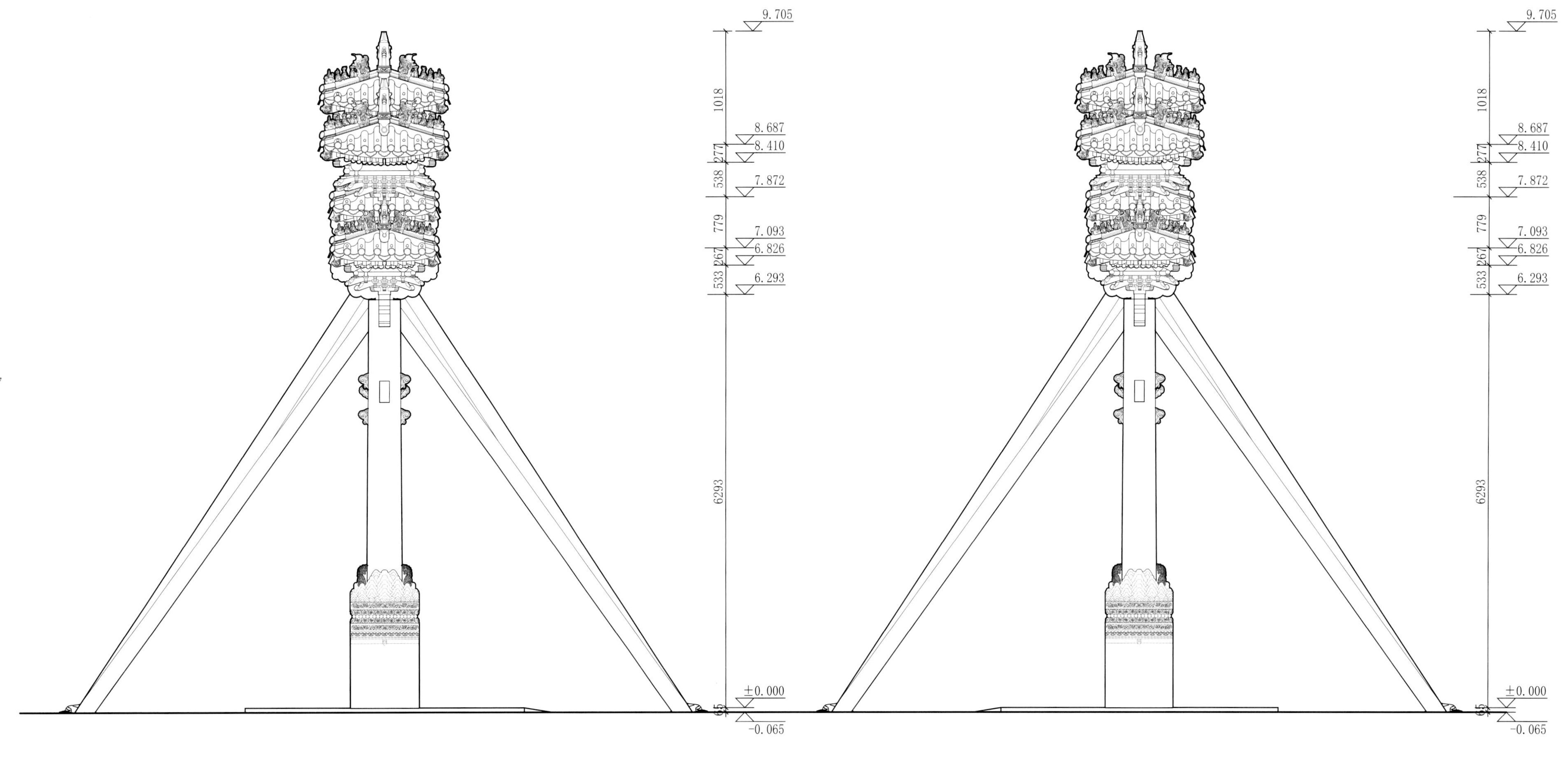

显承无斁昭格惟馨牌坊东立面图
East elevation of Xianchengwuduzhaogeweixin Archway

显承无斁昭格惟馨牌坊西立面图
West elevation of Xianchengwuduzhaogeweixin Archway

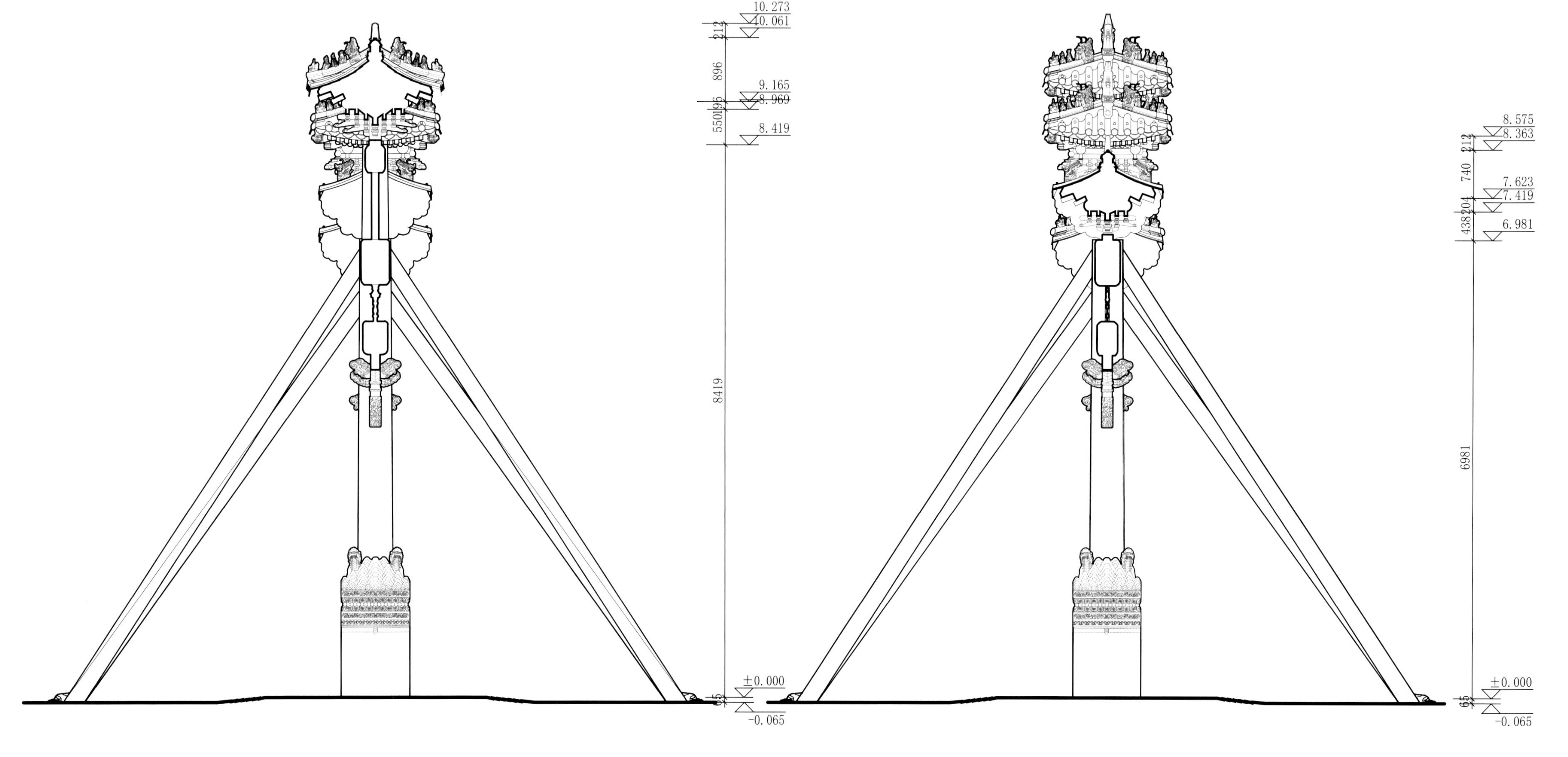

显承无斁昭格惟馨牌坊明间横剖面图
Central bay section of Xianchengwuduzhaogeweixin Archway

显承无斁昭格惟馨牌坊次间横剖面图
Lateral bay section of Xianchengwuduzhaogeweixin Archway

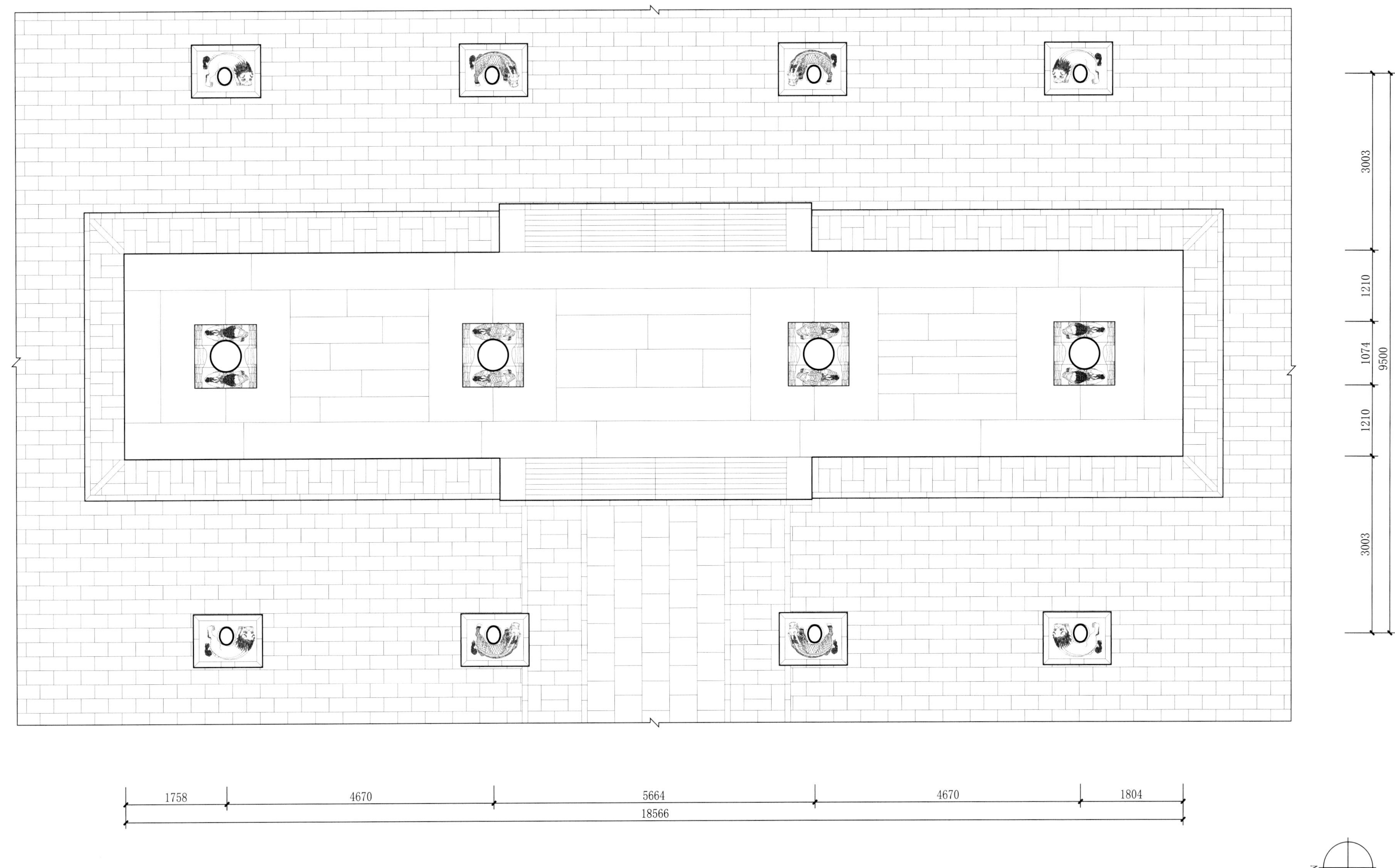

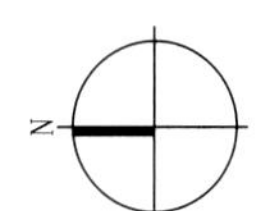

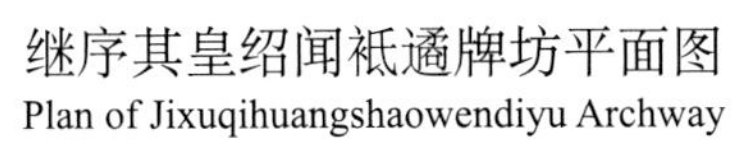

继序其皇绍闻祗遹牌坊平面图
Plan of Jixuqihuangshaowendiyu Archway

继序其皇绍闻祗遹牌坊正立面图

Front elevation of Jixuqihuangshaowendiyu Archway

继序其皇绍闻祗遹牌坊背立面图

Rear elevation of Jixuqihuangshaowendiyu Archway

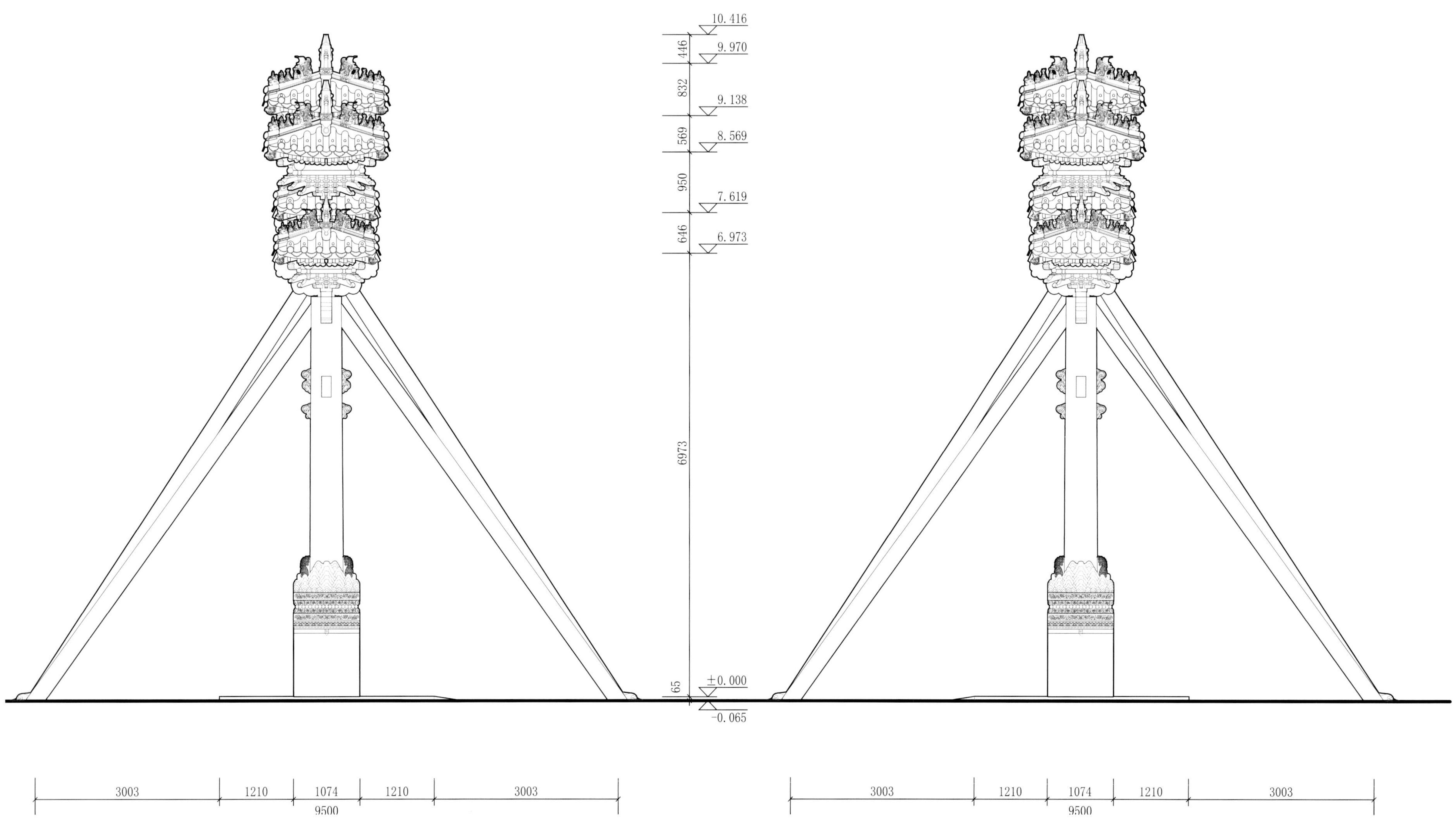

继序其皇绍闻衹遹牌坊北立面图
North elevation of Jixuqihuangshaowendiyu Archway

继序其皇绍闻衹遹牌坊南立面图
South elevation of Jixuqihuangshaowendiyu Archway

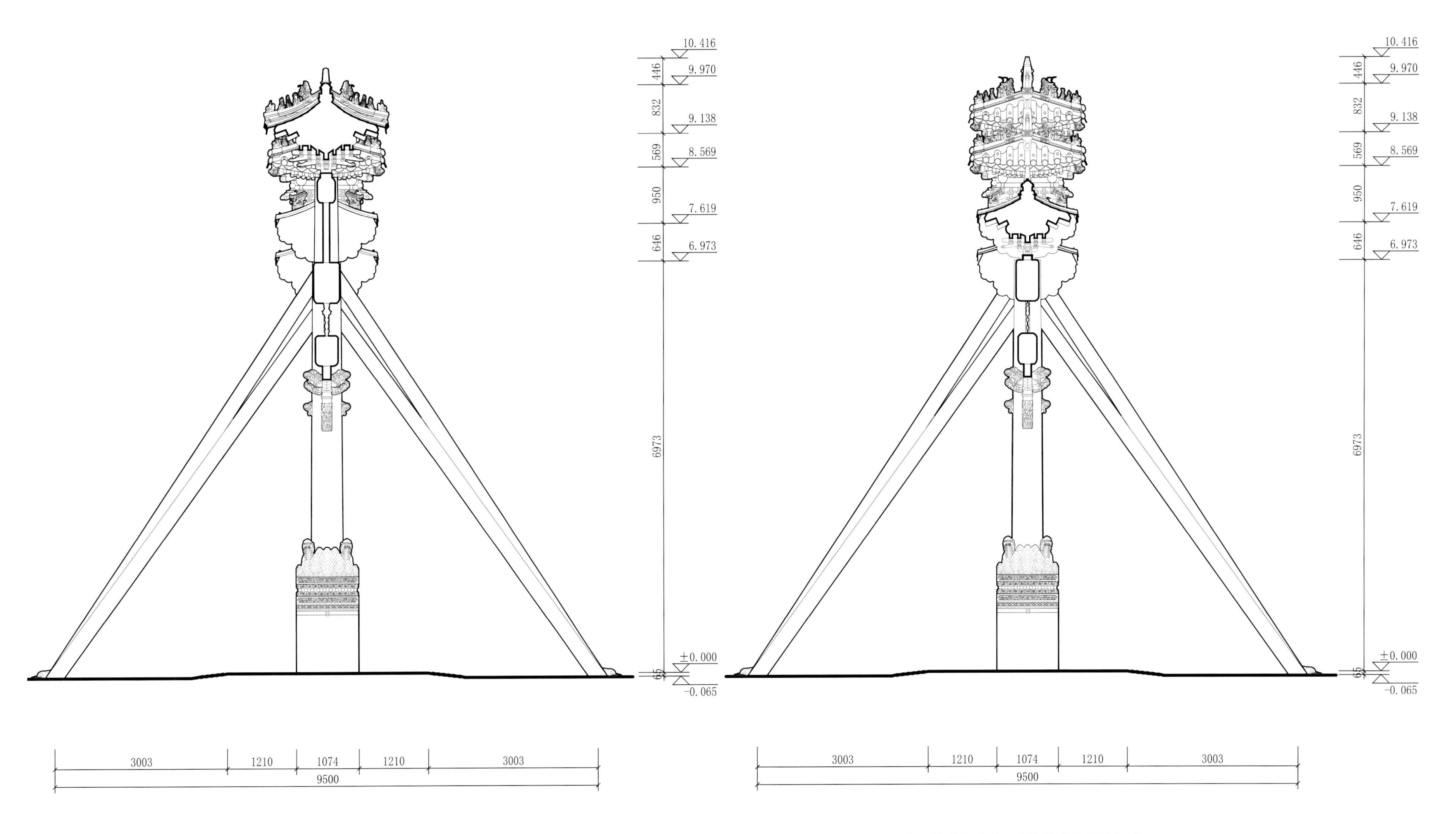

继序其皇绍闻祗遹牌坊明间横剖面图
Central bay section of Jixuqihuangshaowendiyu Archway

继序其皇绍闻祗遹牌坊次间横剖面图
Lateral bay section of Jixuqihuangshaowendiyu Archway

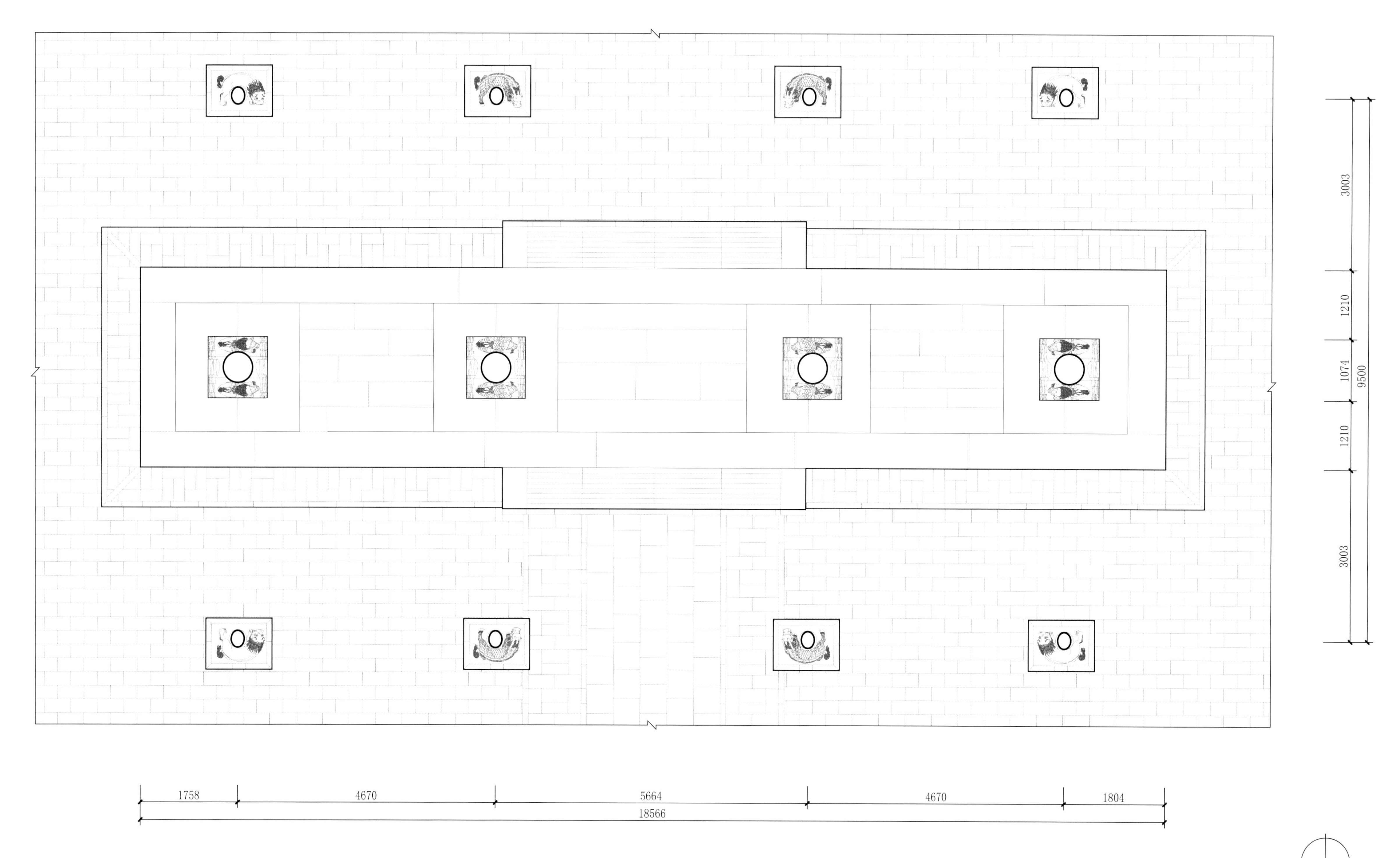

世德作求旧典时式牌坊平面图

Plan of Shidezuoqiujiudianshishi Archway

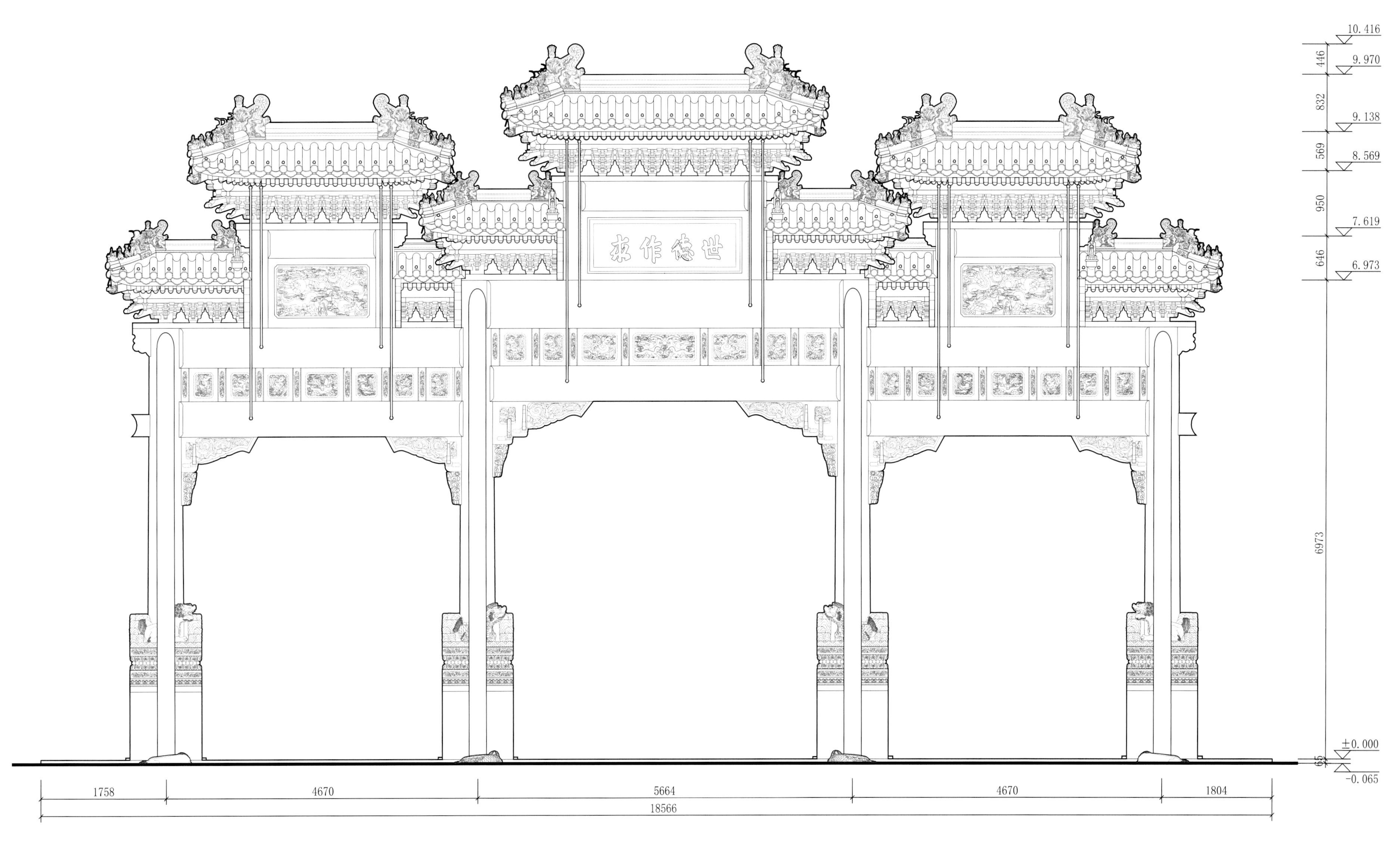

世德作求旧典时式牌坊正立面图

Front elevation of Shidezuoqiujiudianshishi Archway

世德作求旧典时式牌坊背立面图

Rear elevation of Shidezuoqiujiudianshishi Archway

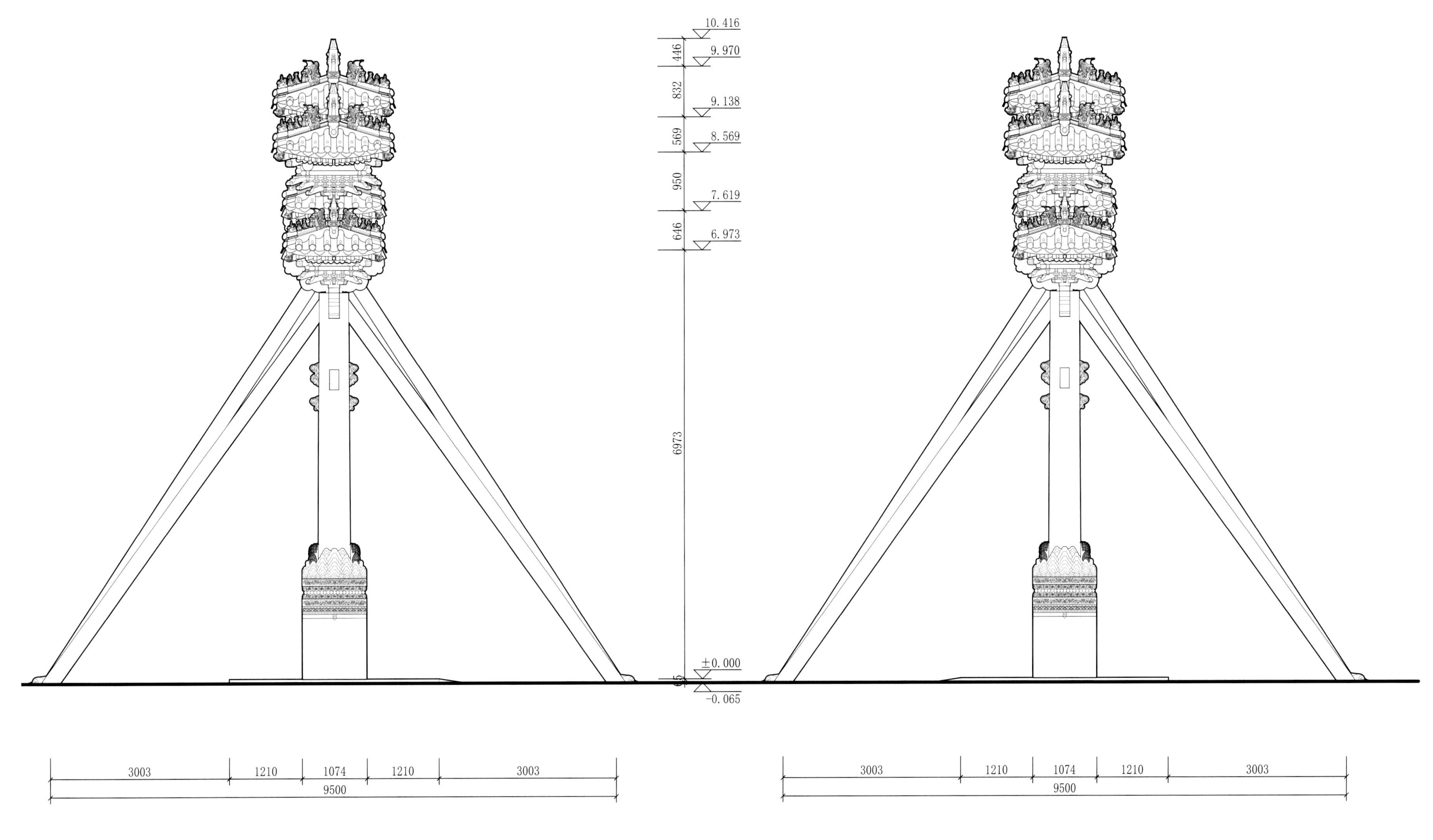

世德作求旧典时式牌坊北立面图
North elevation of Shidezuoqiujiudianshishi Archway

世德作求旧典时式牌坊南立面图
South elevation of Shidezuoqiujiudianshishi Archway

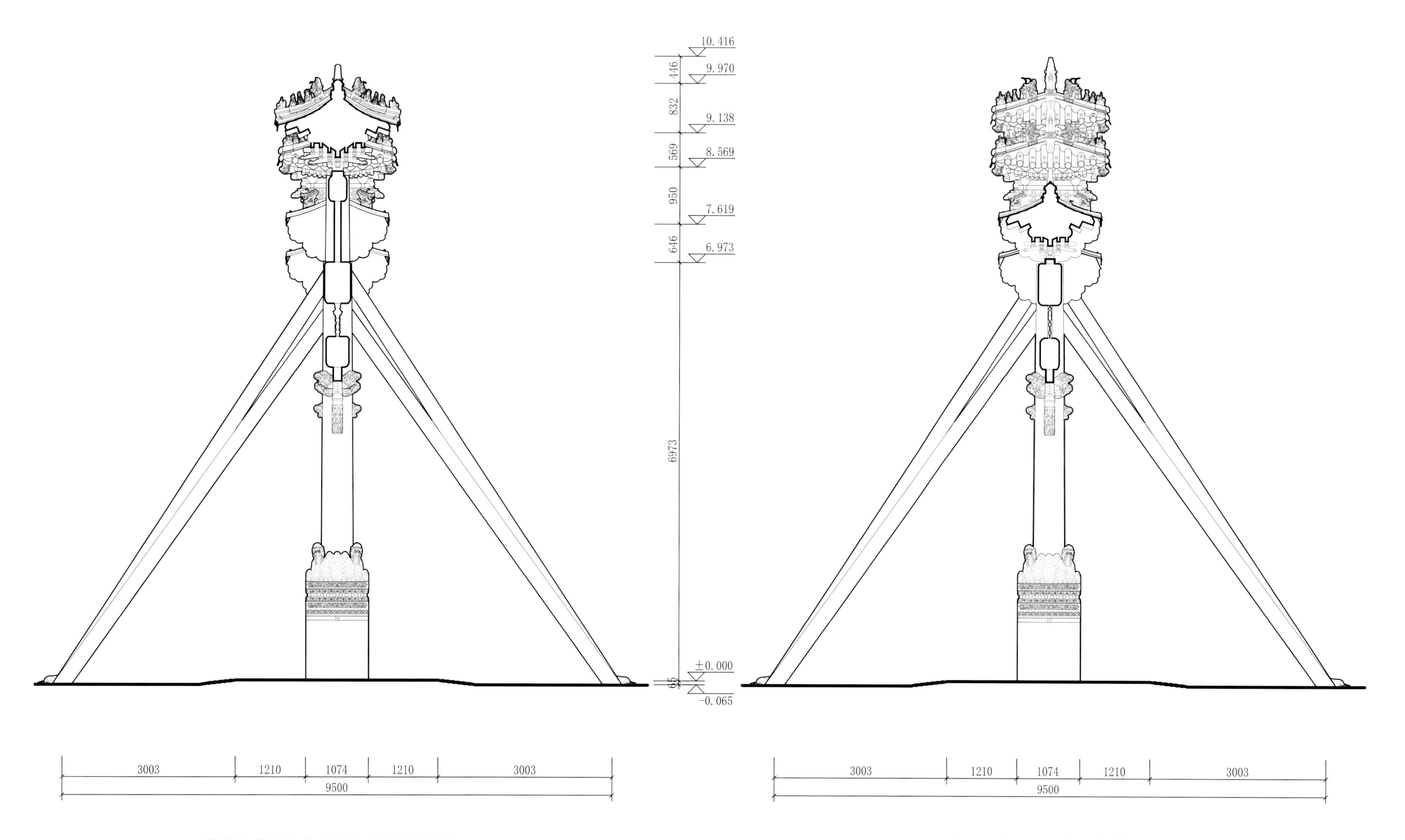

世德作求旧典时式牌坊明间横剖面图
Central bay section of Shidezuoqiujiudianshishi Archway

世德作求旧典时式牌坊次间横剖面图
Lateral bay section of Shidezuoqiujiudianshishi Archway

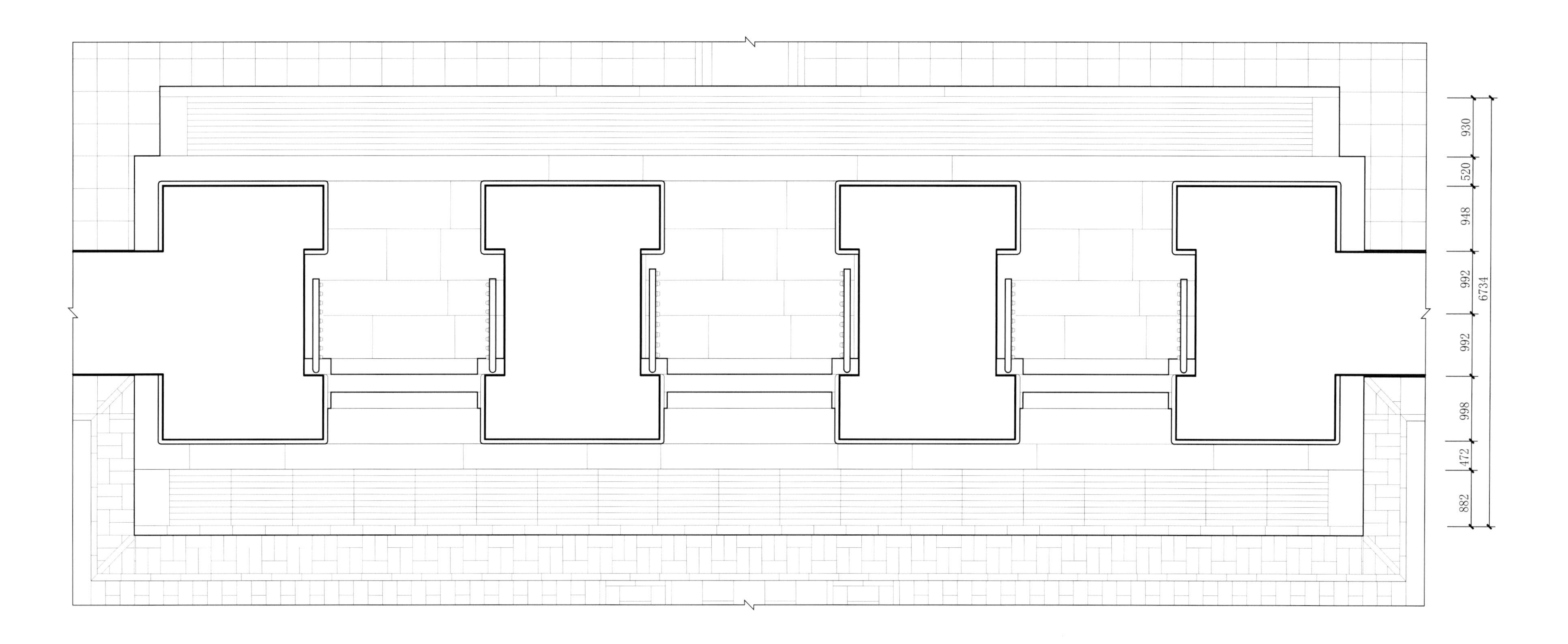

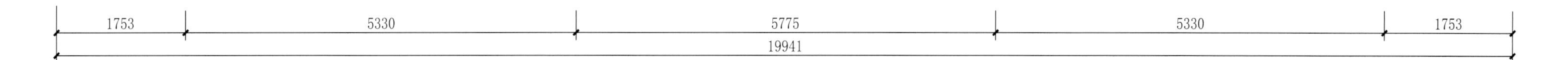

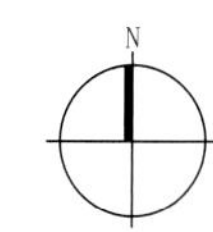

南砖城门平面图
Plan of South Bricked Gate

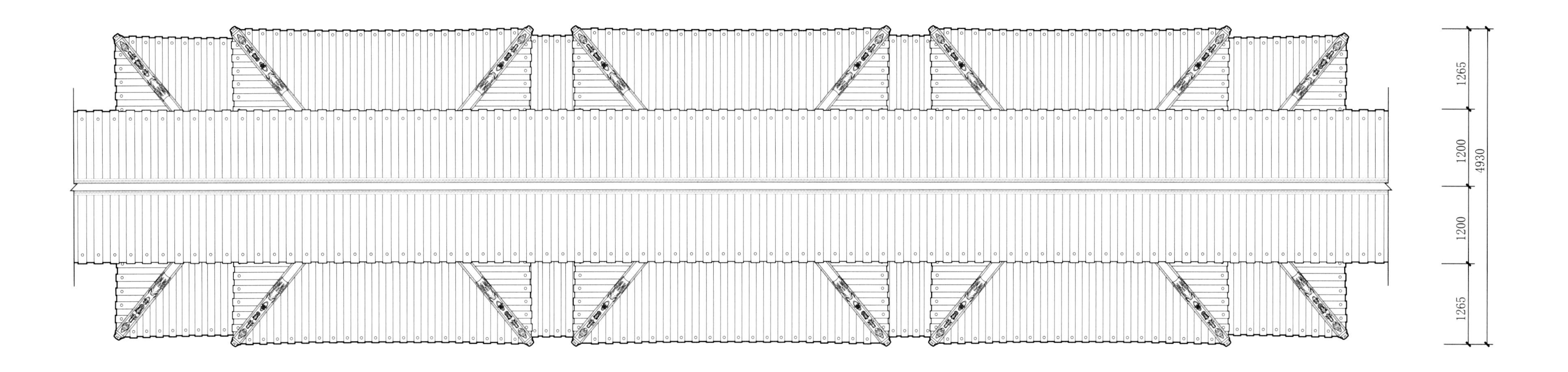

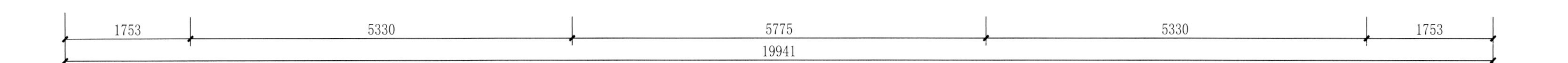

南砖城门屋顶平面图
Roof plan of South Bricked Gate

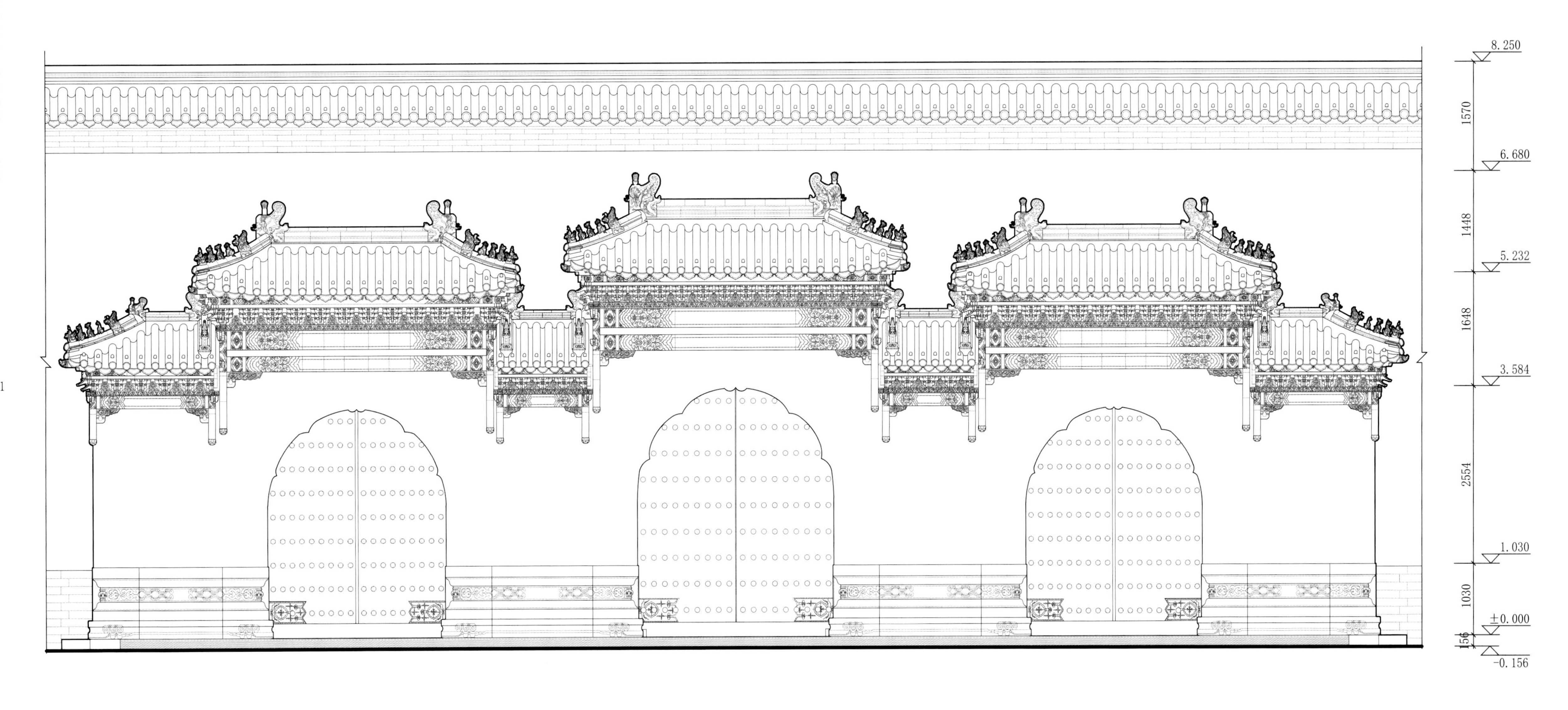

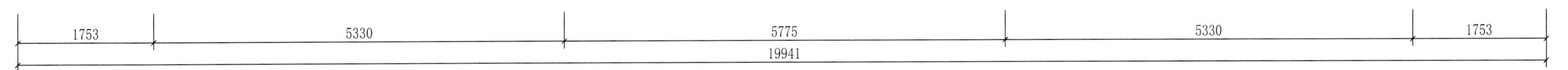

南砖城门正立面图

Front elevation of South Bricked Gate

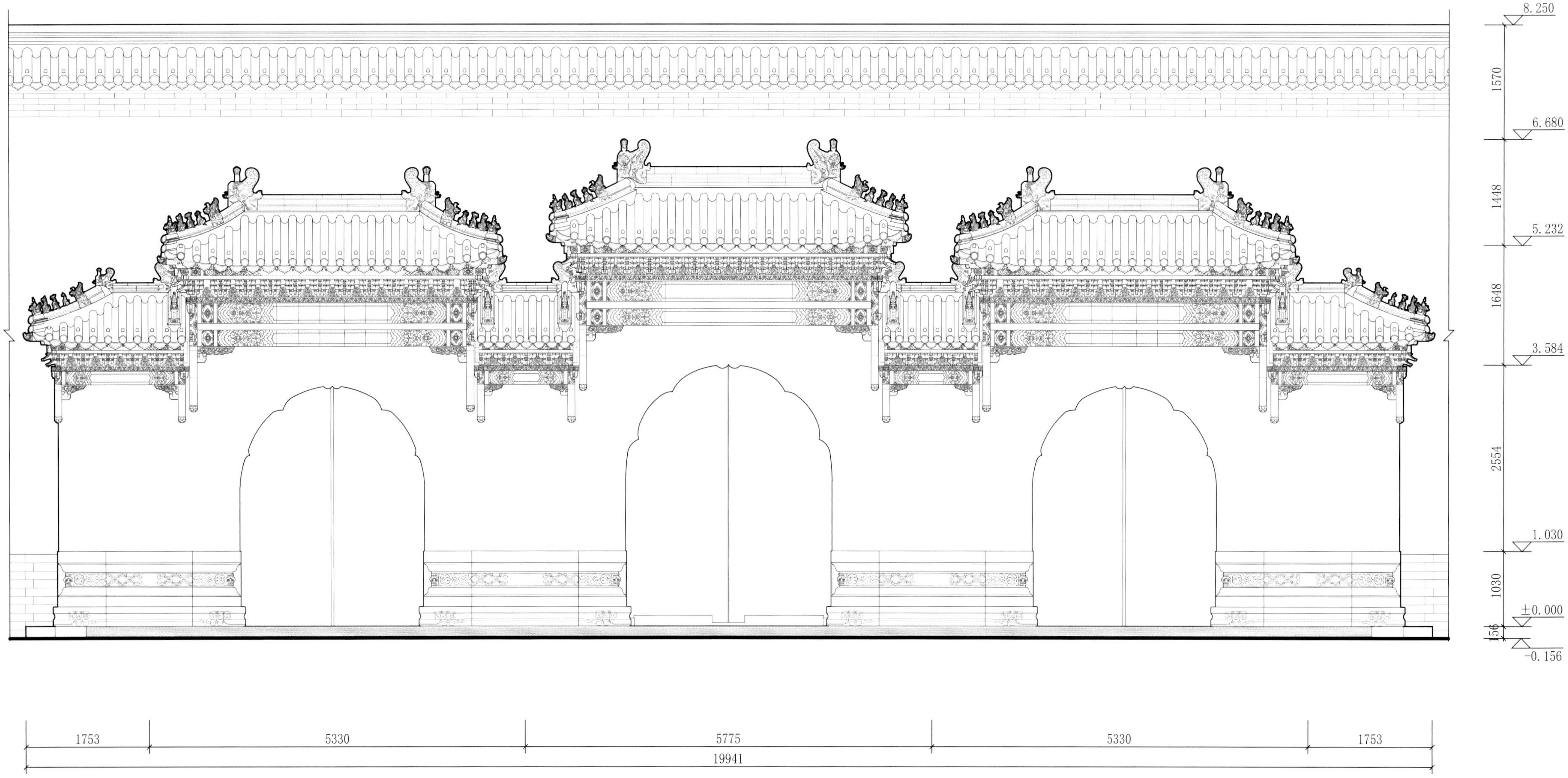

8.250
1570
6.680
1448
5.232
1648
3.584
2554
1.030
1030
±0.000
156
-0.156

1753 | 5330 | 5775 | 5330 | 1753
19941

南砖城门背立面图
Rear elevation of South Bricked Gate

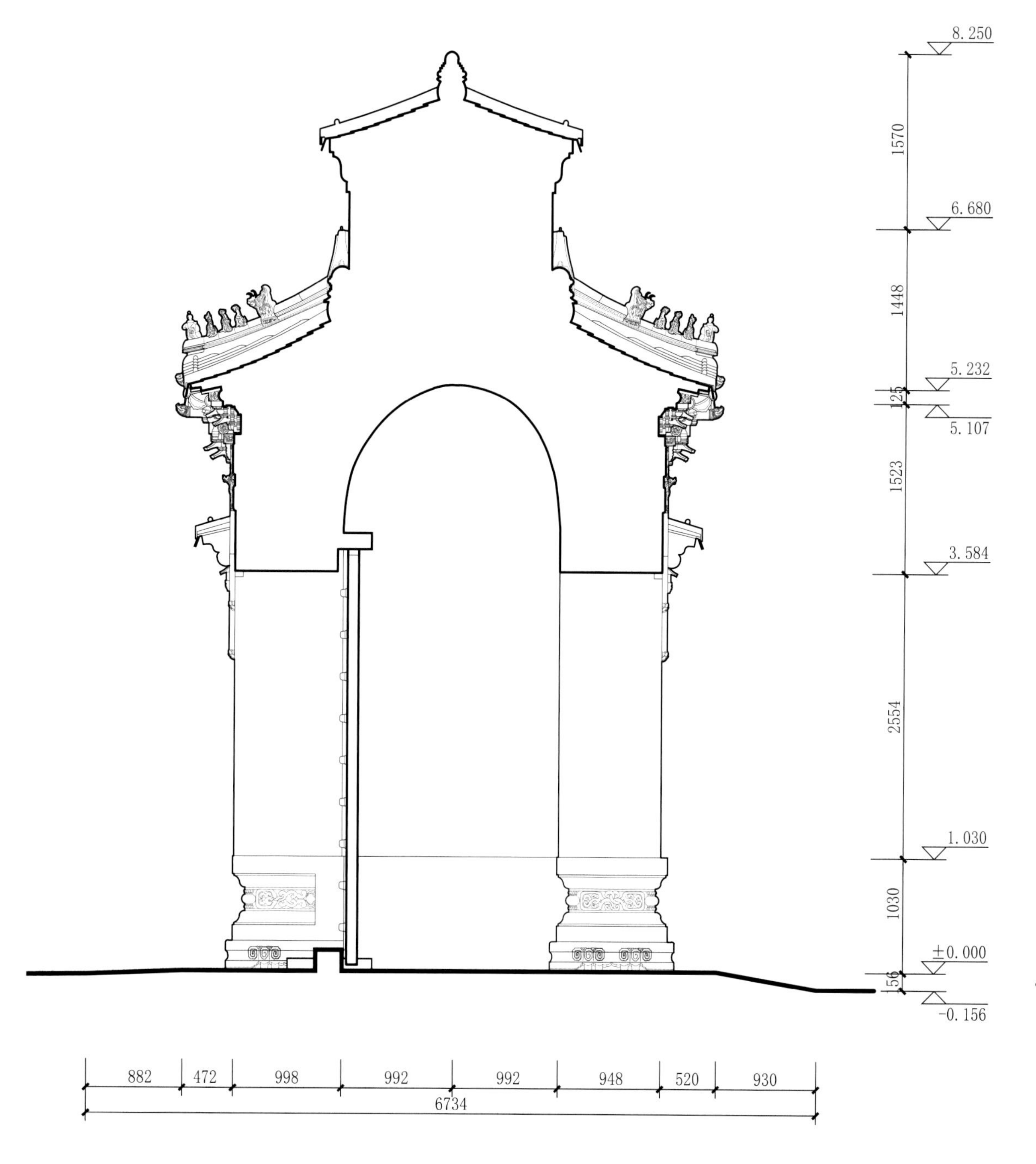

南砖城门明间横剖面图
Central bay section of South Bricked Gate

南砖城门侧立面图
Side elevation of South Bricked Gate

南砖城门东石狮平面图
Plan of east stone lion in front of South Bricked Gate

南砖城门东石狮南立面图
South elevation of east stone lion in front of South Bricked Gate

南砖城门东石狮北立面图
North elevation of east stone lion in front of South Bricked Gate

南砖城门东石狮西立面图
West elevation of east stone lion in front of South Bricked Gate

南砖城门东石狮东立面图
East elevation of east stone lion in front of South Bricked Gate

南砖城门西石狮平面图
Plan of west stone lion in front of South Bricked Gate

南砖城门西石狮南立面图
South elevation of west stone lion in front of South Bricked Gate

南砖城门西石狮北立面图
North elevation of west stone lion in front of South Bricked Gate

南砖城门西石狮西立面图
West elevation of west stone lion in front of South Bricked Gate

南砖城门西石狮东立面图
East elevation of west stone lion in front of South Bricked Gate

南砖城门琉璃纹样大样图

Detail drawing of the texture of Hall of South Bricked Gate

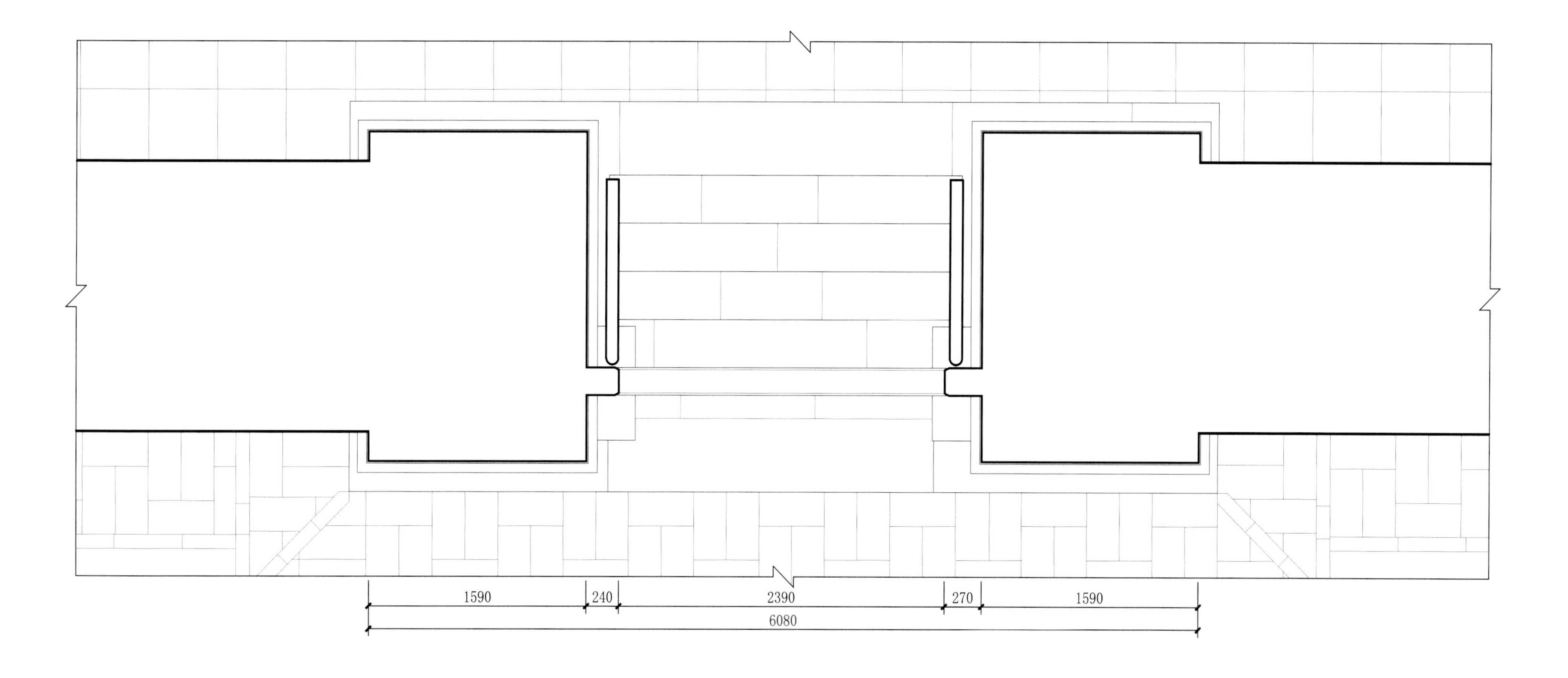

东南砖城门平面图

Plan of Southeast Bricked Gate

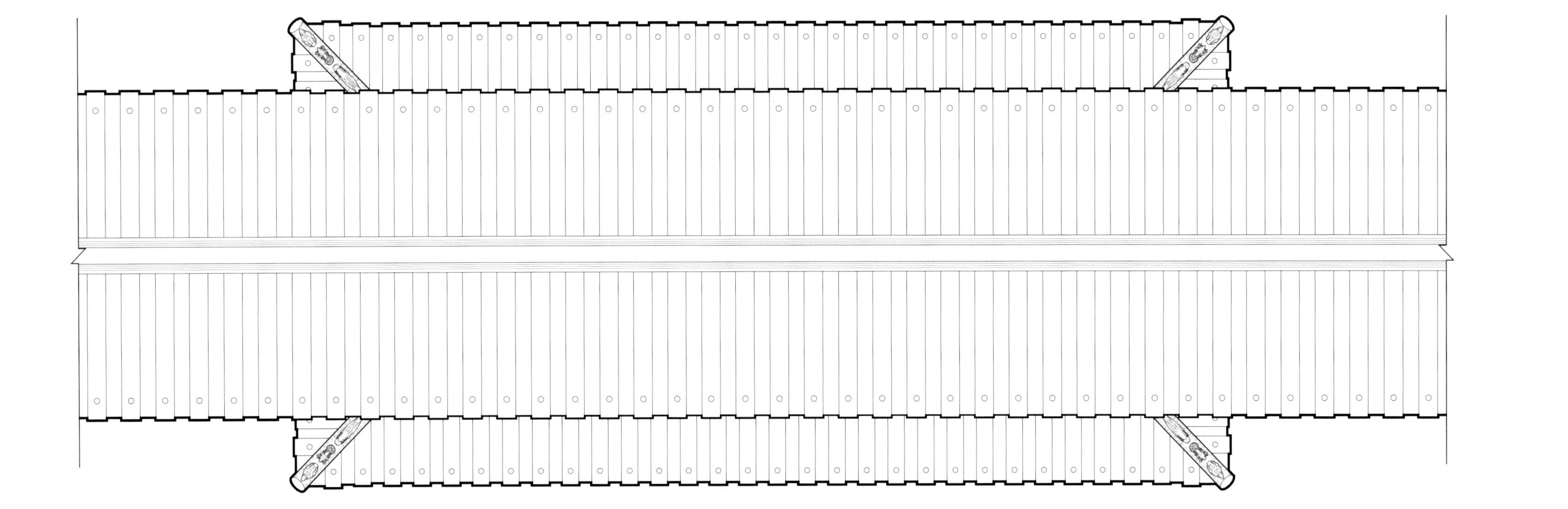

东南砖城门屋顶平面图
Roof plan of Southeast Bricked Gate

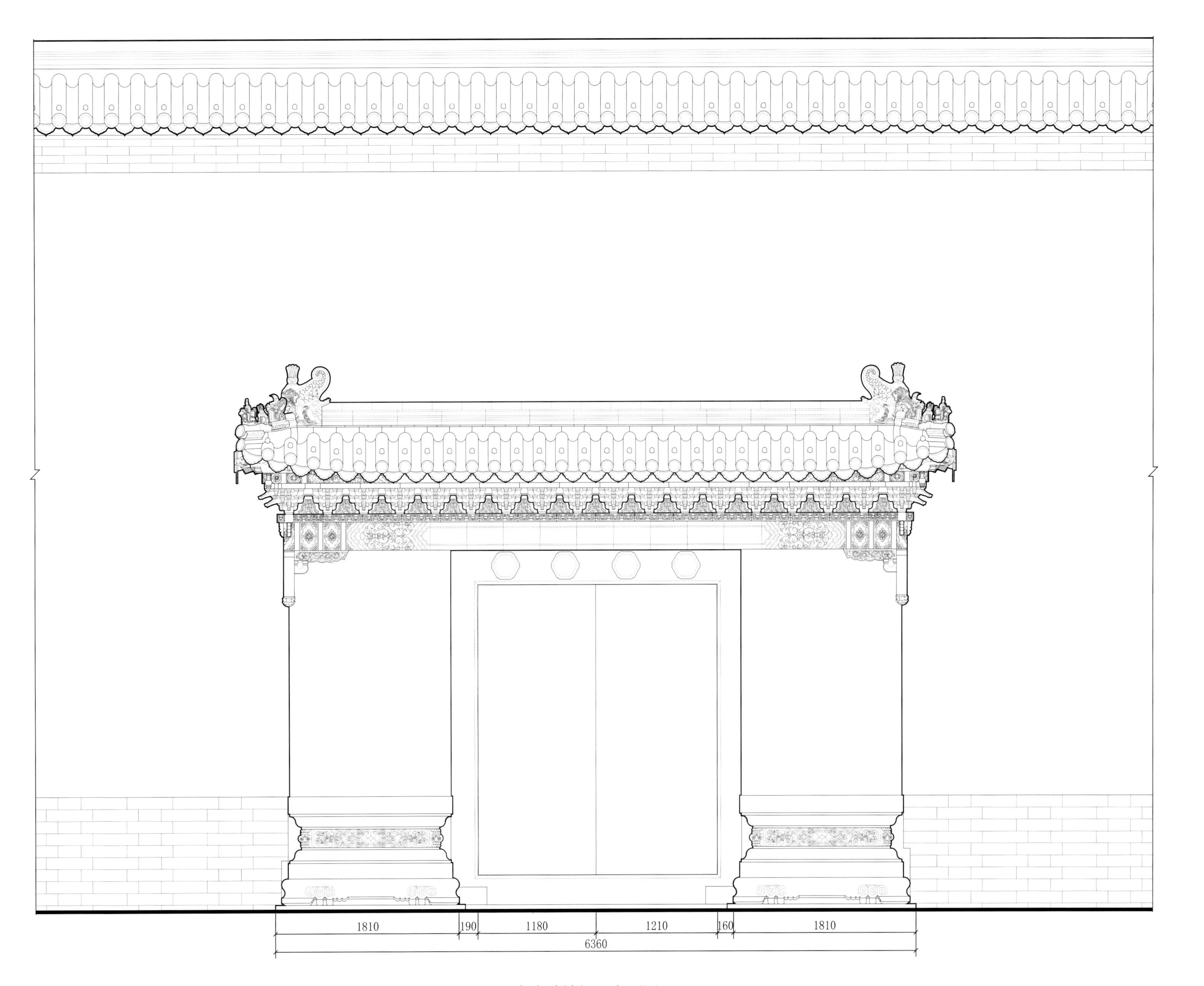

东南砖城门正立面图
Front elevation of Southeast Bricked Gate

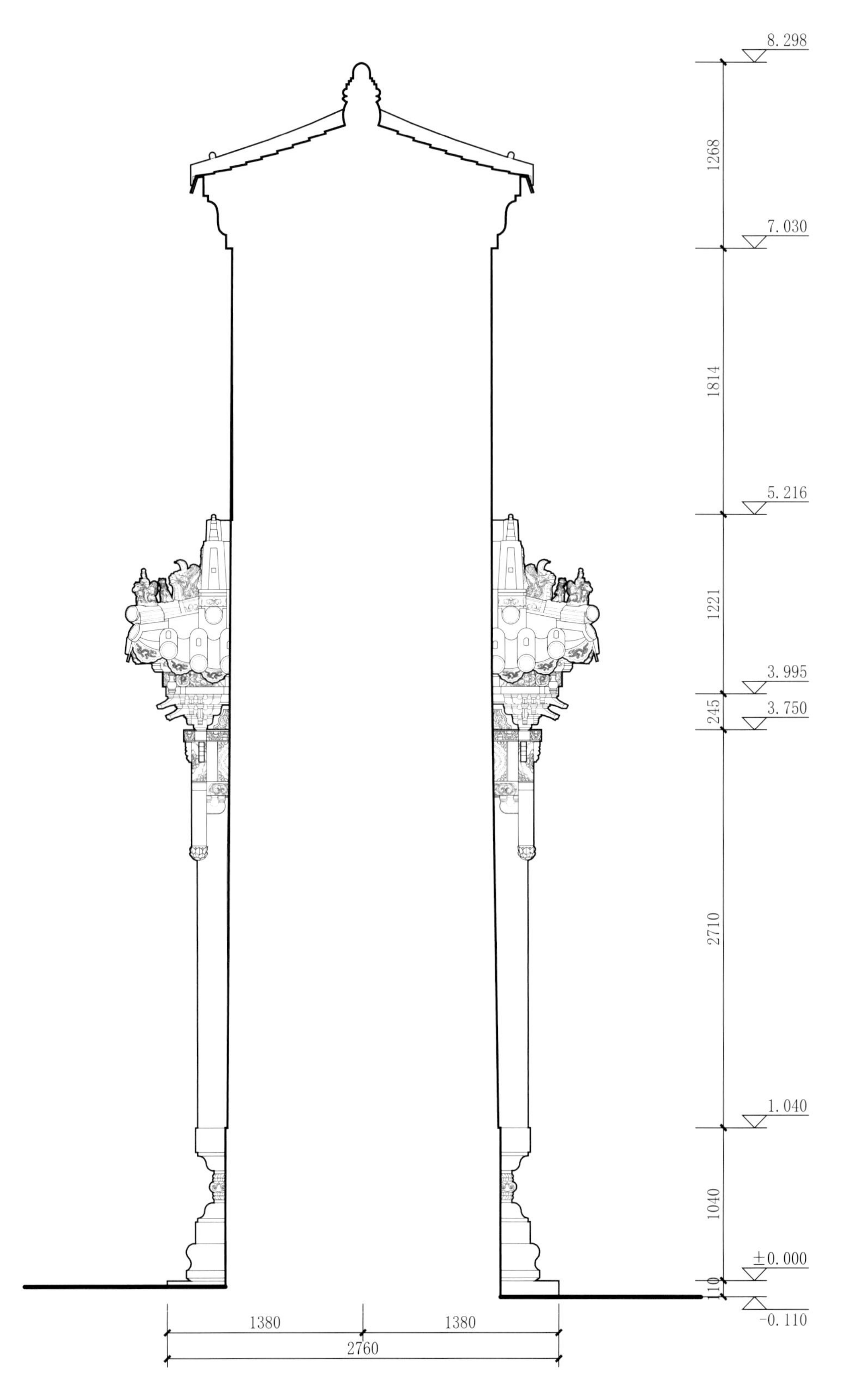

东南砖城门侧立面图
Side elevation of Southeast Bricked Gate

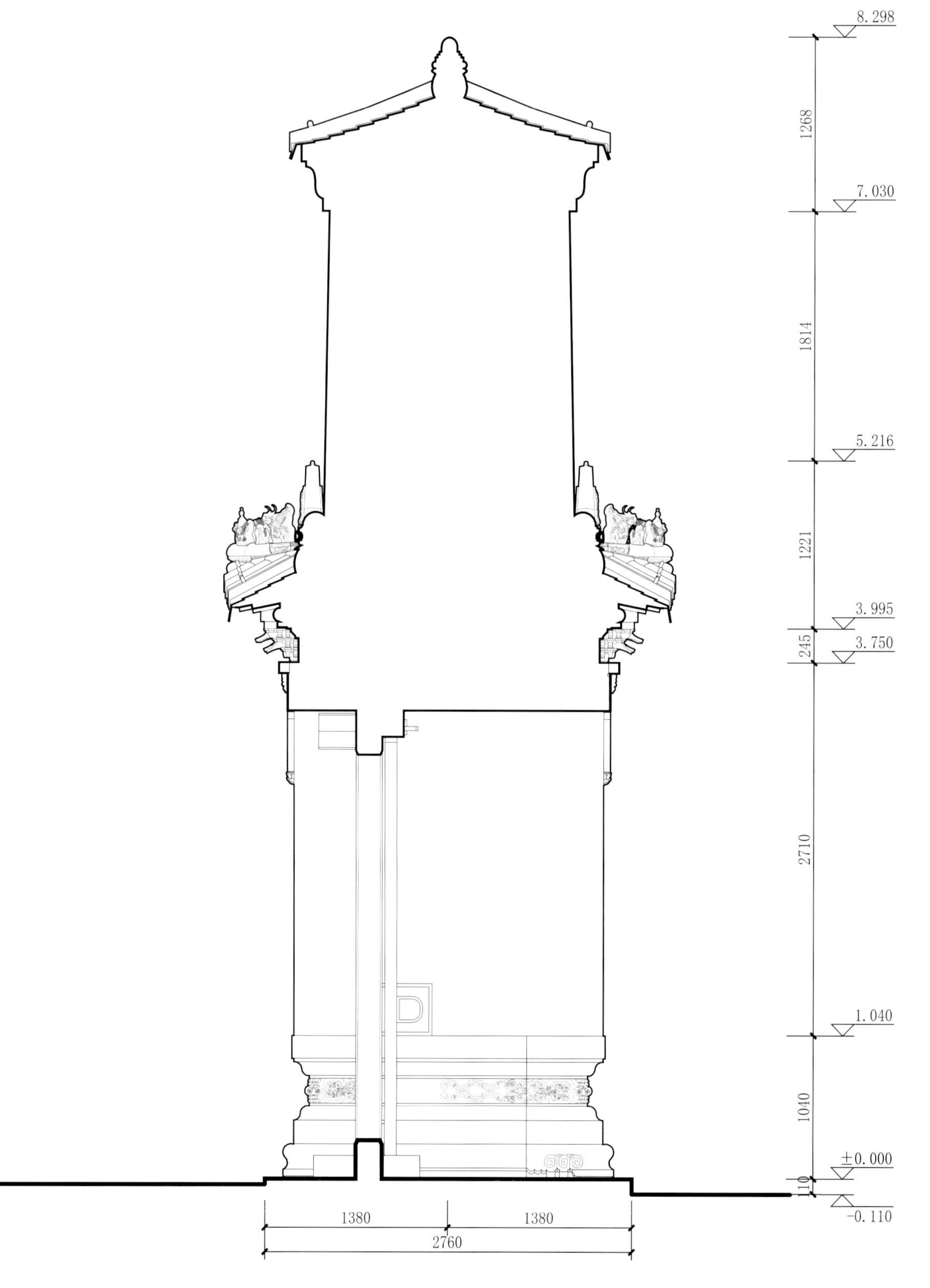

东南砖城门剖面图
Section of Southeast Bricked Gate

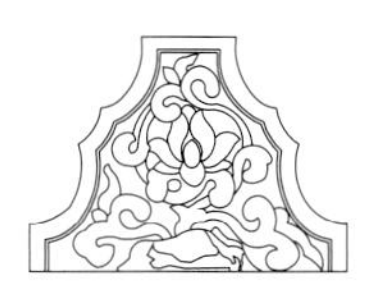

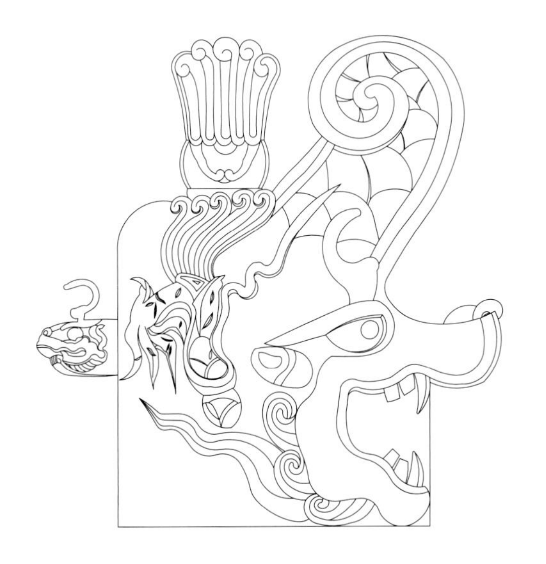

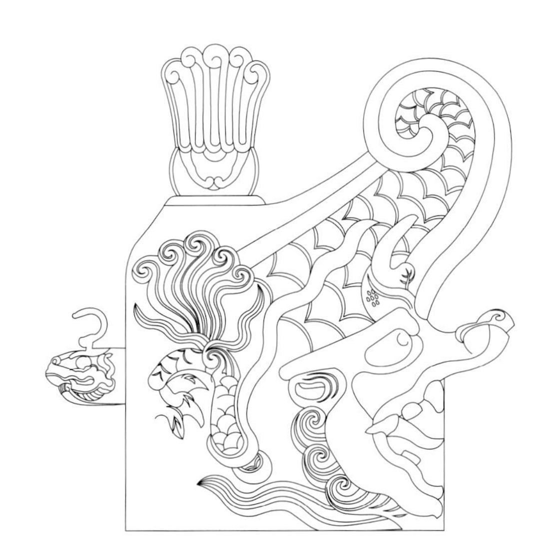

东南砖城门琉璃纹大样图

Detail drawing of the texture of Hall of Southeast Bricked Gate

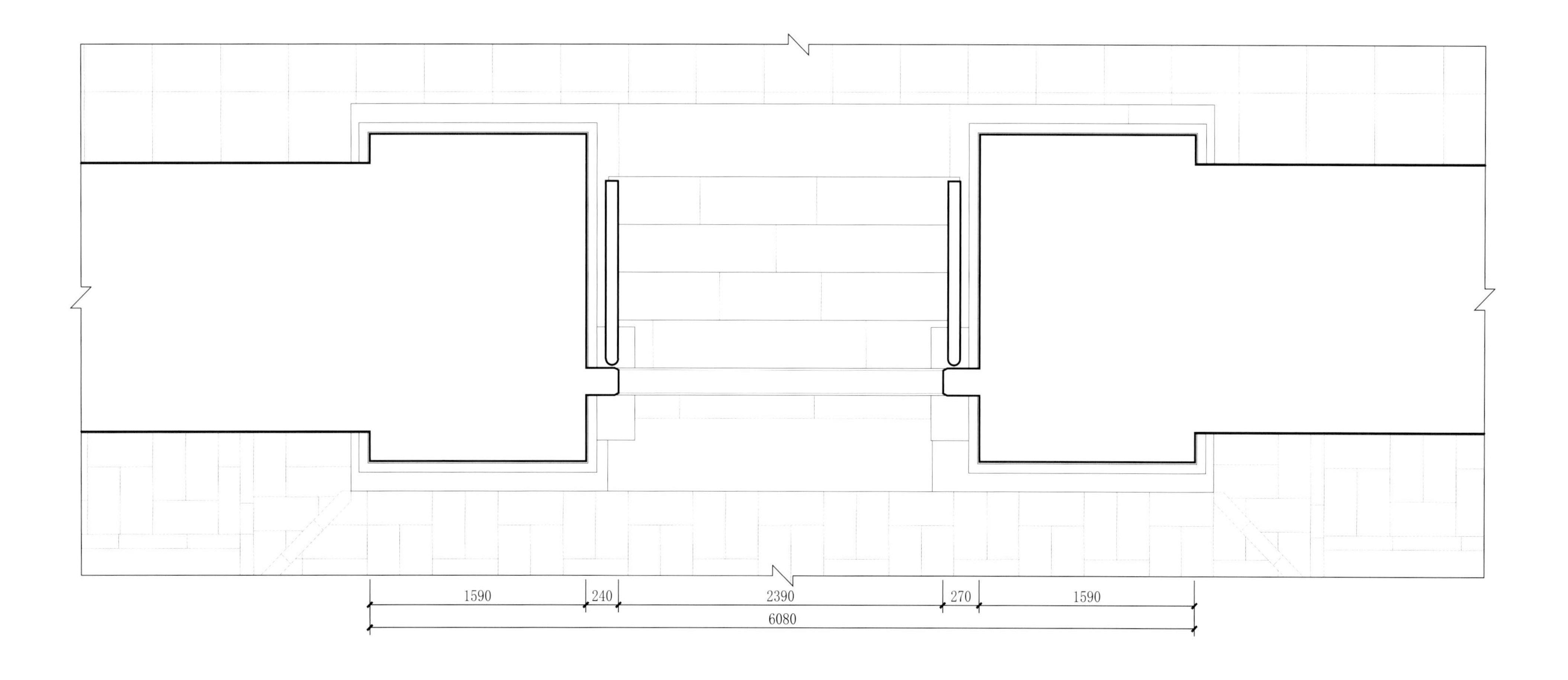

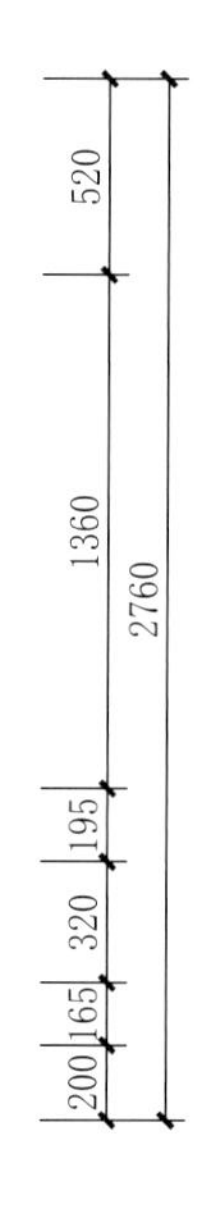

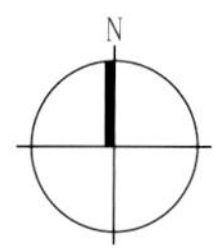

西南砖城门平面图
Plan of Southwest Bricked Gate

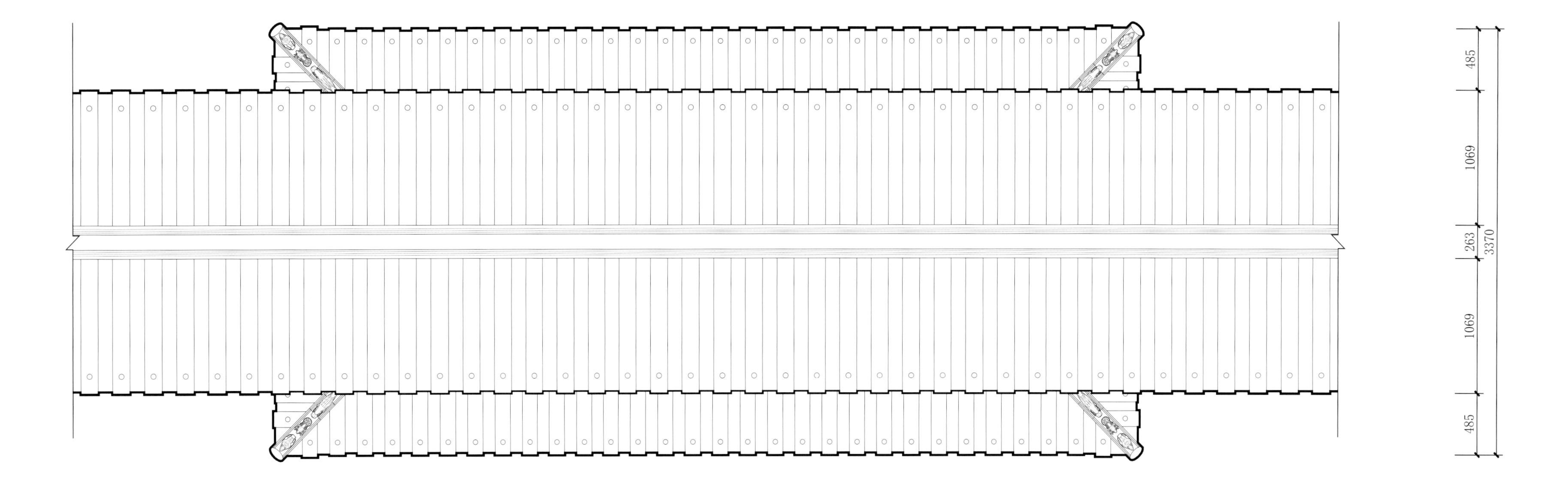

西南砖城门屋顶平面图
Roof plan of Southwest Bricked Gate

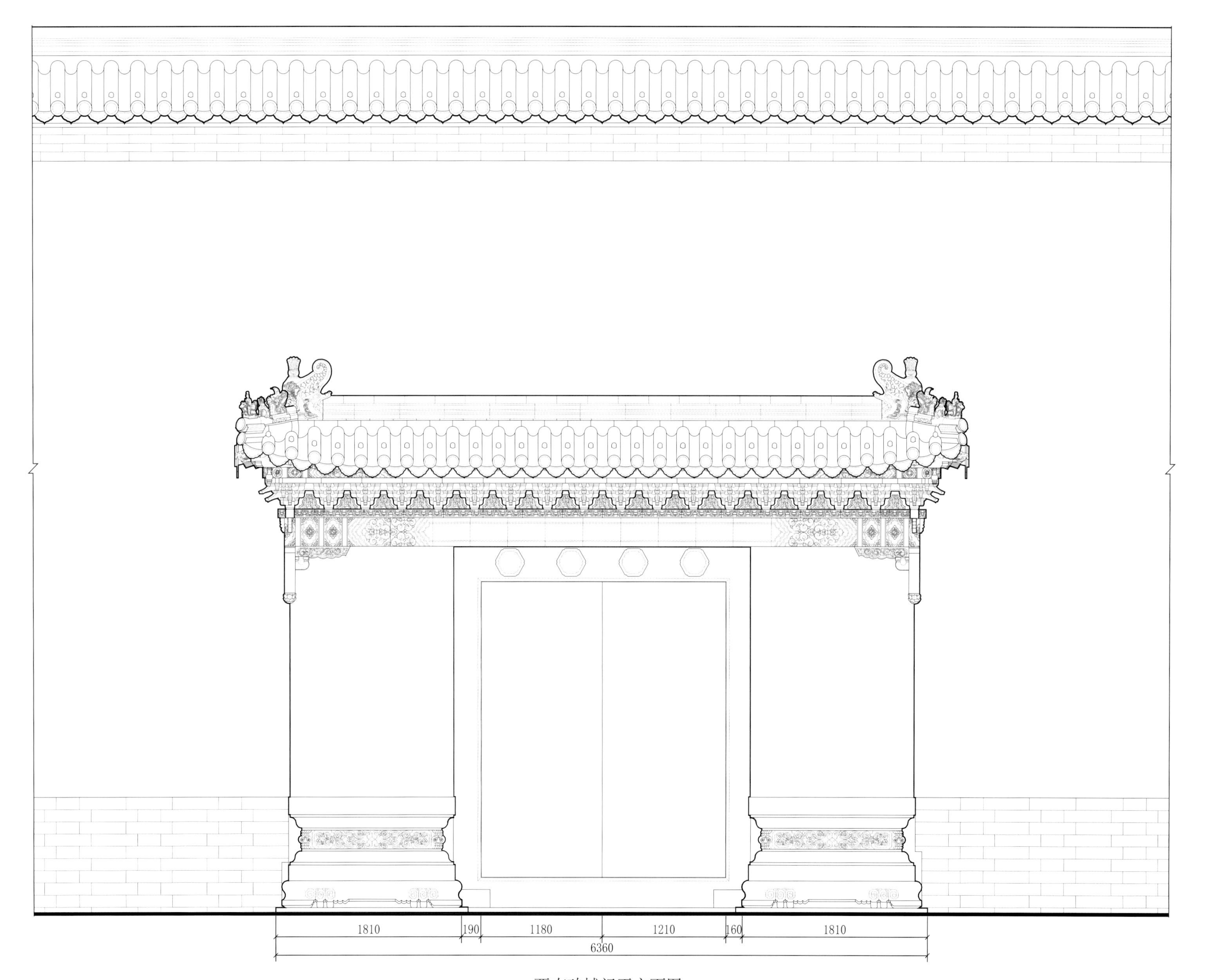

西南砖城门正立面图
Front elevation of Southwest Bricked Gate

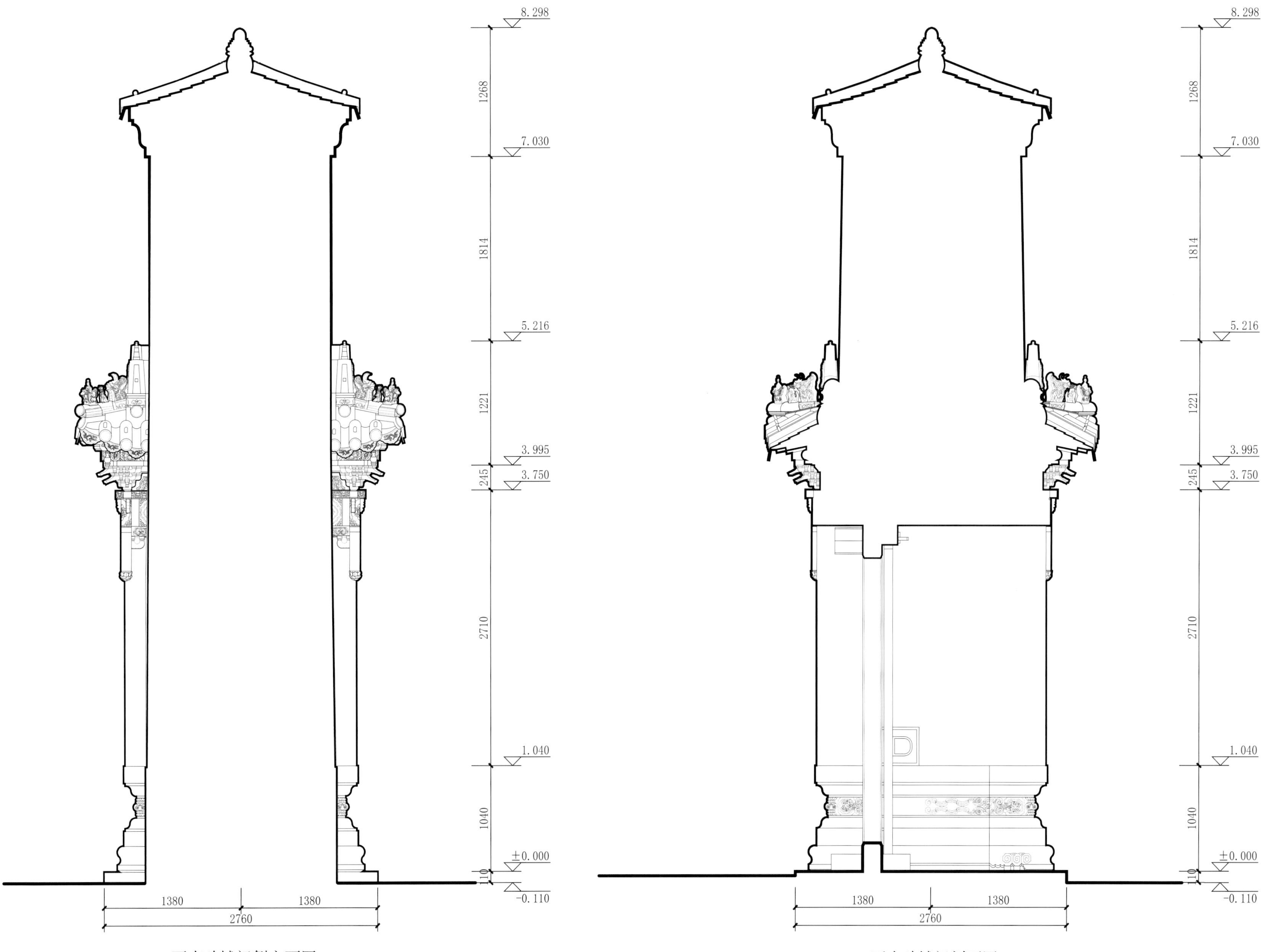

西南砖城门侧立面图
Side elevation of Southwest Bricked Gate

西南砖城门剖面图
Section of Southwest Bricked Gate

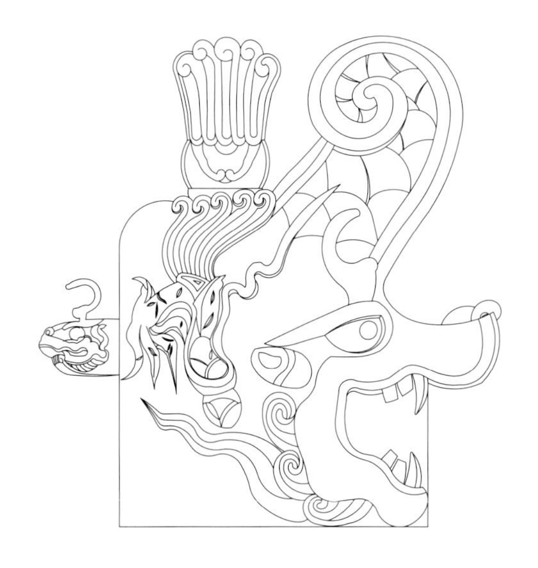

西南砖城门琉璃纹大样图
Detail drawing of the texture of Hall of Southwest Bricked Gate

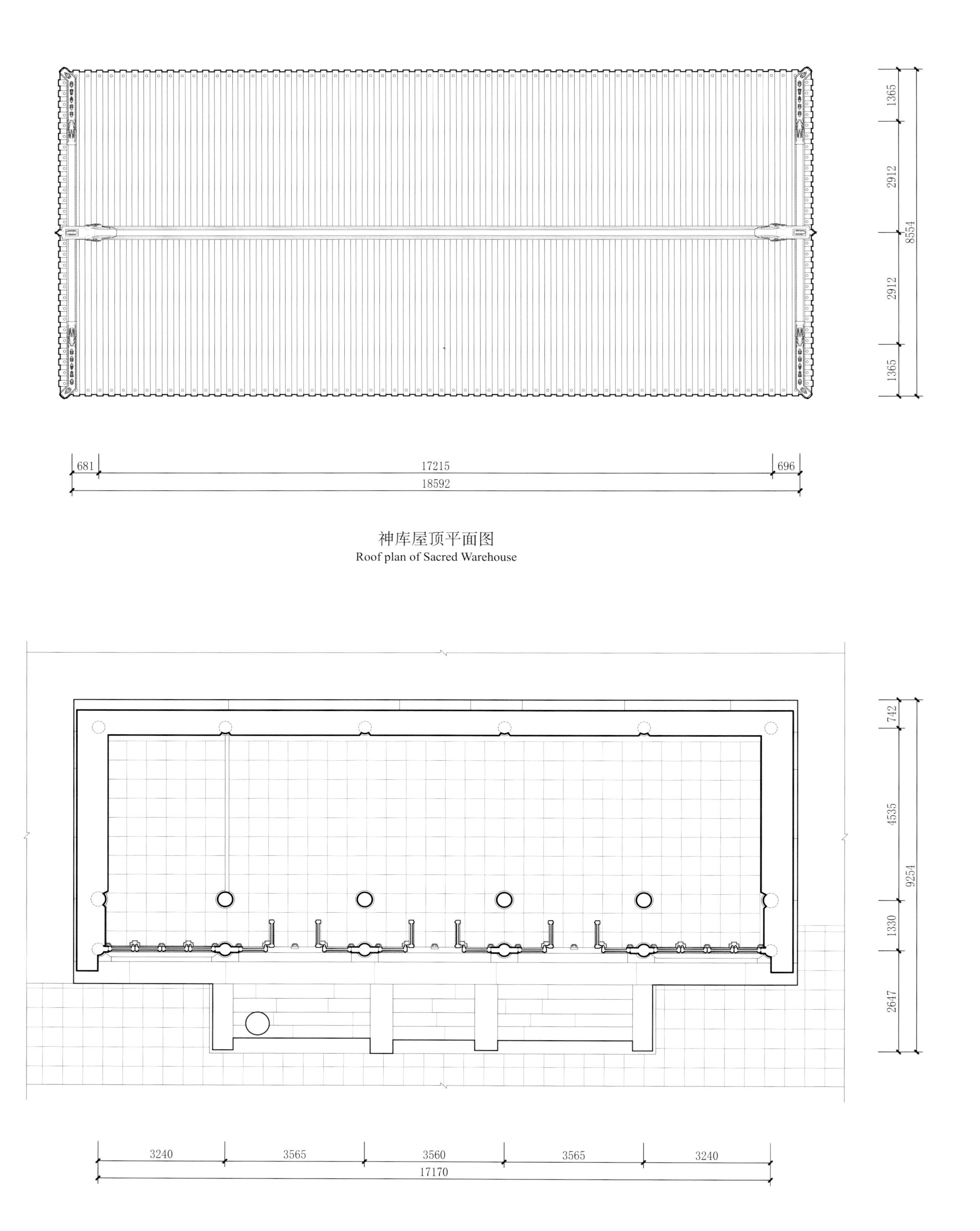

神库屋顶平面图
Roof plan of Sacred Warehouse

神库平面图
Plan of Sacred Warehouse

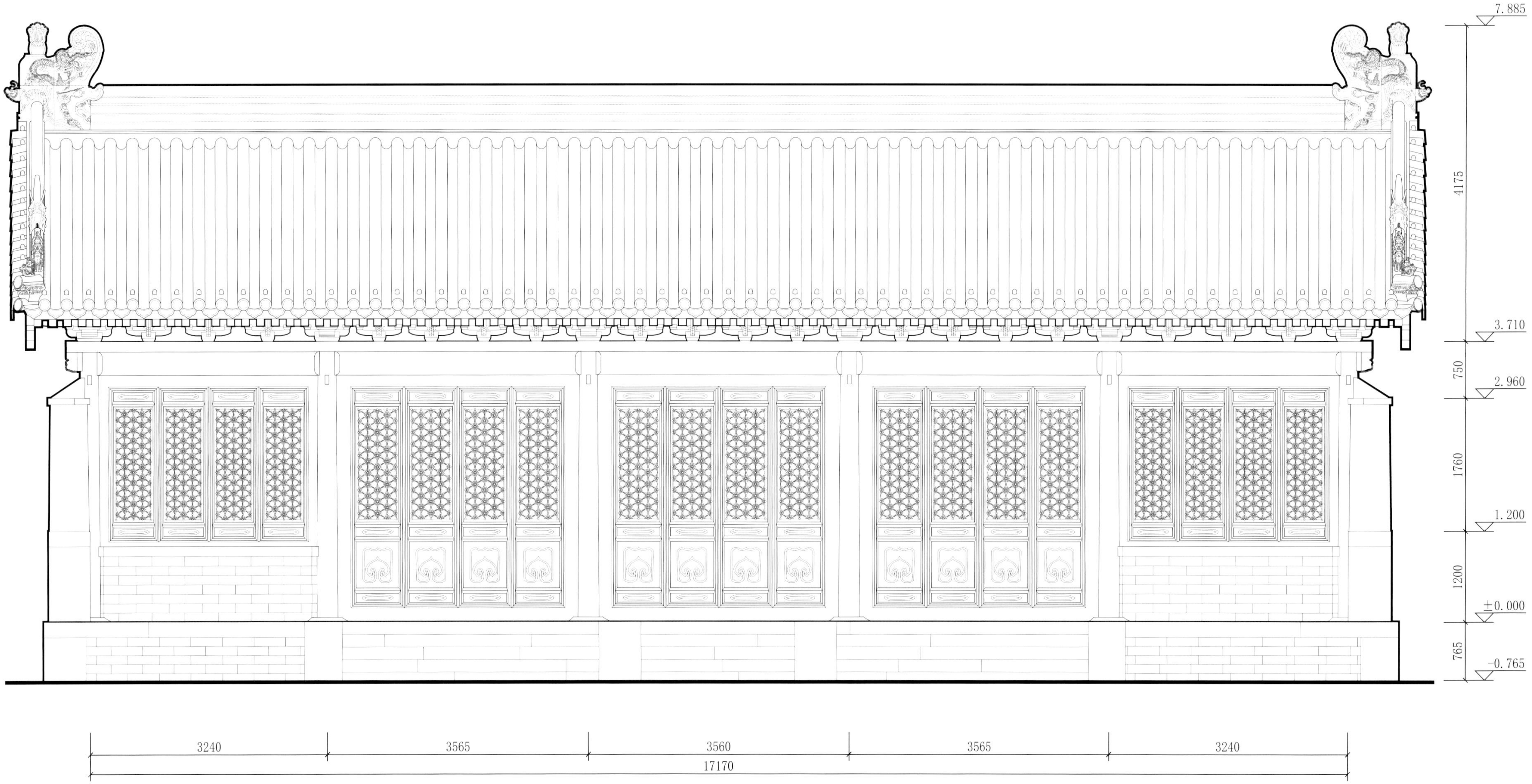

7.885
4175
3.710
750
2.960
1760
1.200
1200
±0.000
765
-0.765

3240 3565 3560 3565 3240
17170

神库正立面图
Front elevation of Sacred Warehouse

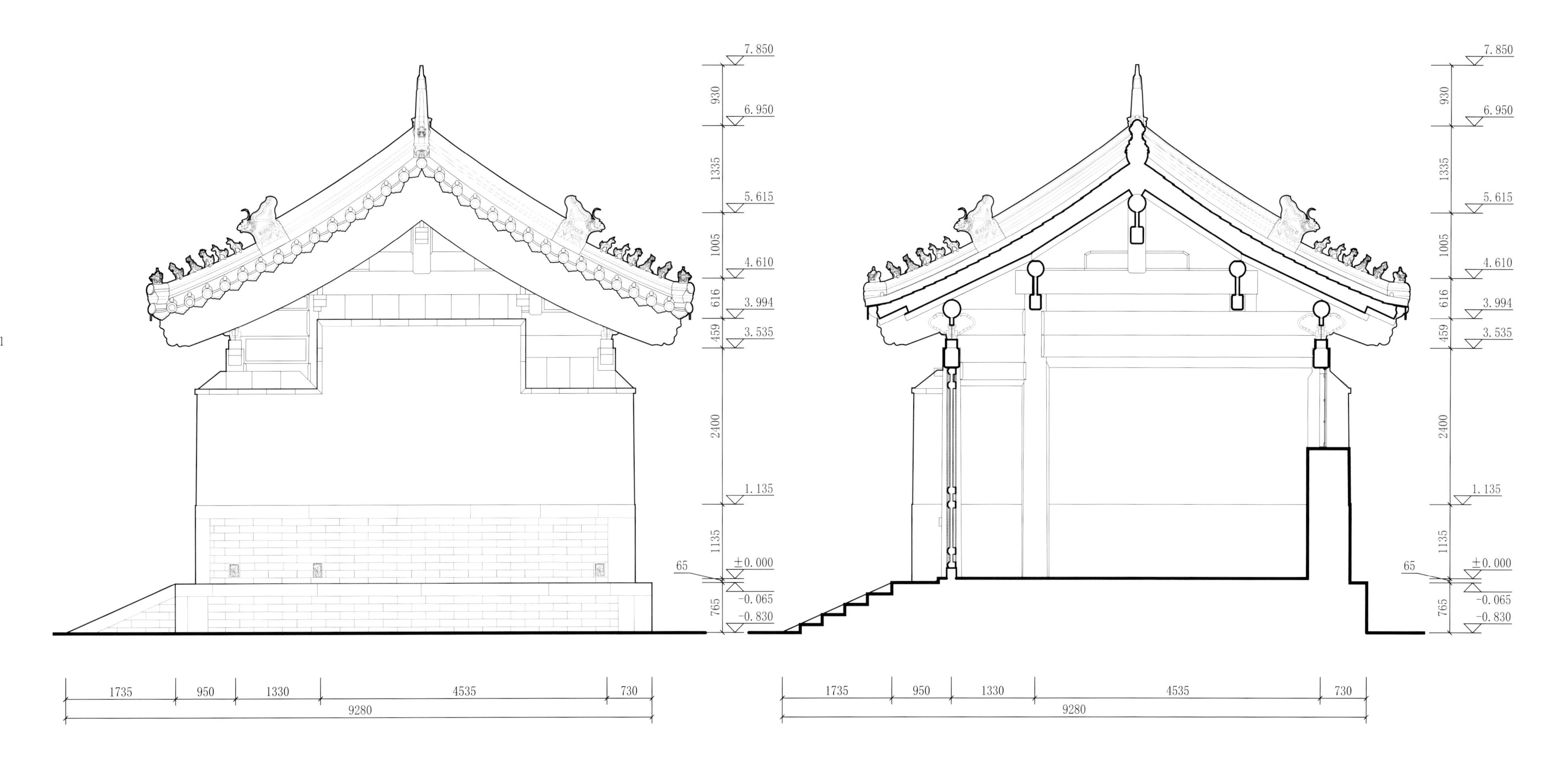

神库侧立面图

Side elevation of Sacred Warehouse

神库横剖面图

Transverse section of Sacred Warehouse

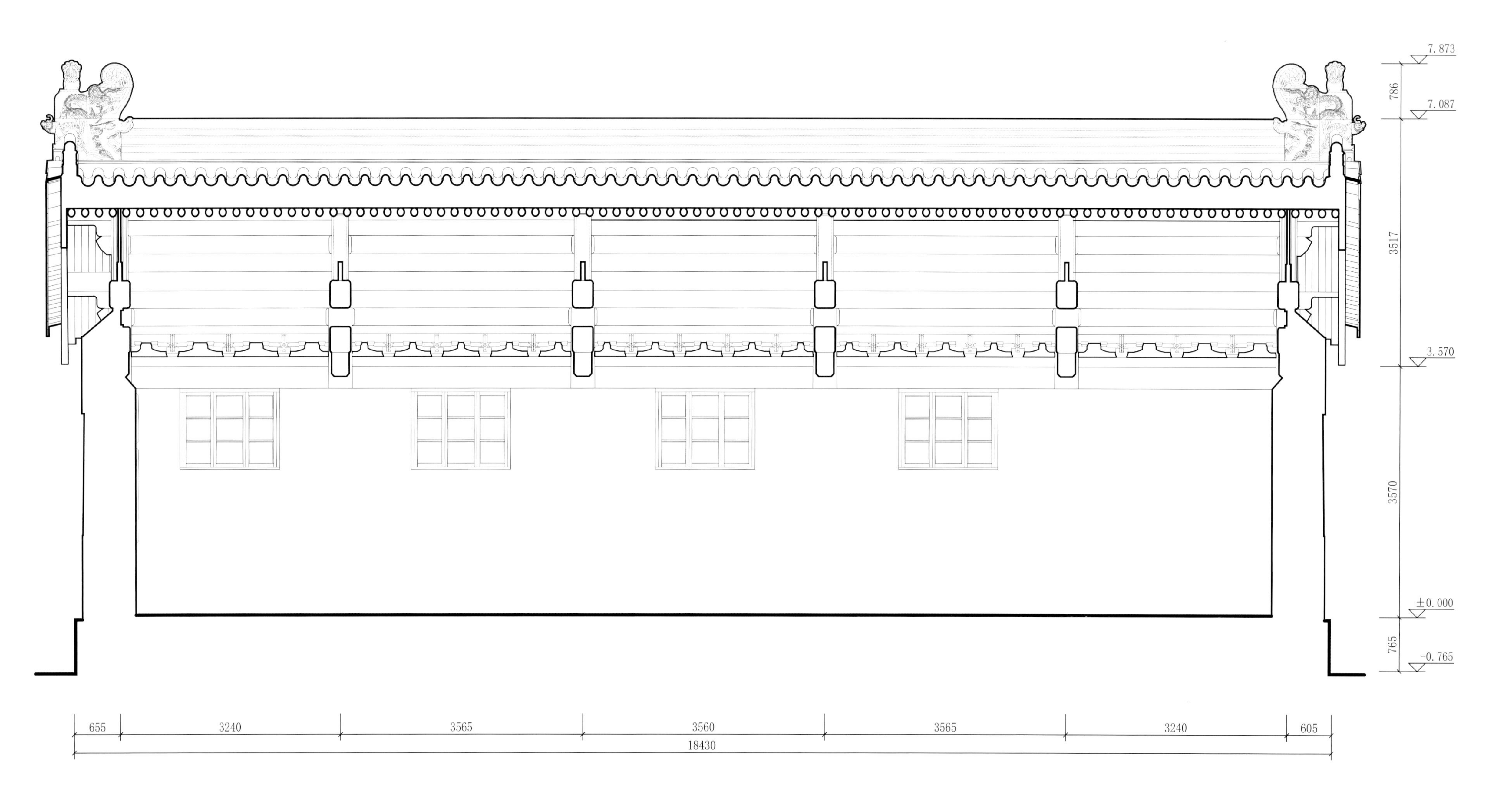

神库纵剖面图
Longitudinal section of Sacred Warehouse

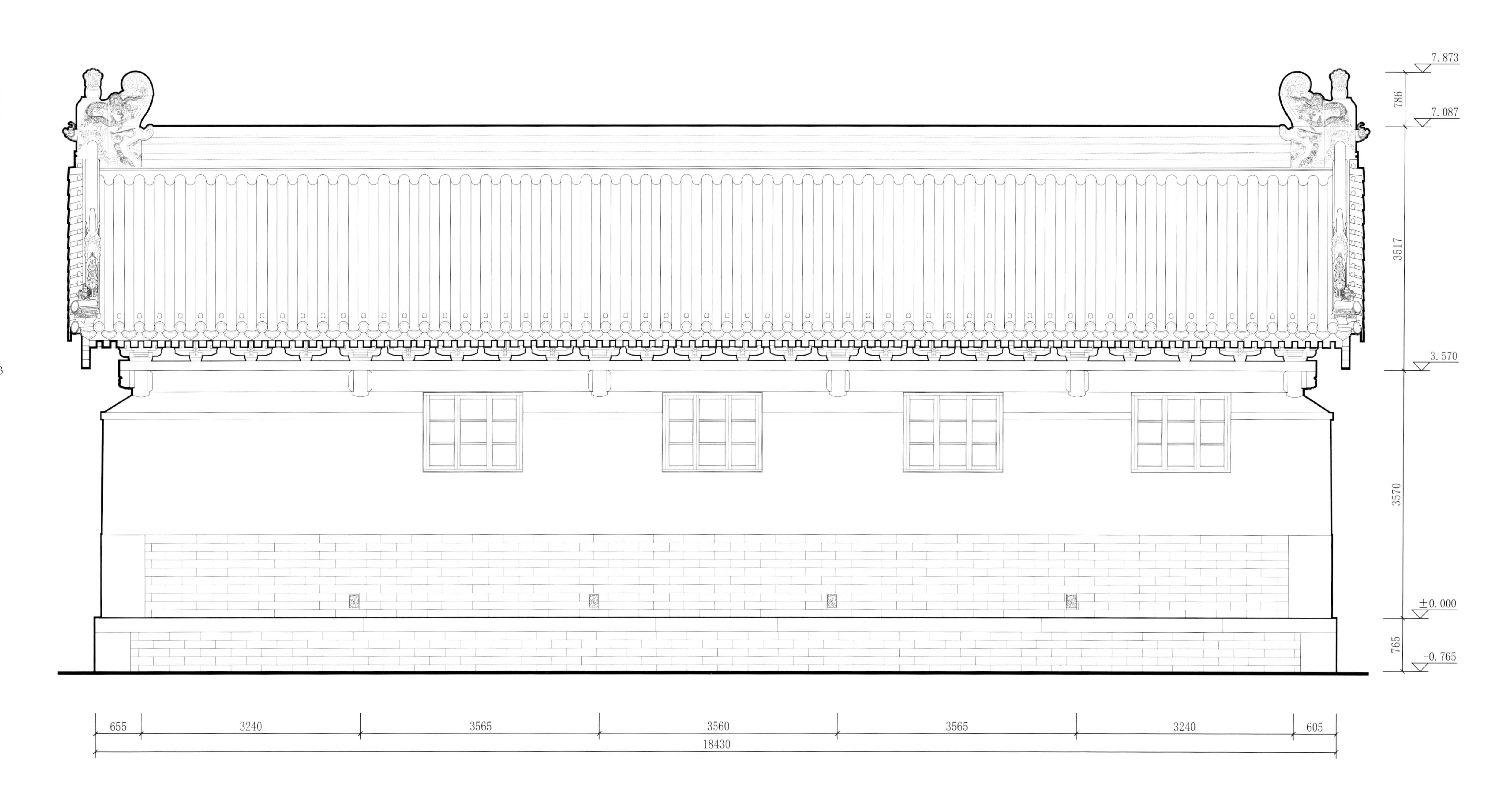

神库背立面图
Rear elevation of Sacred Warehouse

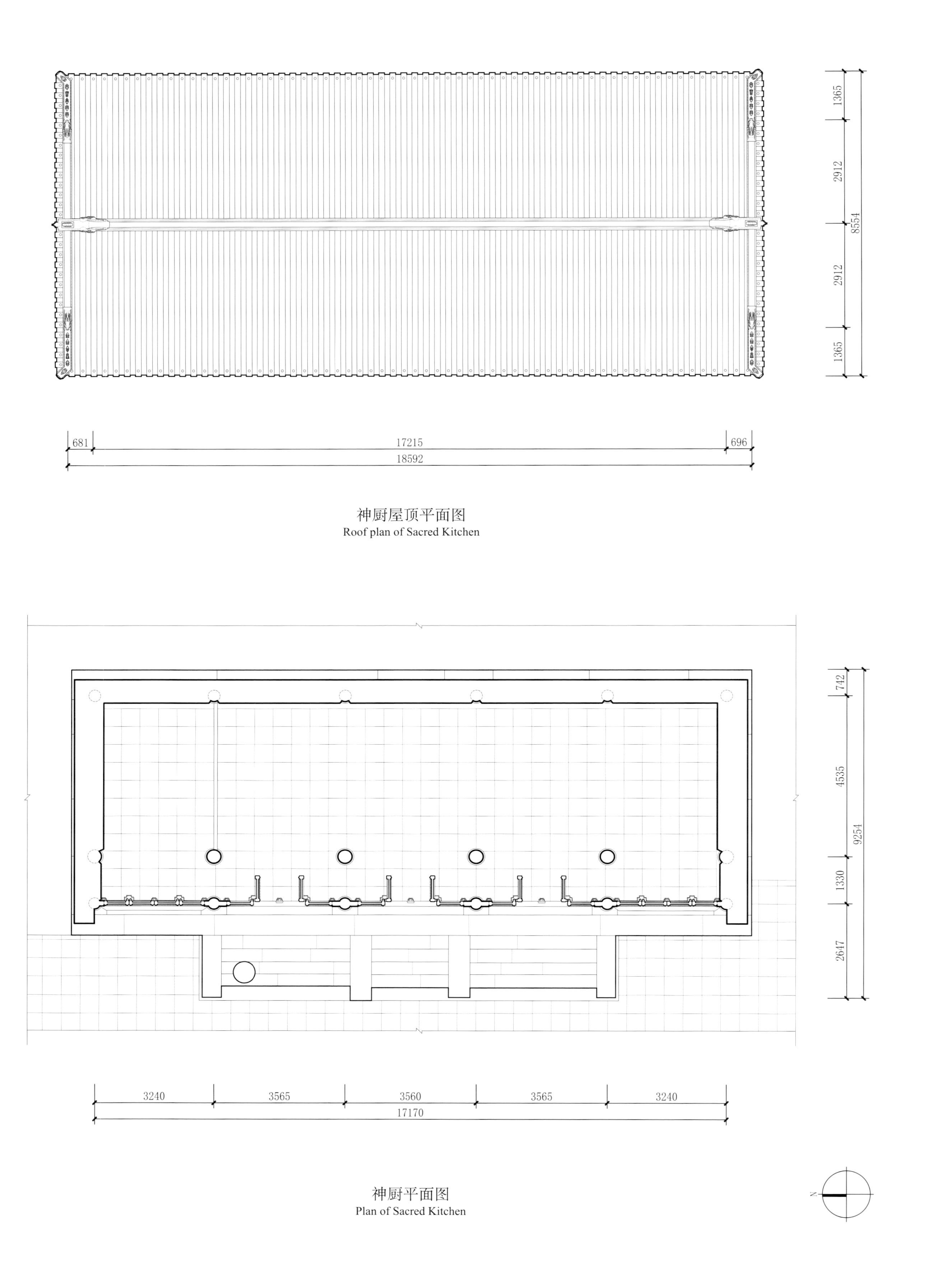

神厨屋顶平面图
Roof plan of Sacred Kitchen

神厨平面图
Plan of Sacred Kitchen

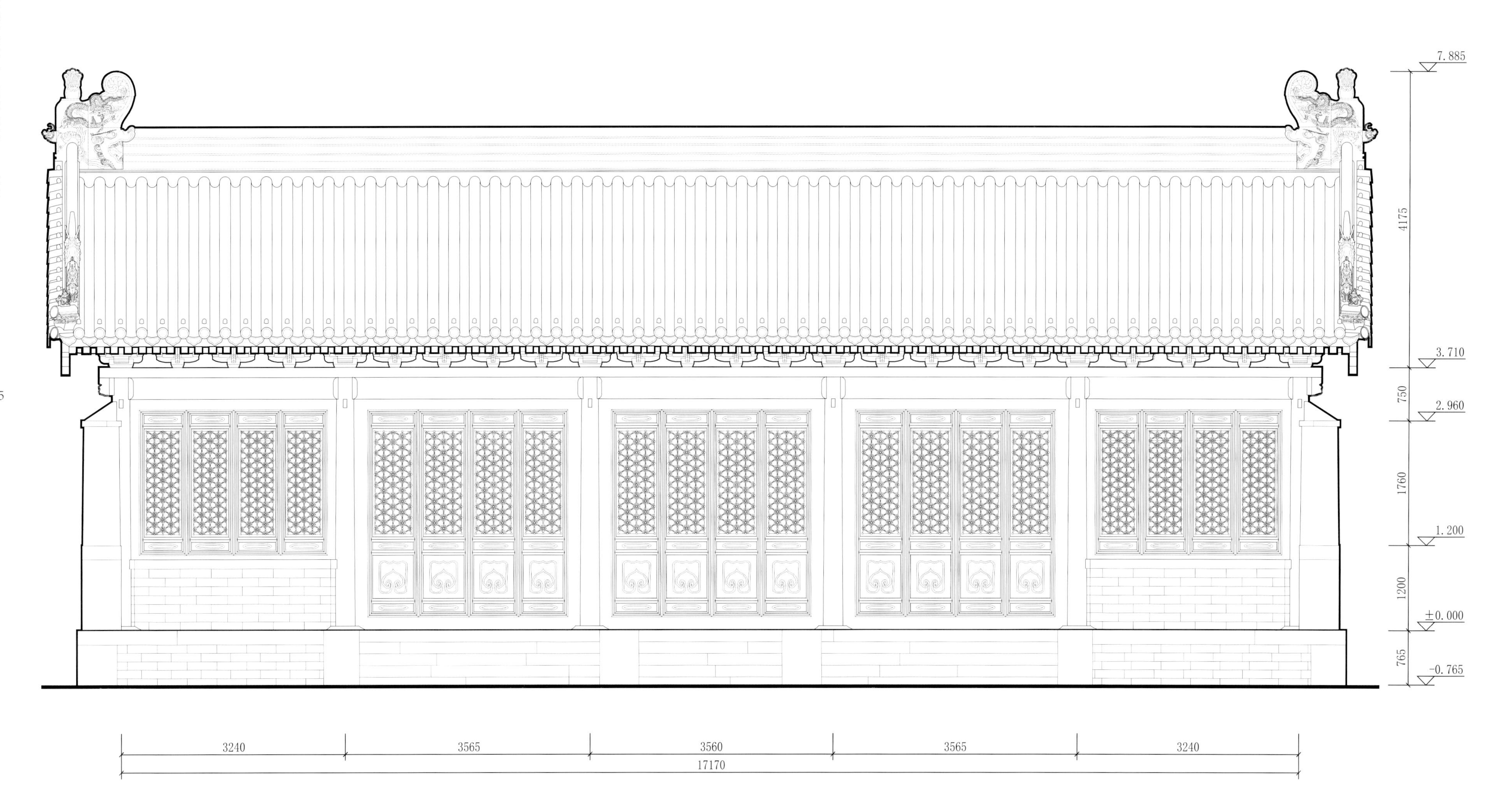

神厨正立面图
Front elevation of Sacred Kitchen

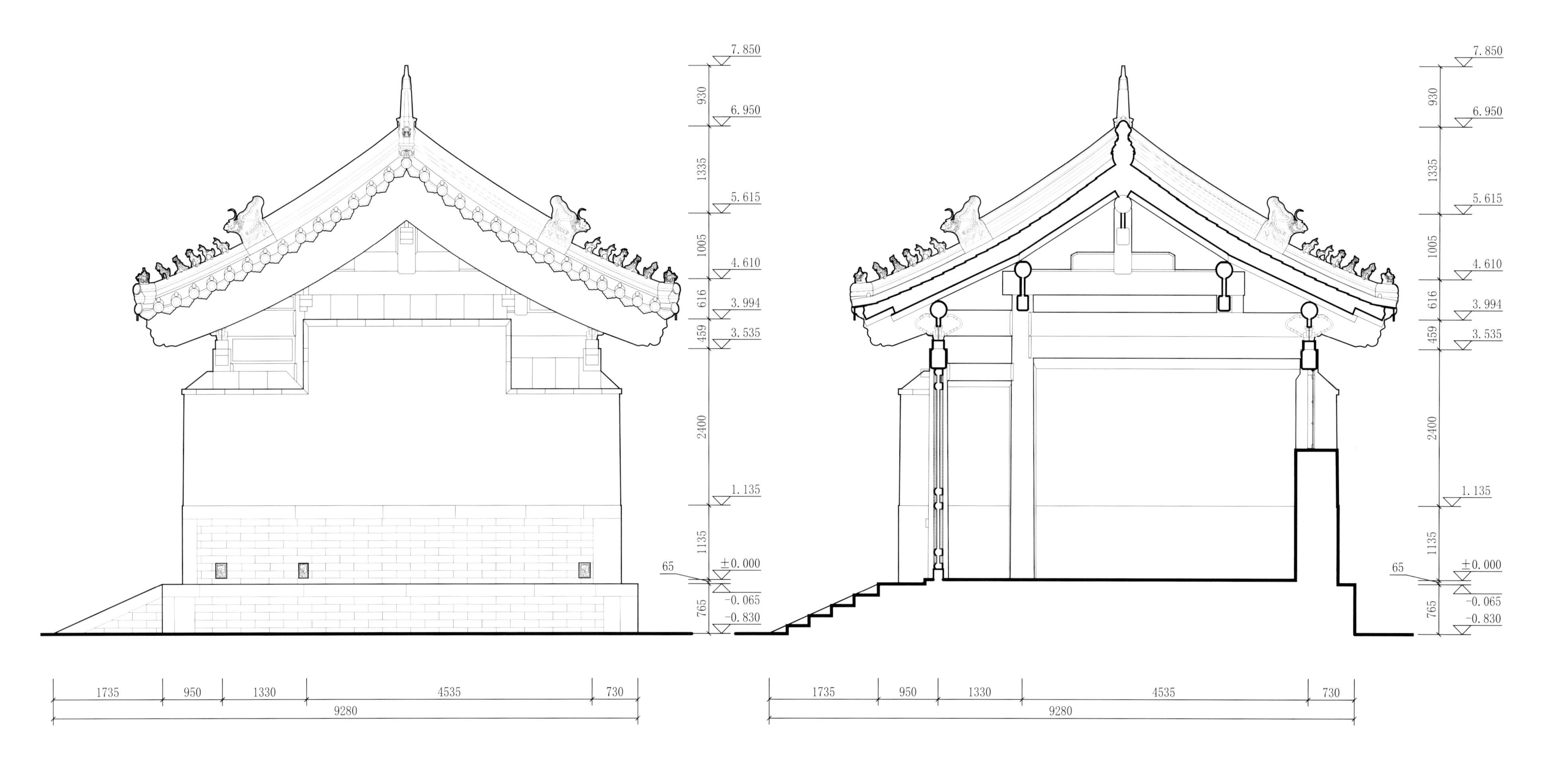

神厨侧立面图
Side elevation of Sacred Kitchen

神厨横剖面图
Transverse section of Sacred Kitchen

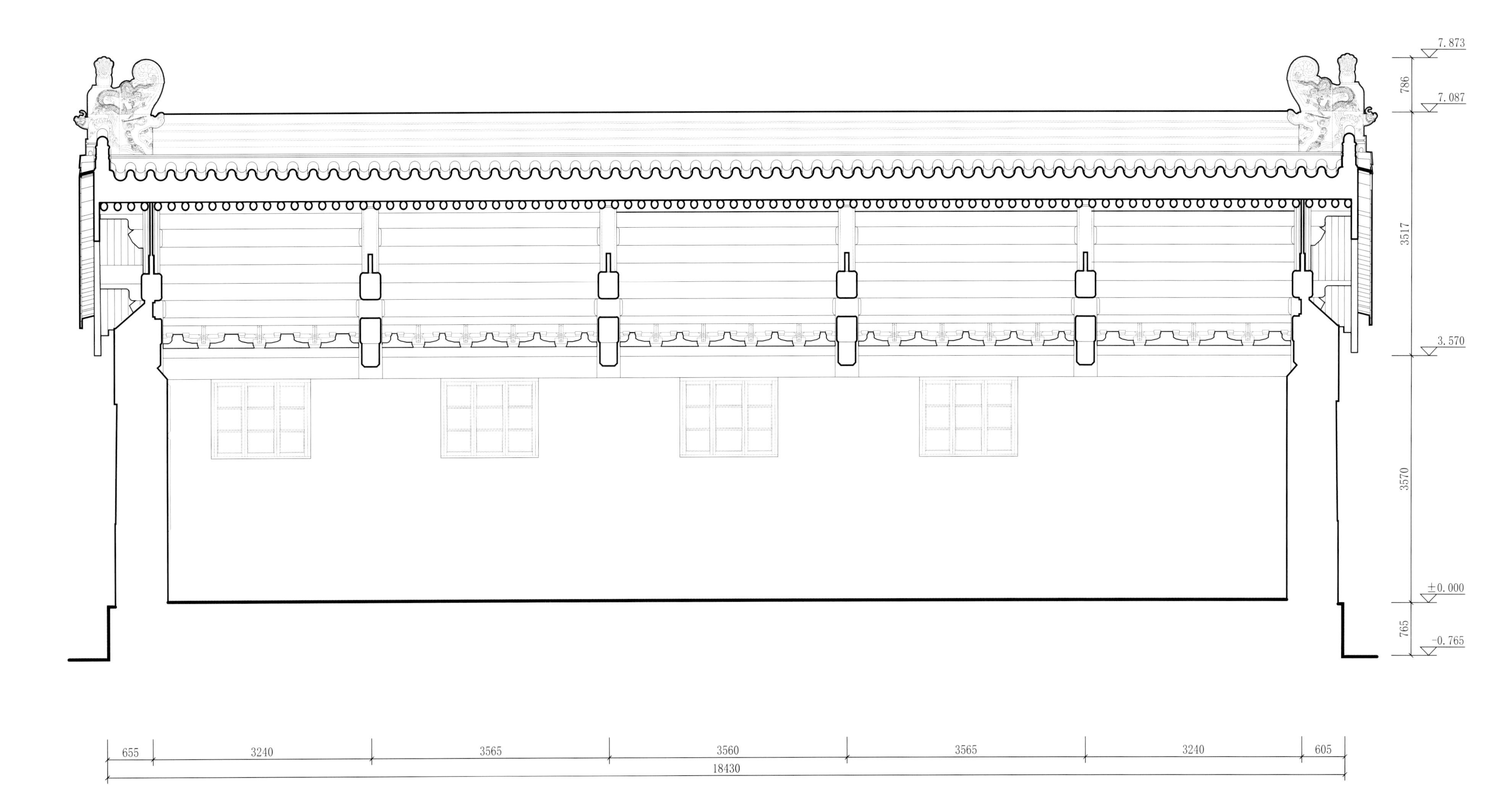

神厨纵剖面图

Longitudinal section of Sacred Kitchen

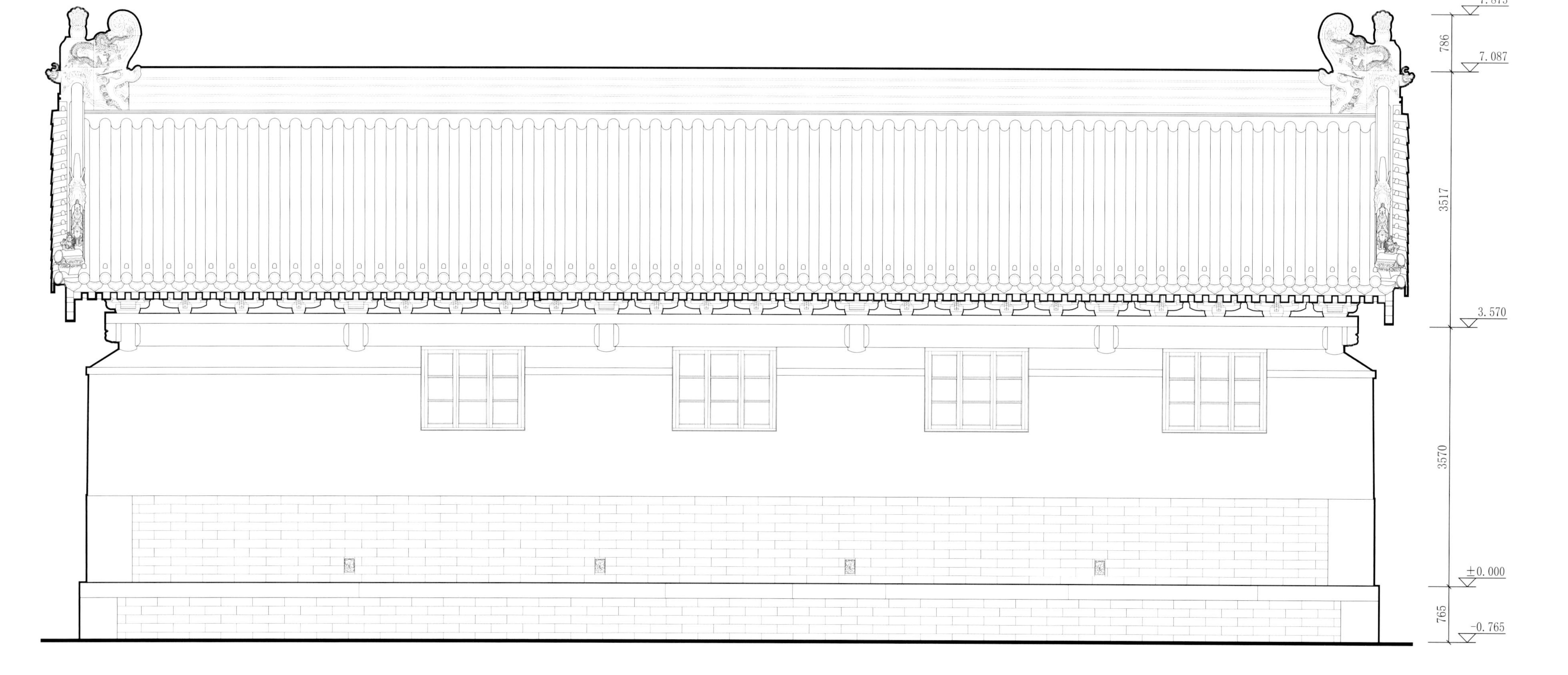

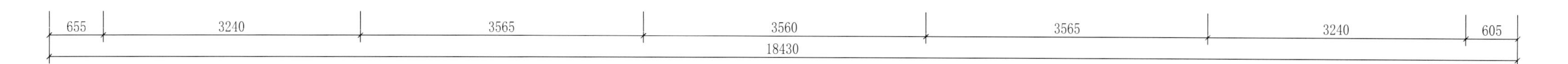

神厨背立面图

Rear elevation of Sacred Kitchen

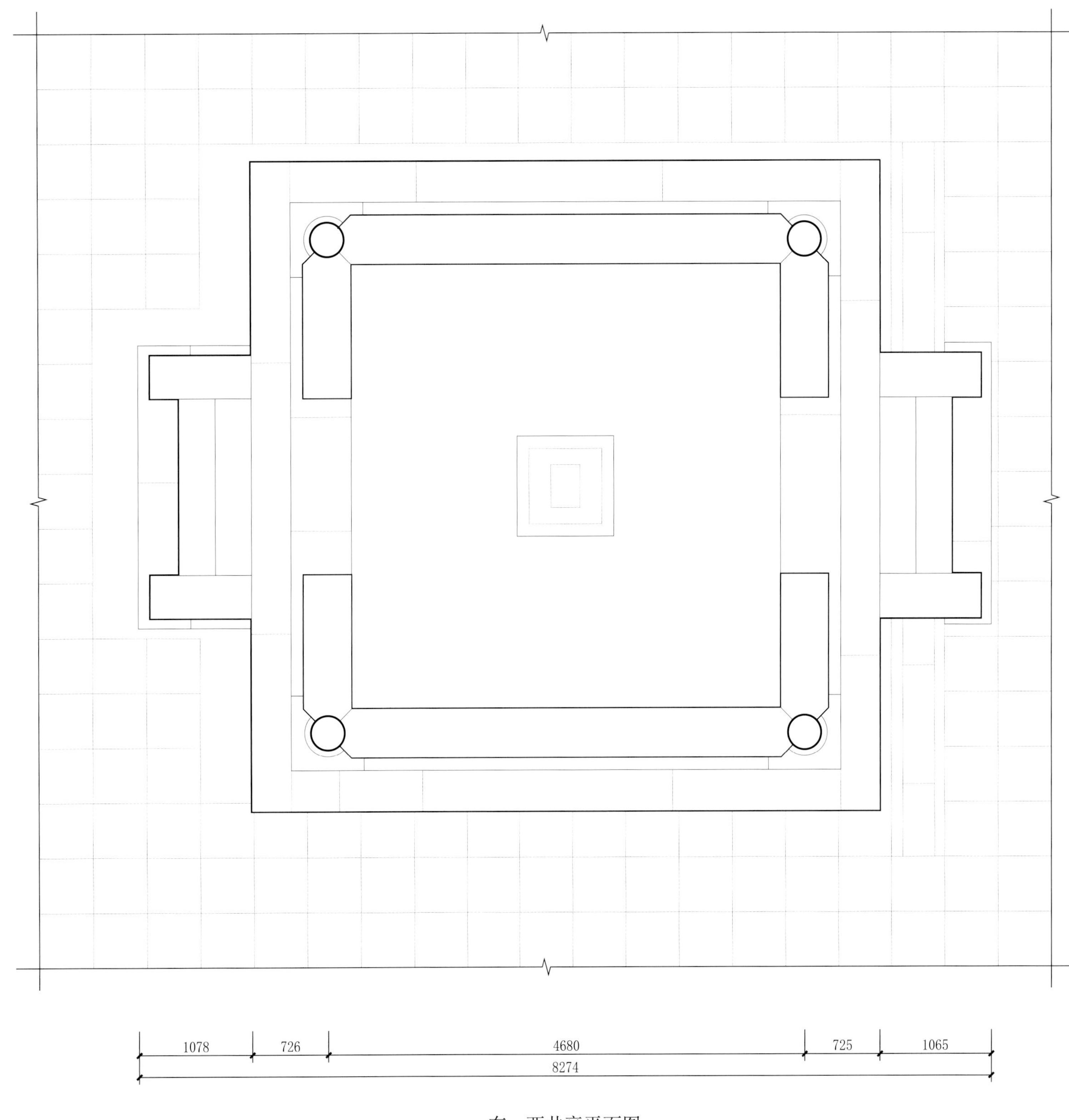

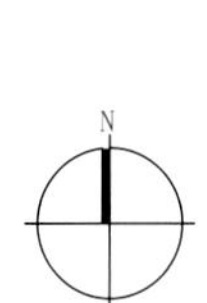

东、西井亭平面图
Plan of East or West Well Pavilion

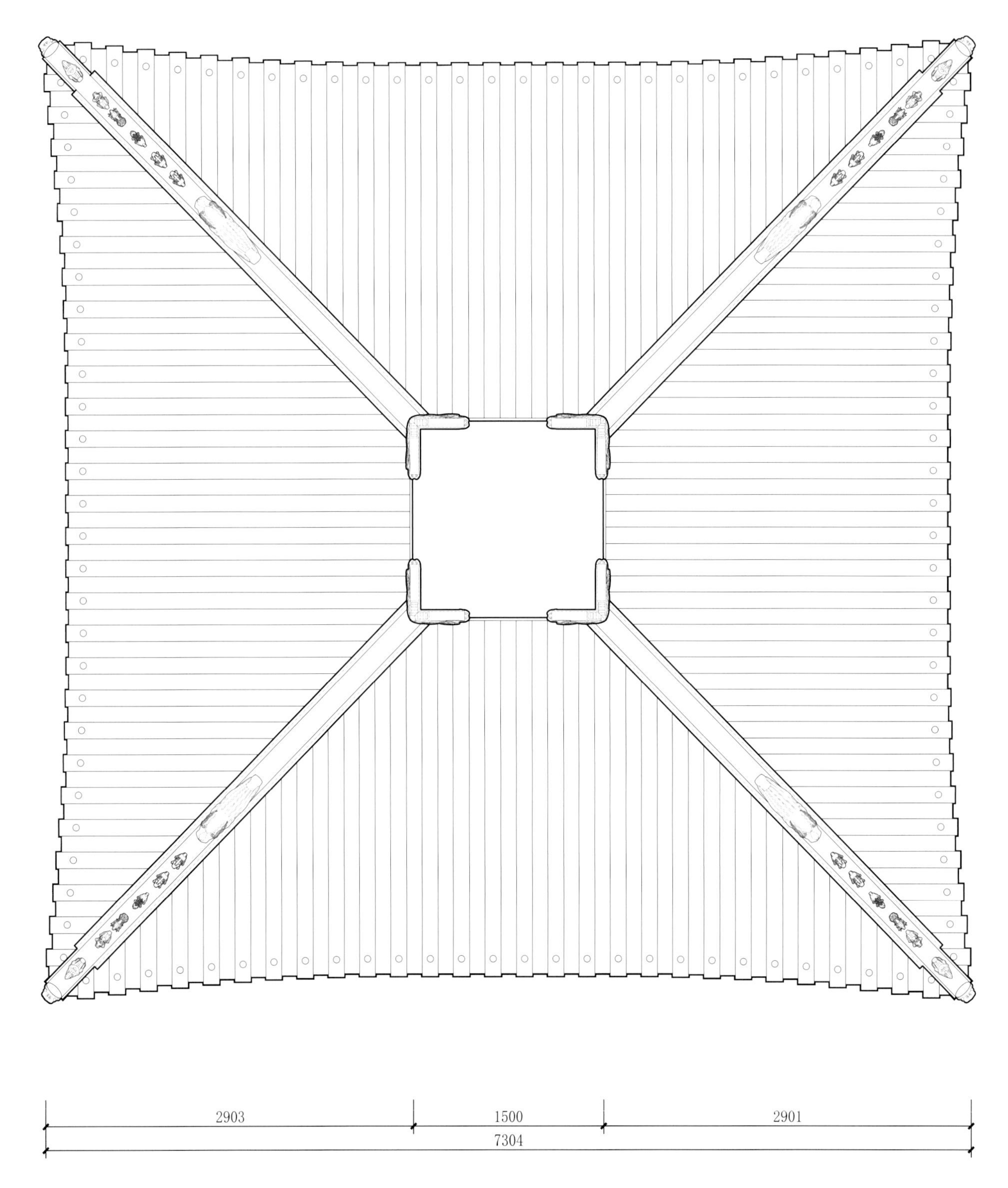

东、西井亭屋顶平面图
Roof Plan of East or West Well Pavilion

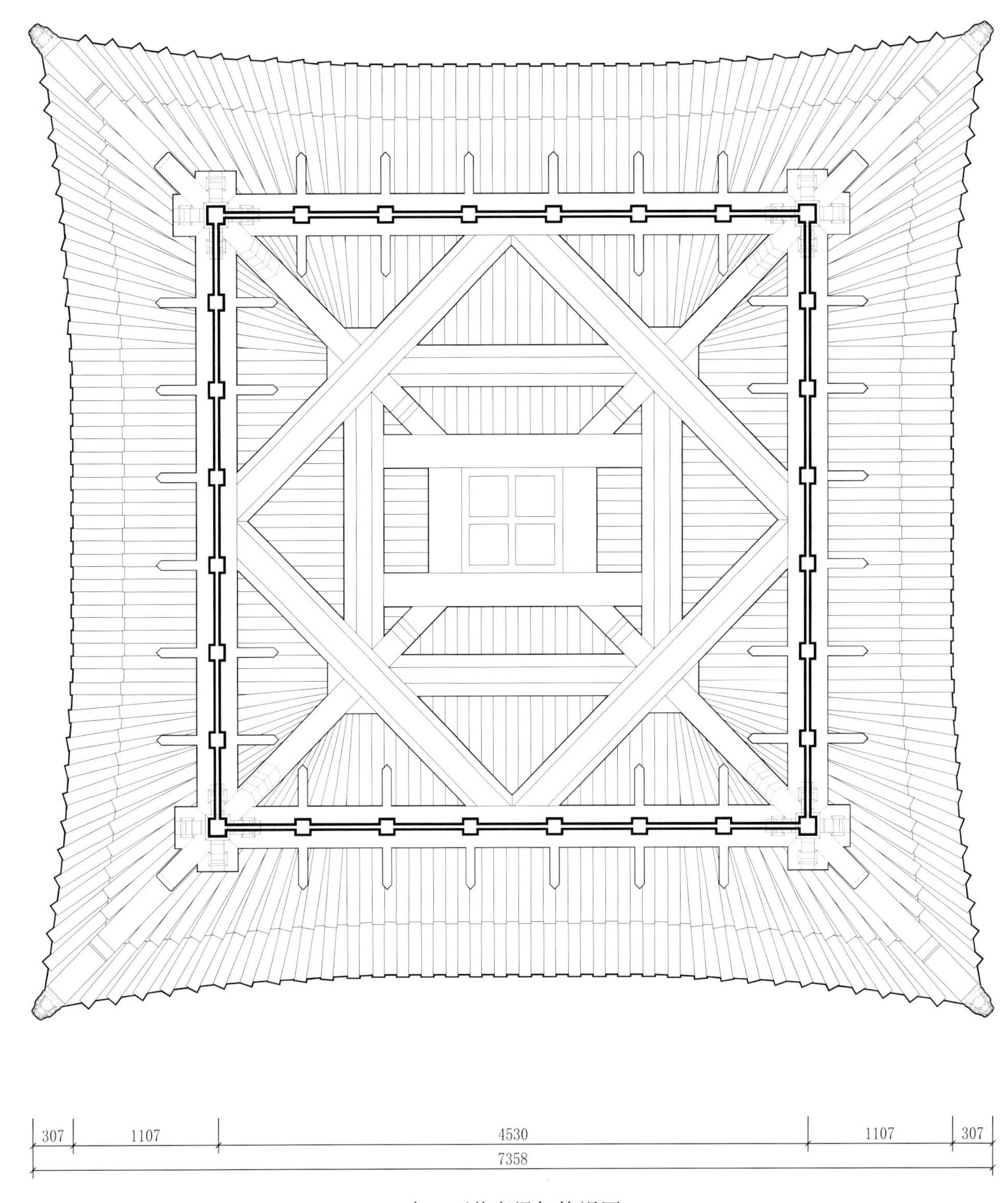

东、西井亭梁架仰视图

Bottom view of beams of East or West Well Pavilion

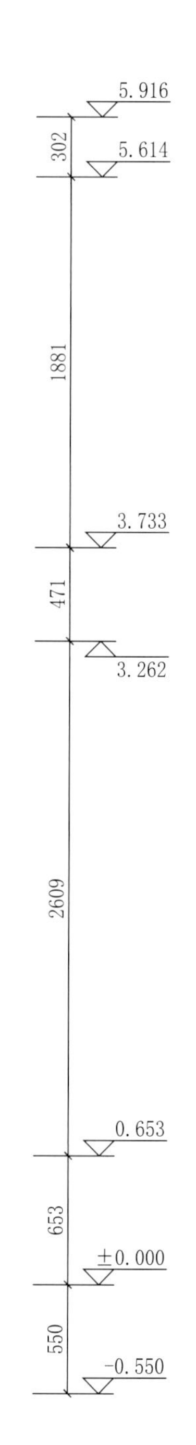

725 4580 725
6030

东、西井亭正立面图

Front elevation of East or West Well Pavilion

东、西井亭侧立面图
Side elevation of East or West Well Pavilion

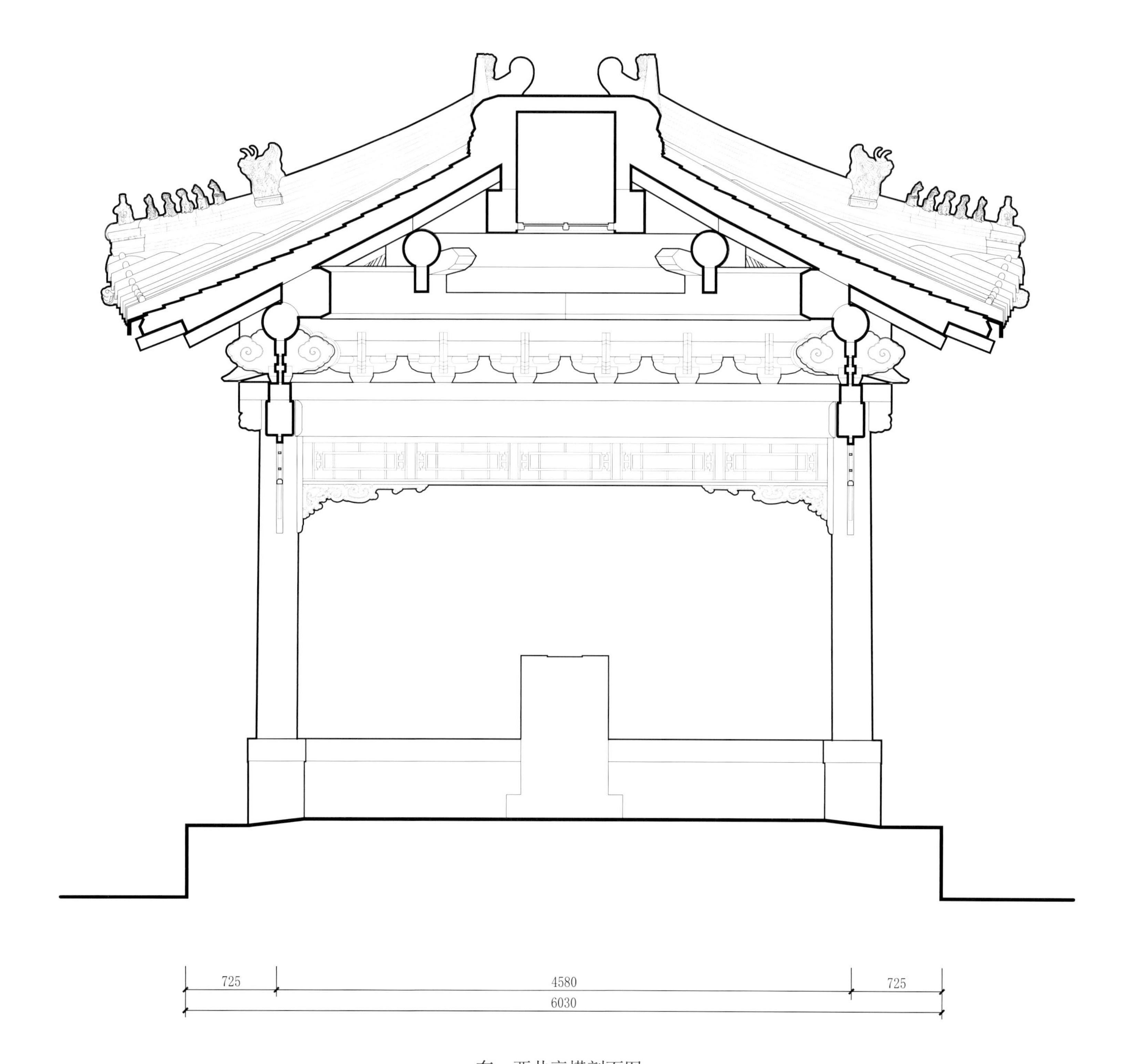

东、西井亭横剖面图
Section of East or West Well Pavilion

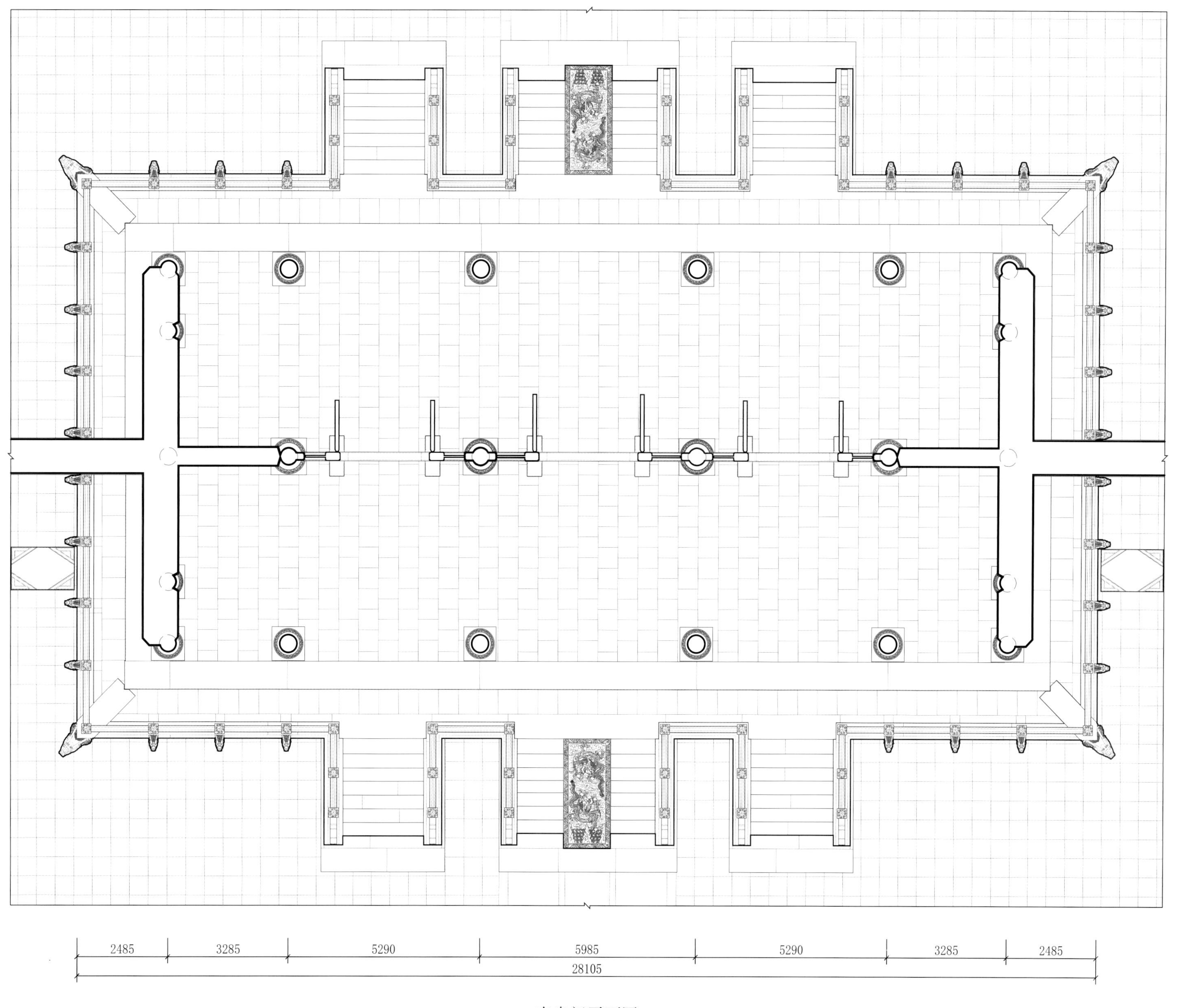

寿皇门平面图
Plan of Gate of Shouhuang

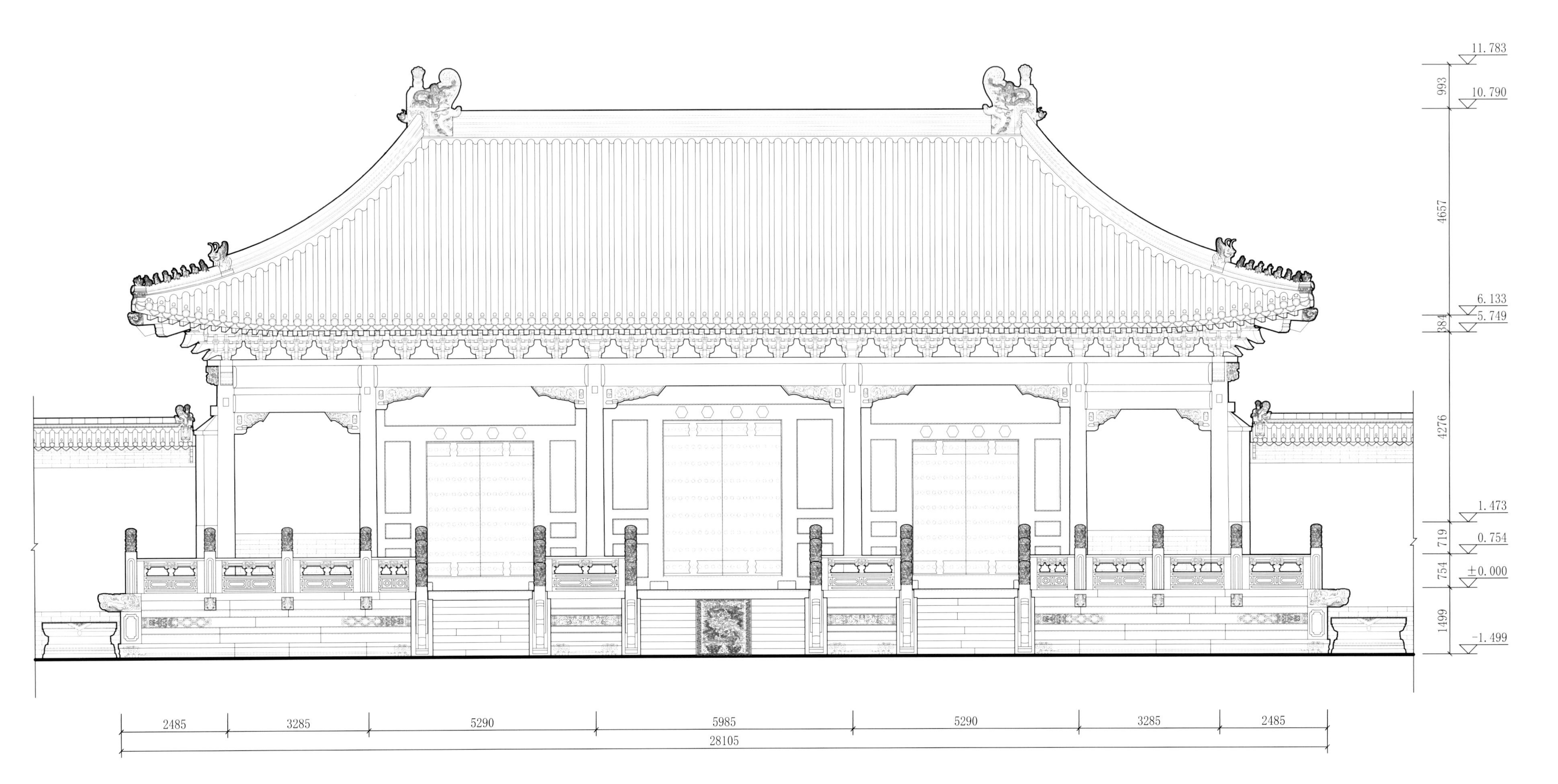

寿皇门南立面图

South elevation of Gate of Shouhuang

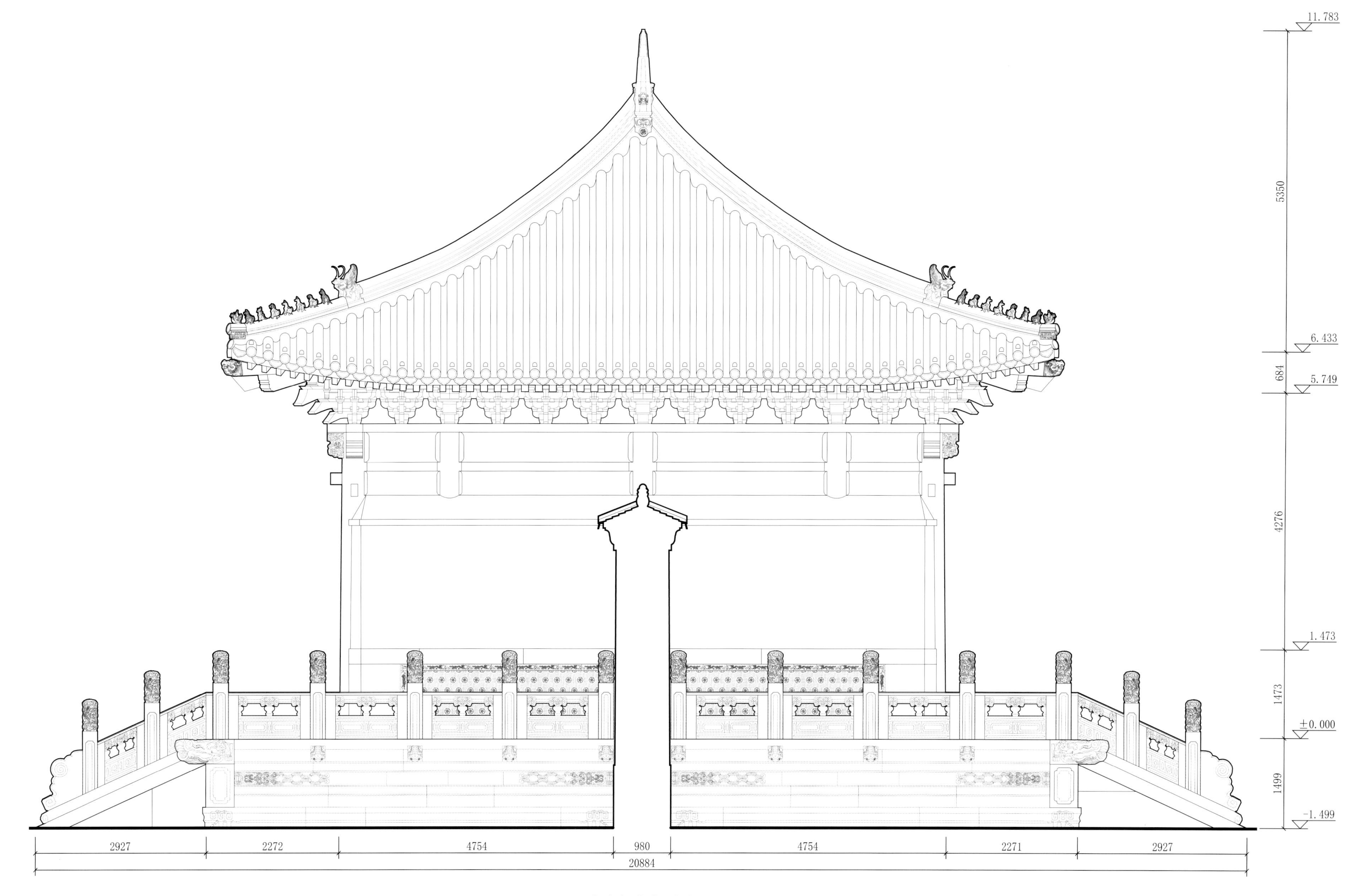

寿皇门东立面图

East elevation of Gate of Shouhuang

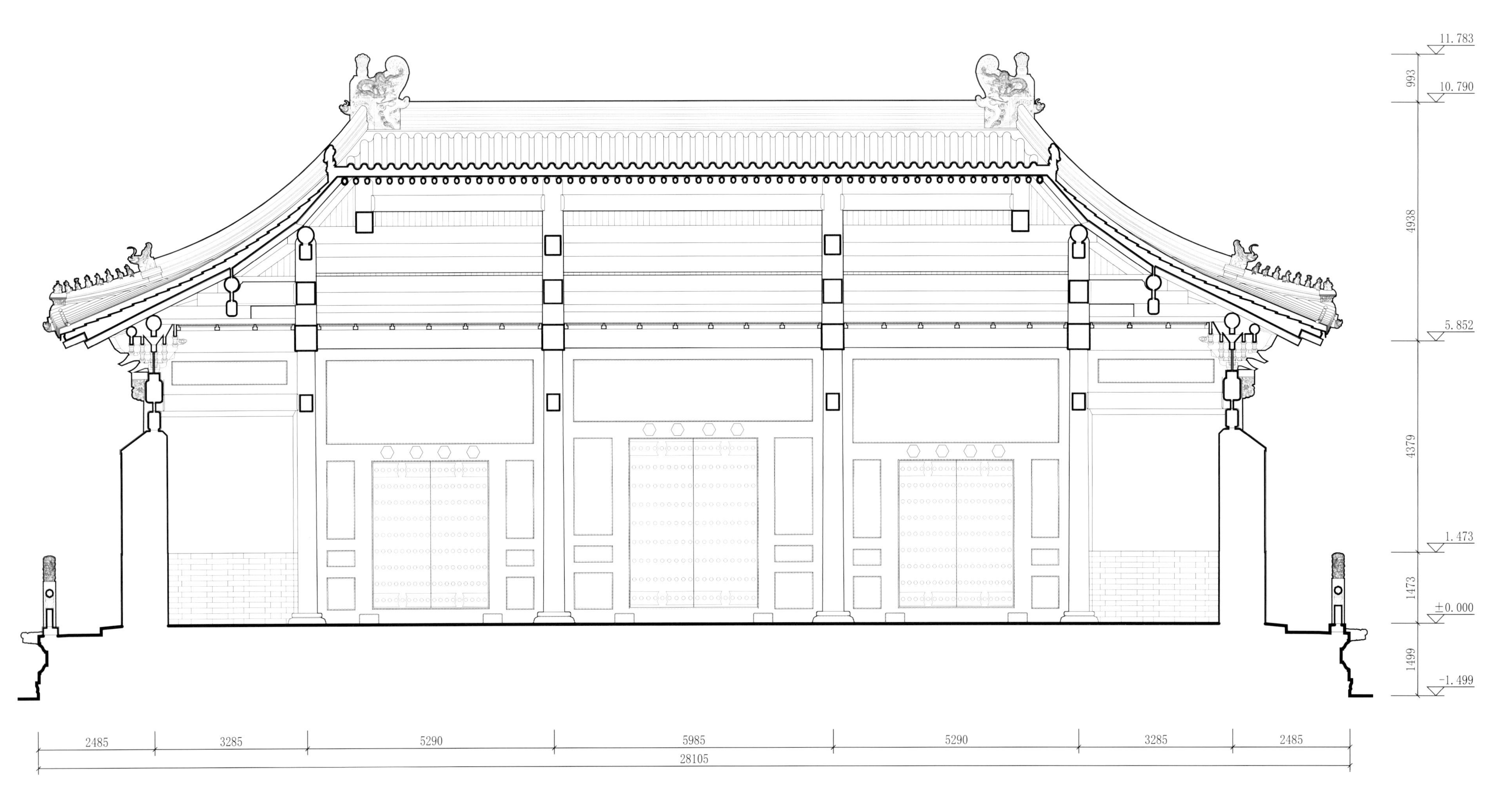

寿皇门纵剖面图

Longitudinal section of Gate of Shouhuang

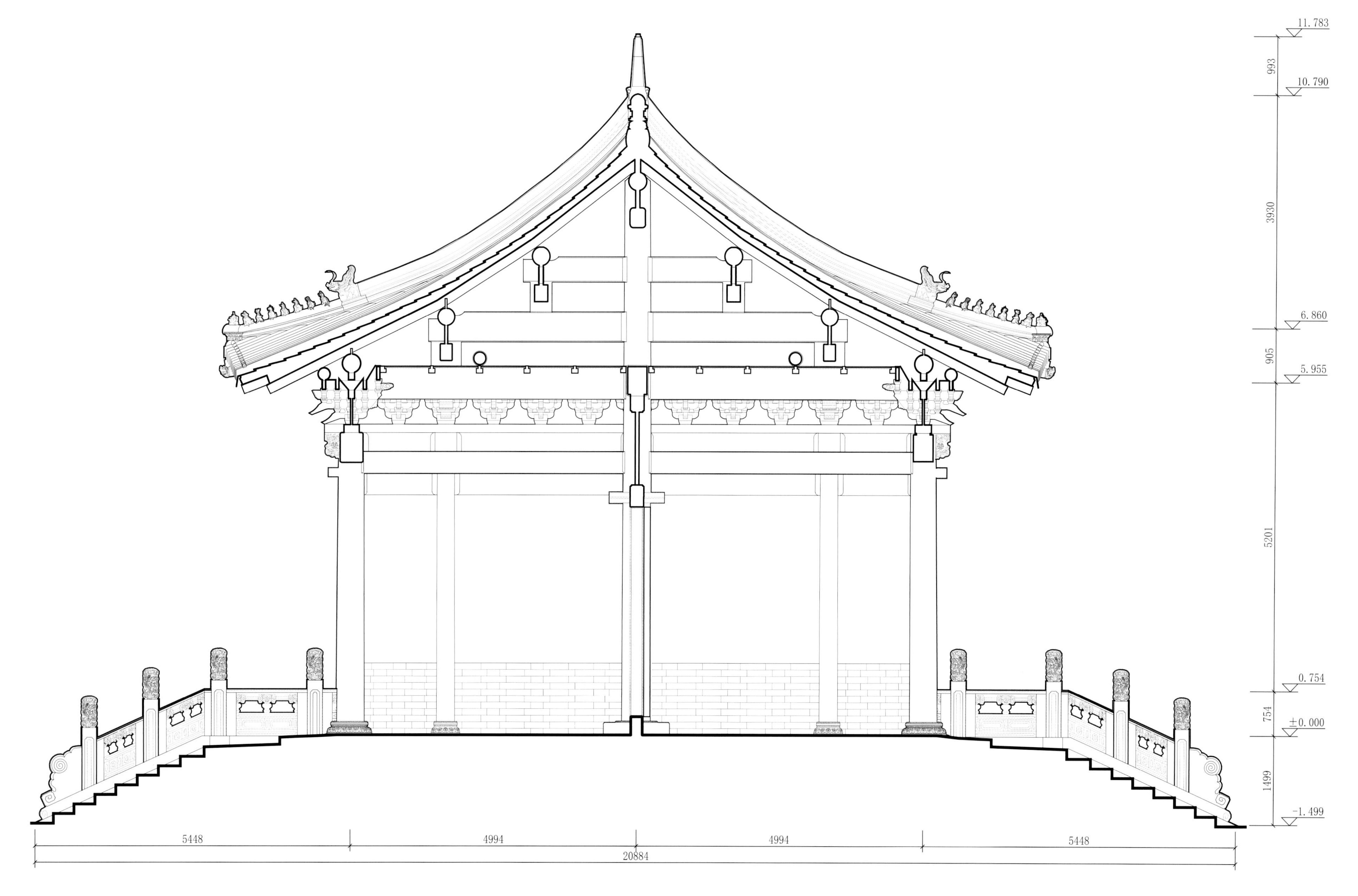

寿皇门明间横剖面图
Central bay transverse section of Gate of Shouhuang

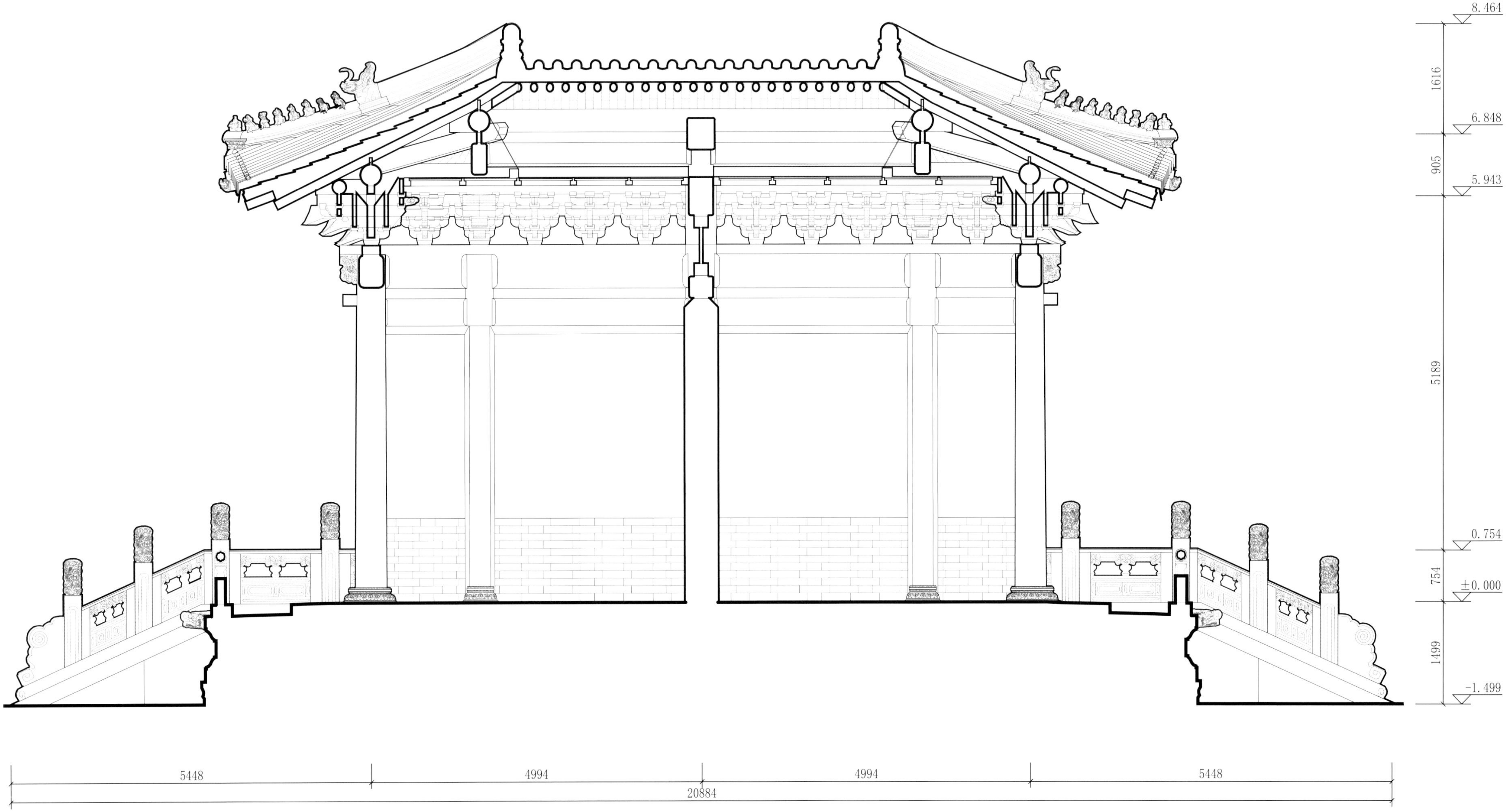

寿皇门梢间横剖面图
Lateral bay transverse section of Gate of Shouhuang

寿皇门东石狮正立面图
Front elevation of the east stone lion of Gate of Shouhuang

寿皇门东石狮背立面图
Back elevation of the east stone lion of Gate of Shouhuang

寿皇门东石狮平面图
Plan of the east stone lion of Gate of Shouhuang

寿皇门东石狮西立面图
West elevation of the east stone lion of Gate of Shouhuang

寿皇门东石狮东立面图
East elevation of the east stone lion of Gate of Shouhuang

寿皇门西石狮正立面图
Front elevation of the west stone lion of Gate of Shouhuang

寿皇门西石狮背立面图
Back elevation of the west stone lion of Gate of Shouhuang

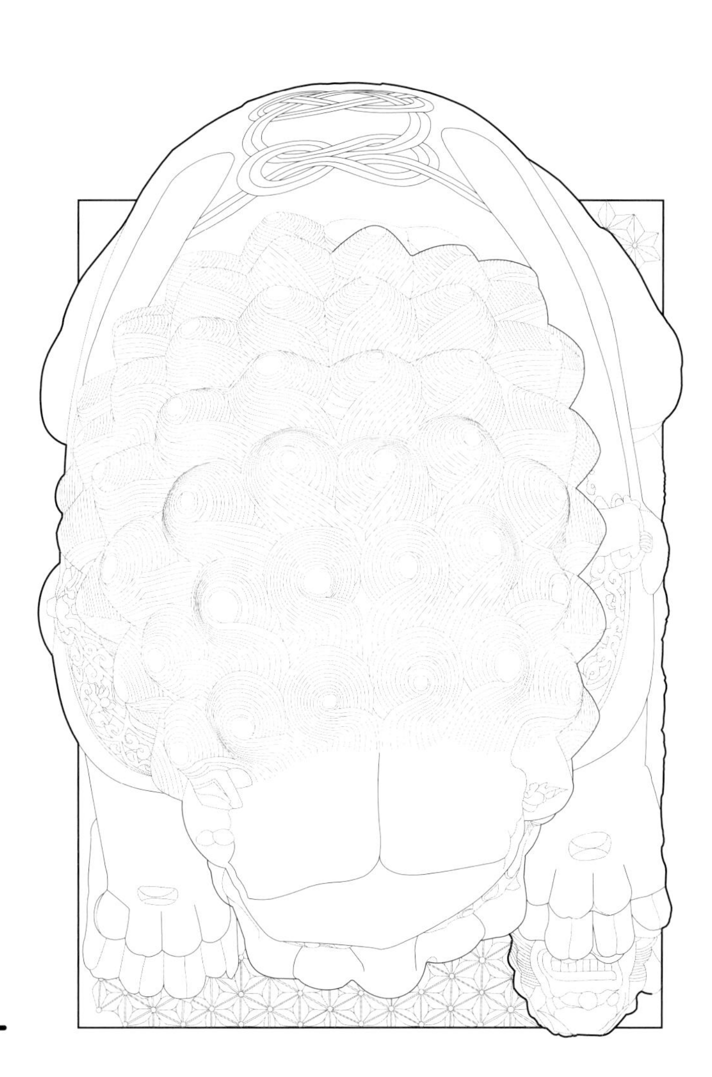

寿皇门西石狮平面图
Plan of the west stone lion of Gate of Shouhuang

寿皇门西石狮西立面图
West elevation of the west stone lion of Gate of Shouhuang

寿皇门西石狮东立面图
East elevation of the west stone lion of Gate of Shouhuang

寿皇门丹陛大样图

Detail drawing of danby of Gate of Shouhuang

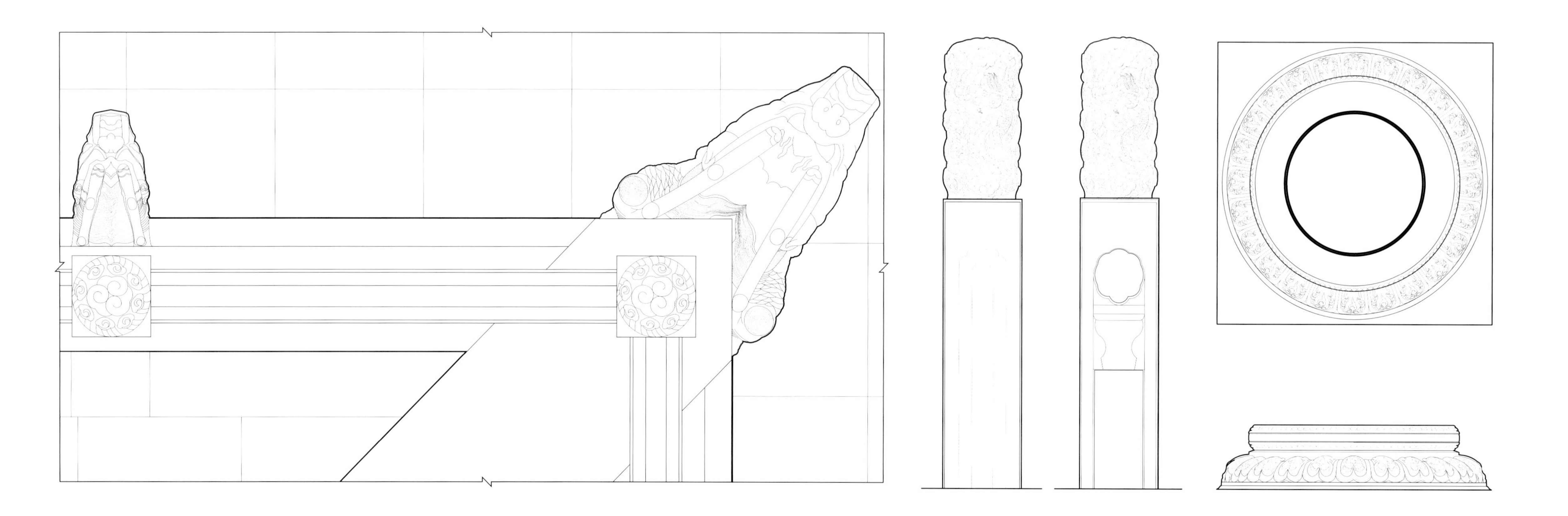

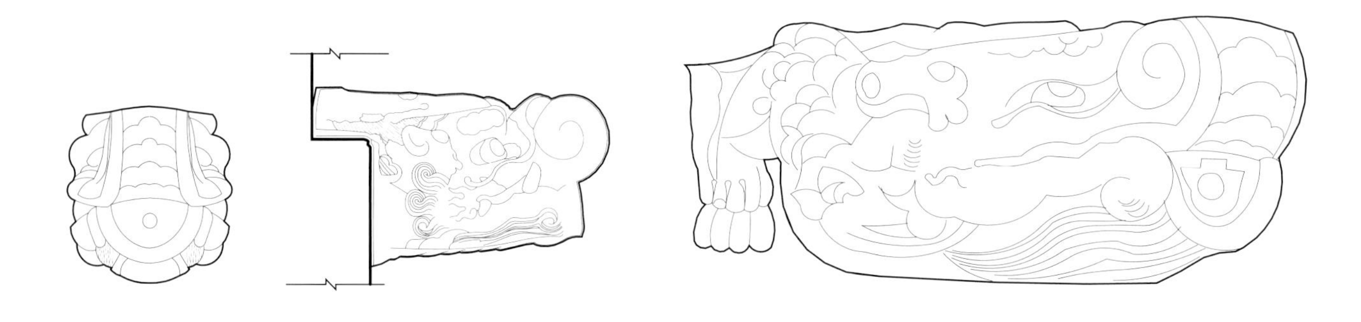

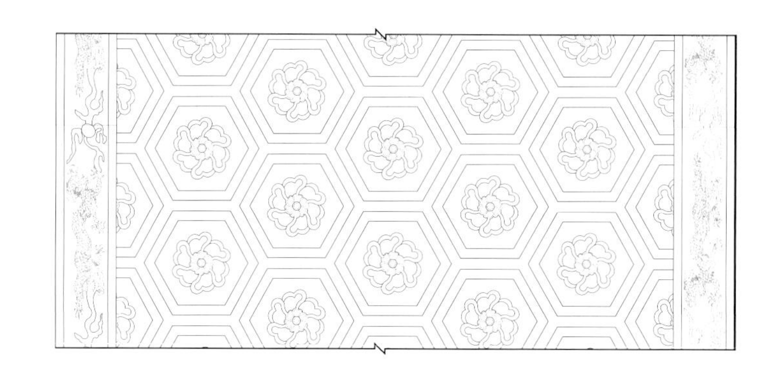

寿皇门须弥座望柱栏板大样图
Detail drawing of stone Xumizuo stylobate of Gate of Shouhuang

寿皇门露陈座平面图

Plan of Luchenzuo of Gate of Shouhuang

寿皇门露陈座正立面图
Front elevation of Luchenzuo of Gate of Shouhuang

寿皇门露陈座侧立面图
Side elevation of Luchenzuo of Gate of Shouhuang

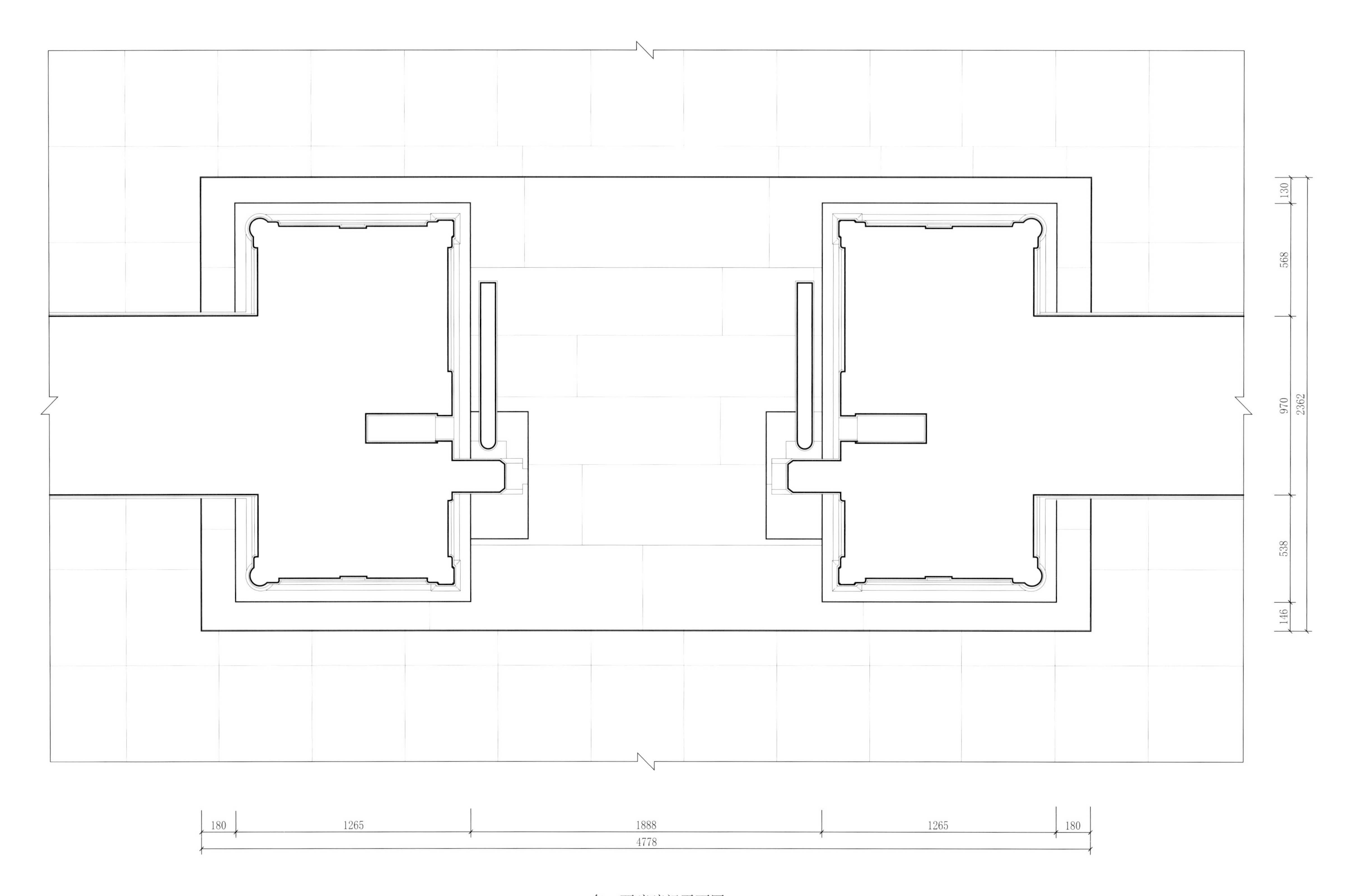

东、西琉璃门平面图
Plan of East or West Colored-glaze Gate

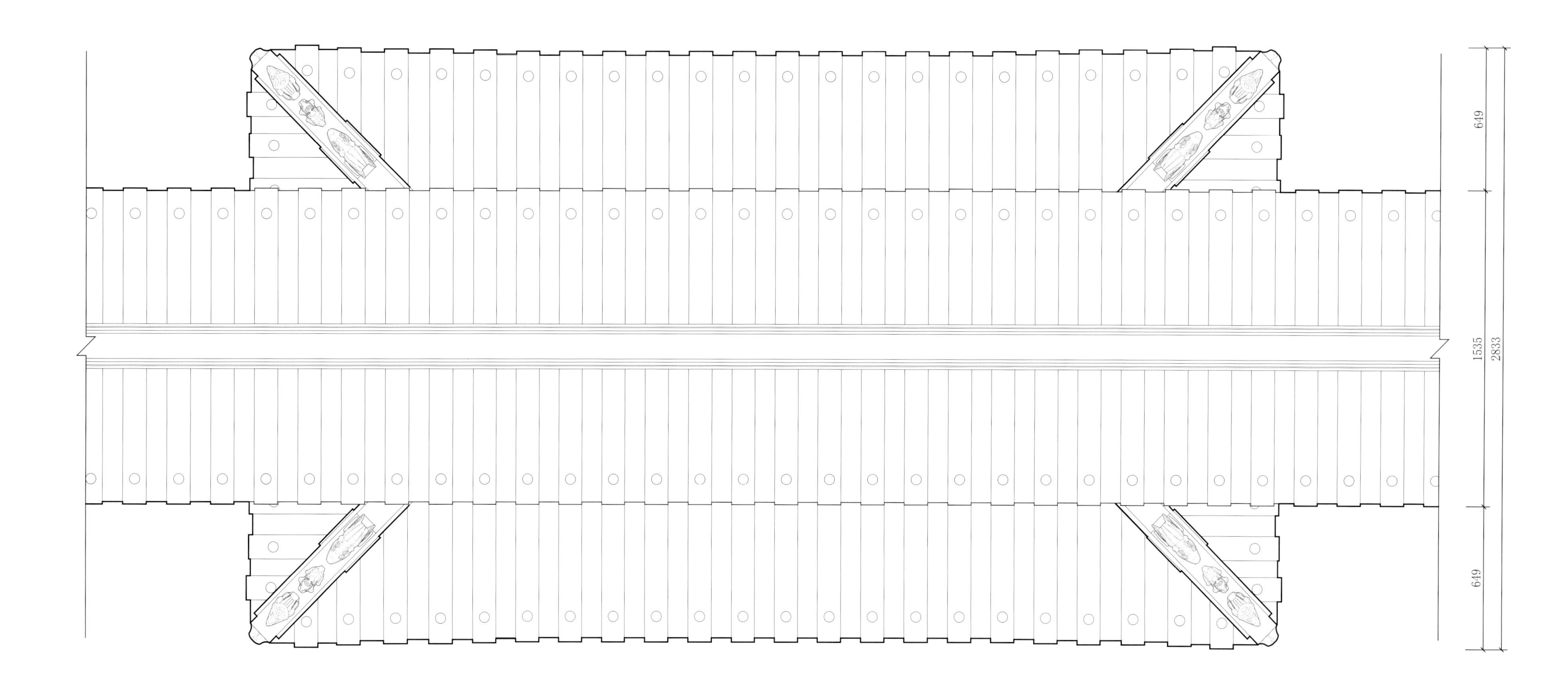

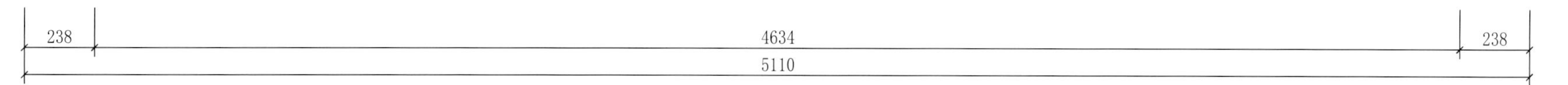

东、西琉璃门屋顶平面图
Roof plan of East or West Colored-glaze Gate

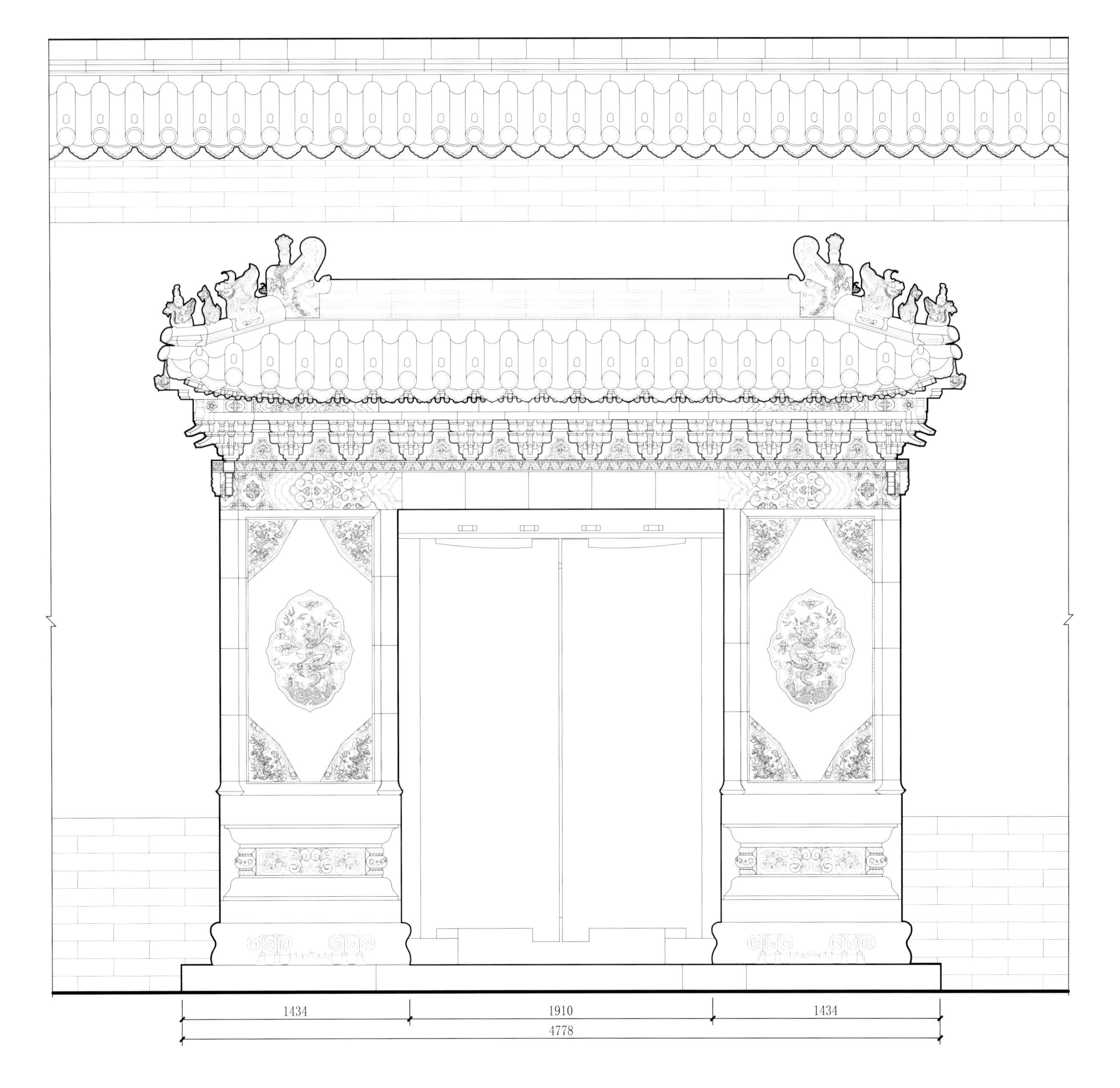

东、西琉璃门正立面图

Front elevation of East or West Colored-glaze Gate

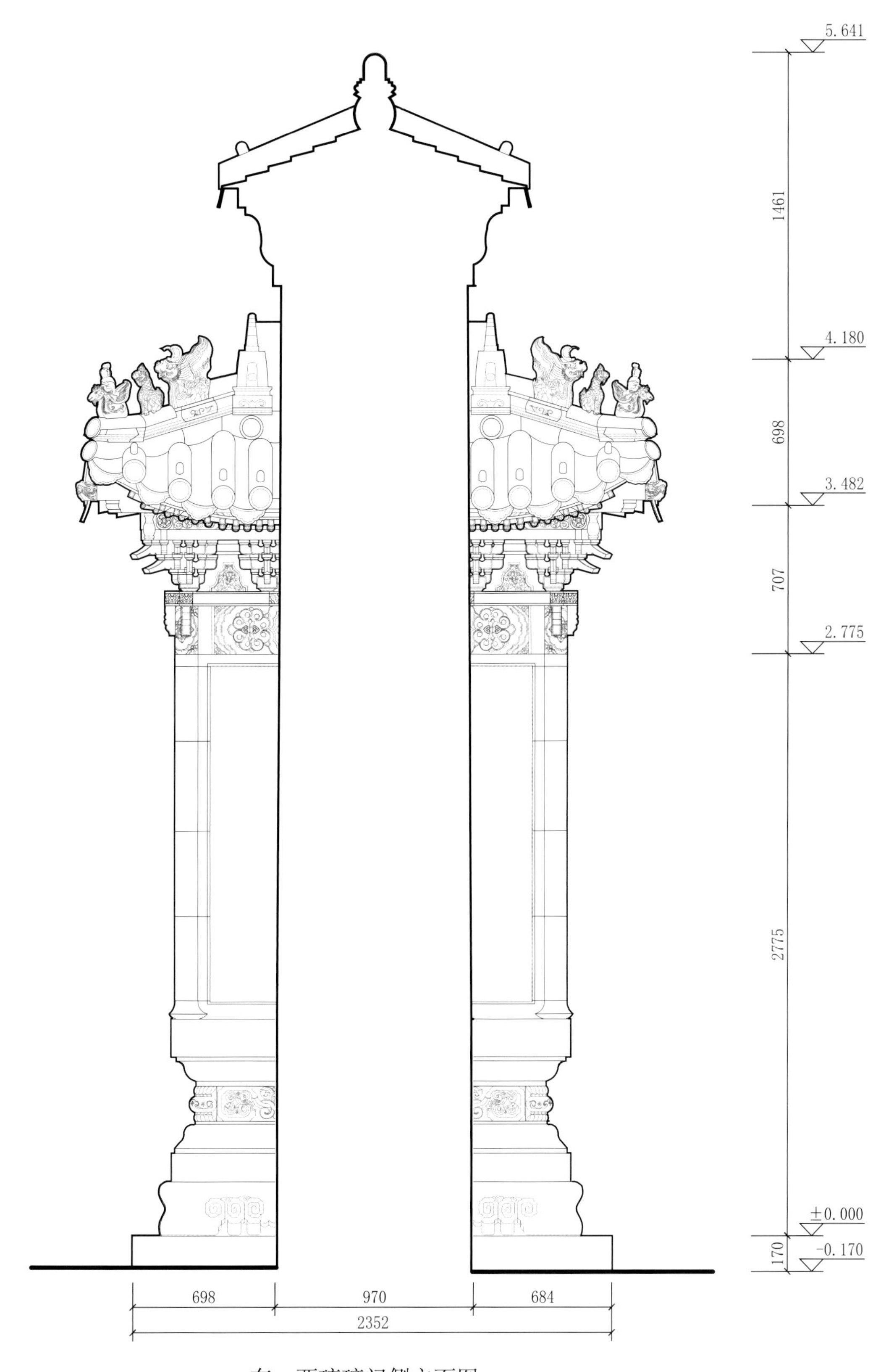

东、西琉璃门侧立面图

Side elevation of East or West Colored-glaze Gate

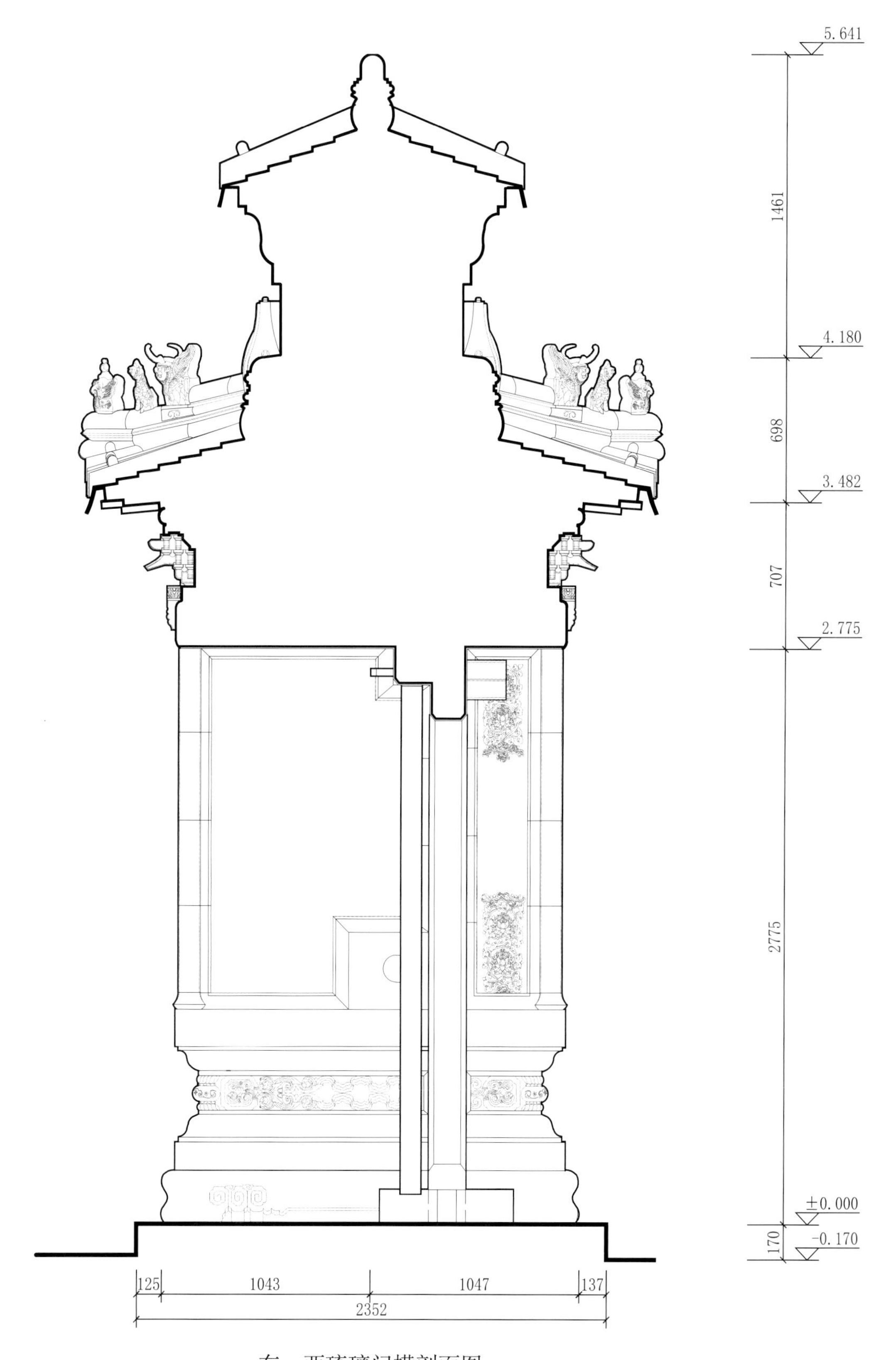

东、西琉璃门横剖面图

Transverse section of East or West Colored-glaze Gate

东、西琉璃门琉璃纹样大样图

Detail drawing of the texture of East or West Colored-glaze Gate

东、西琉璃门琉璃纹样大样图
Detail drawing of the texture of East or West Colored-glaze Gate

东、西琉璃门琉璃纹样大样图
Detail drawing of the texture of East or West Colored-glaze Gate

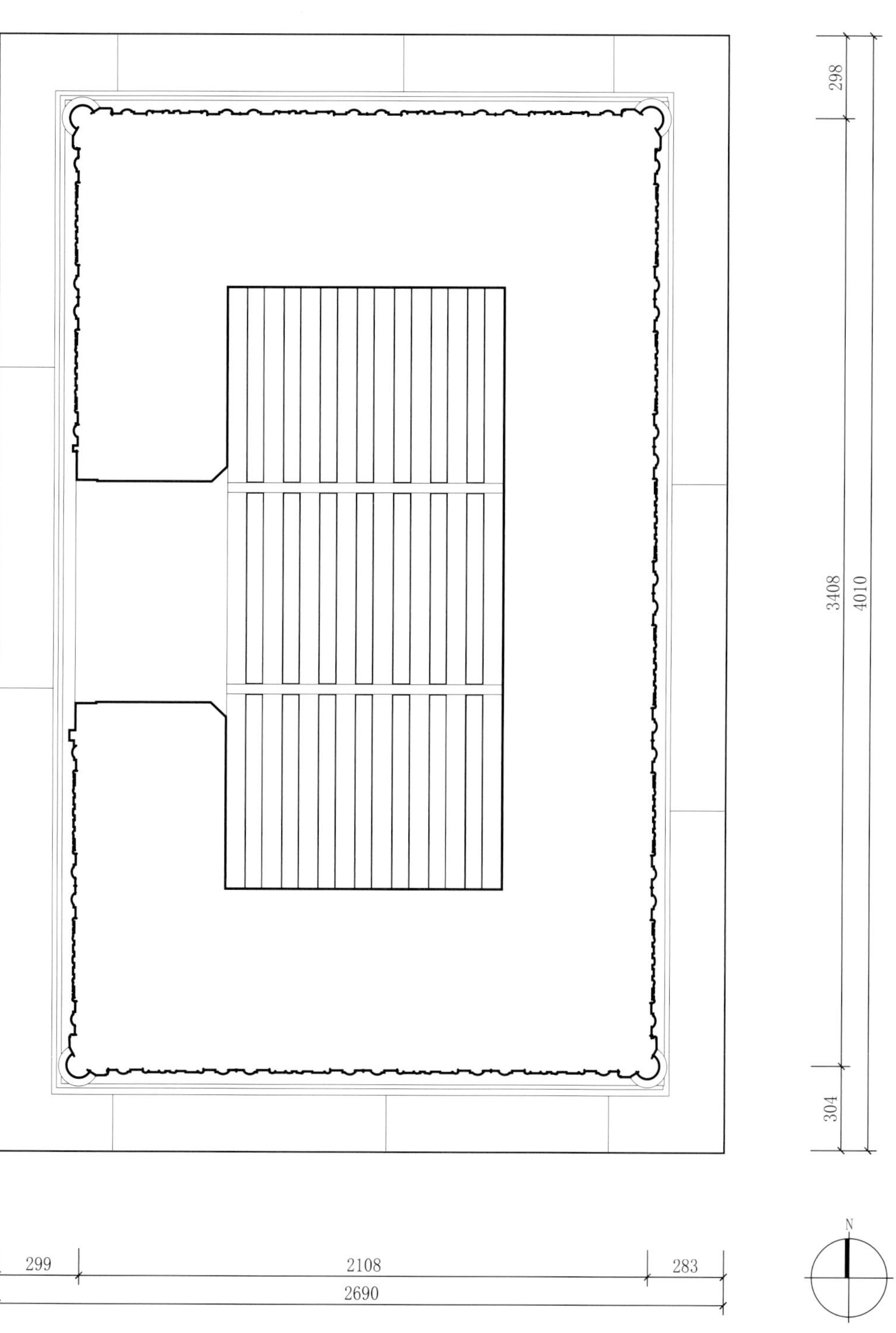

东、西燎炉平面图
Plan of East or West Flame Burner

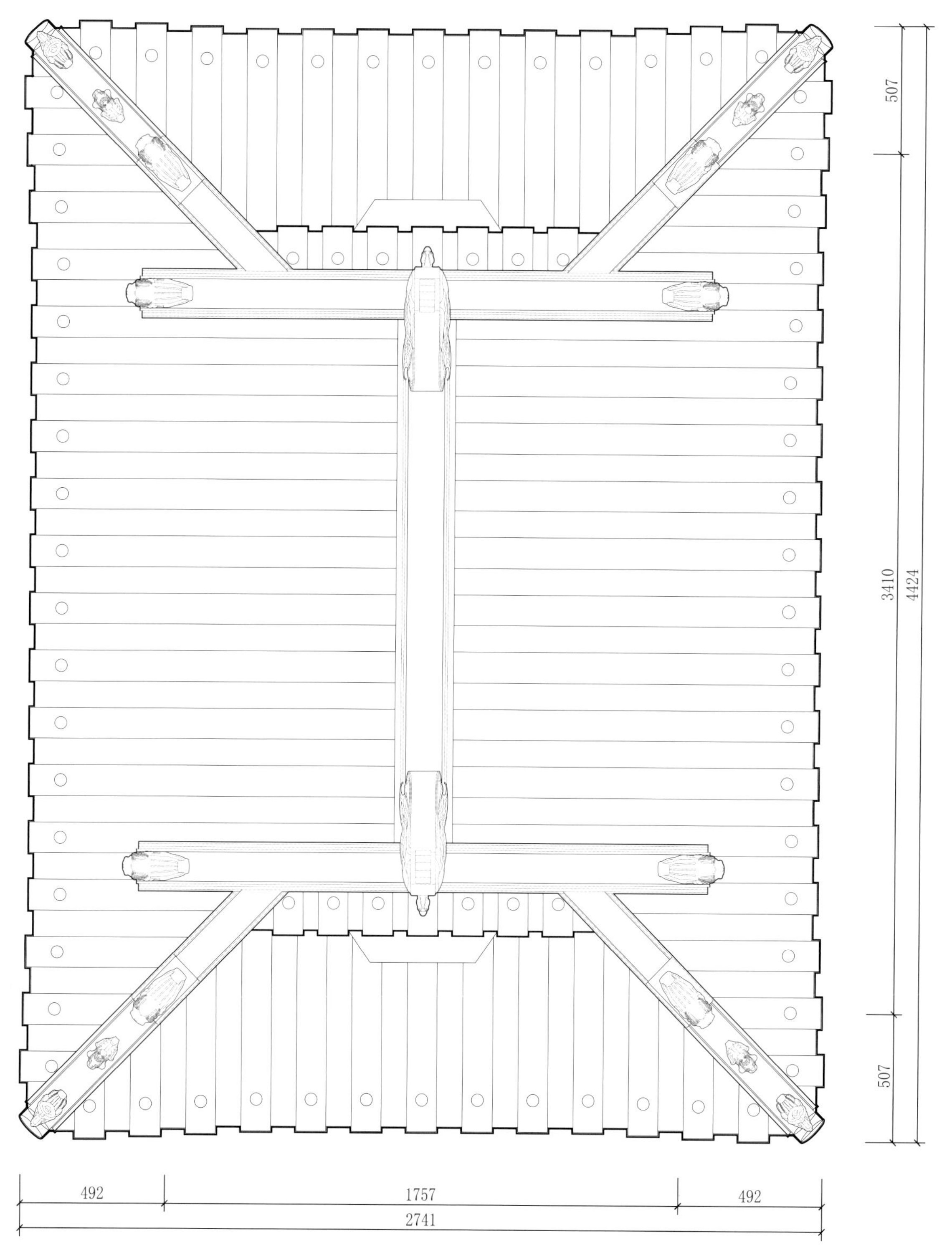

东、西燎炉屋顶平面图
Roof plan of East or West Flame Burner

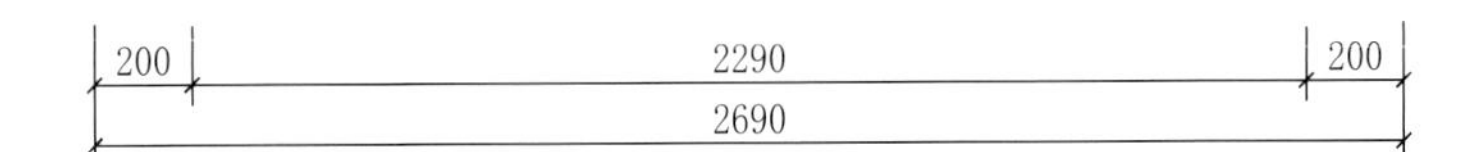

东、西燎炉侧立面图
Side elevation of East or West Flame Burner

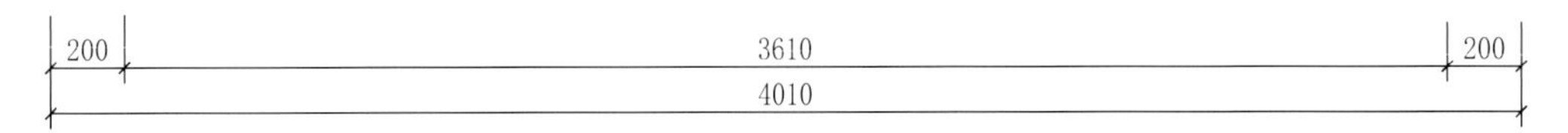

东、西燎炉正立面图
Front elevation of East or West Flame Burner

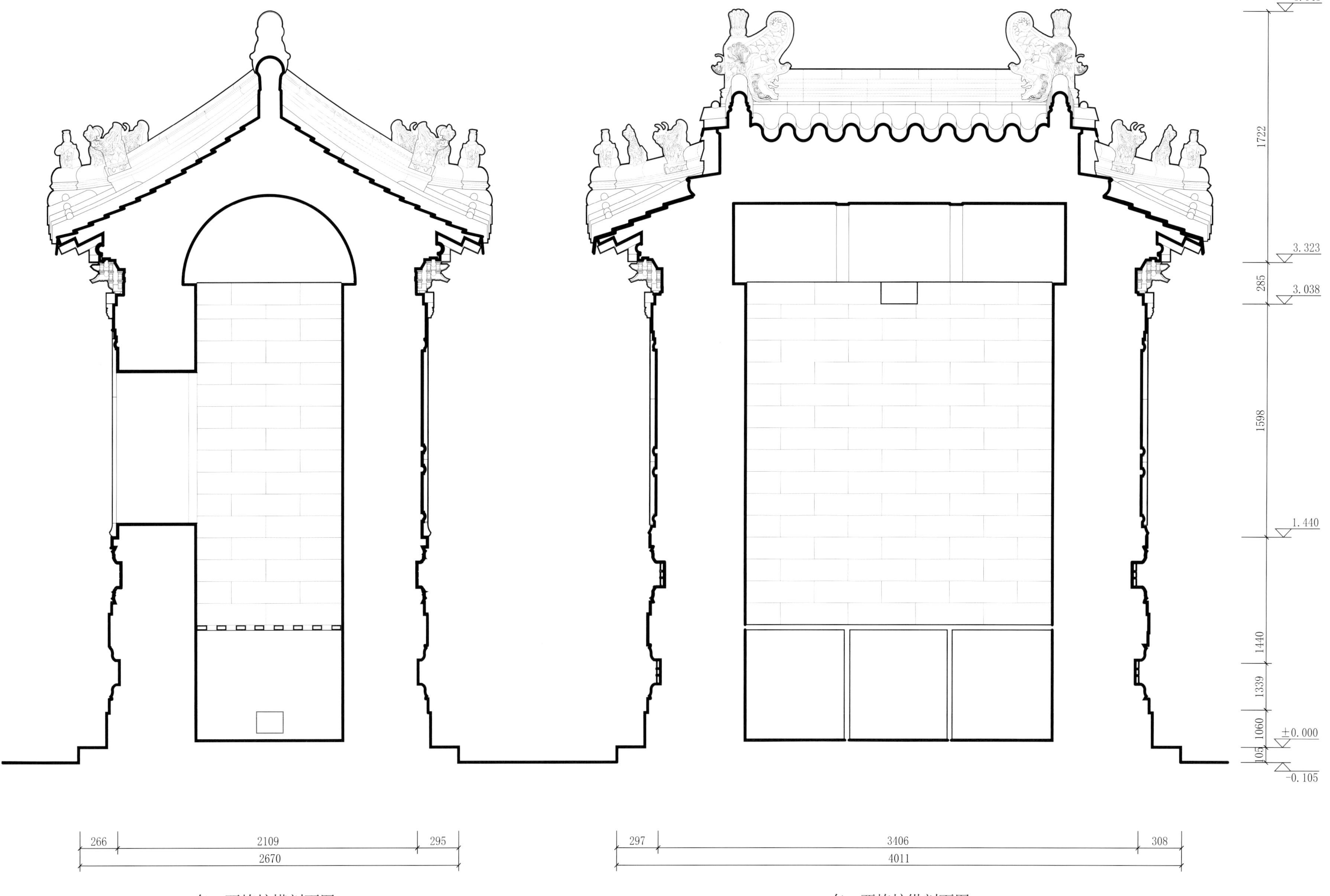

东、西燎炉横剖面图
Transverse section of East or West Flame Burner

东、西燎炉纵剖面图
Longitudinal section of East or West Flame Burner

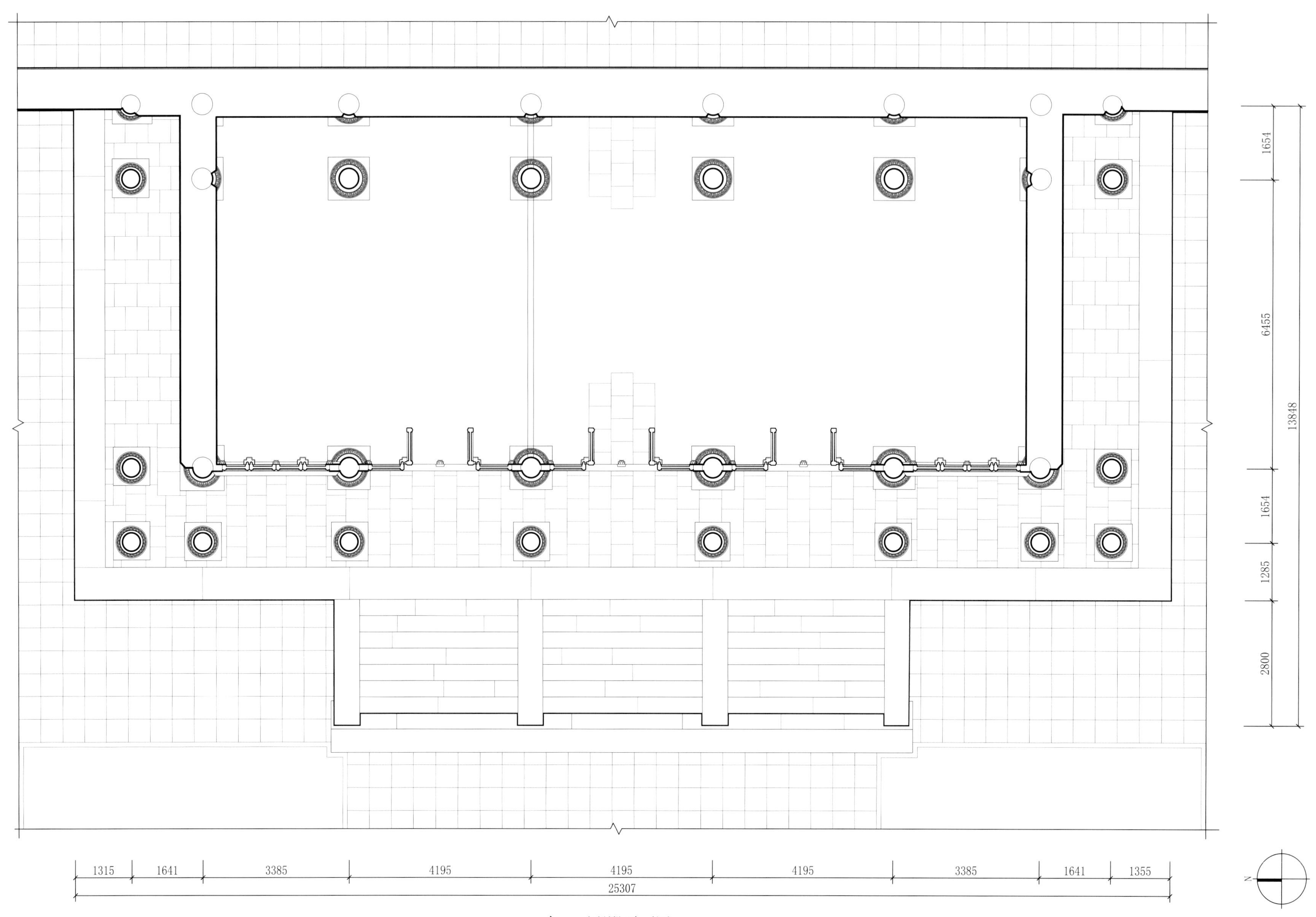

东、西配殿平面图
Plan of East or West Side Hall

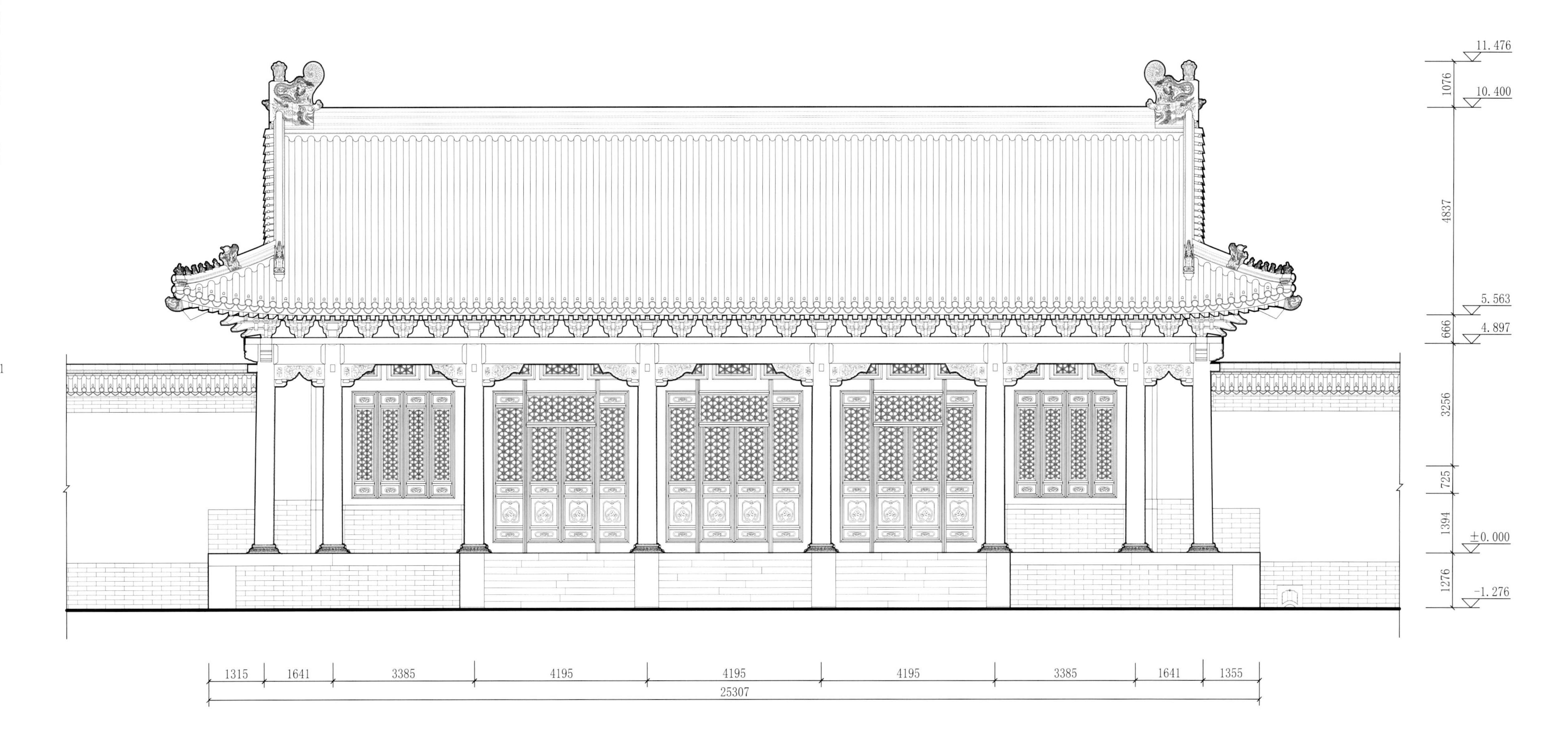

东、西配殿正立面图
Front elevation of East or West Side Hall

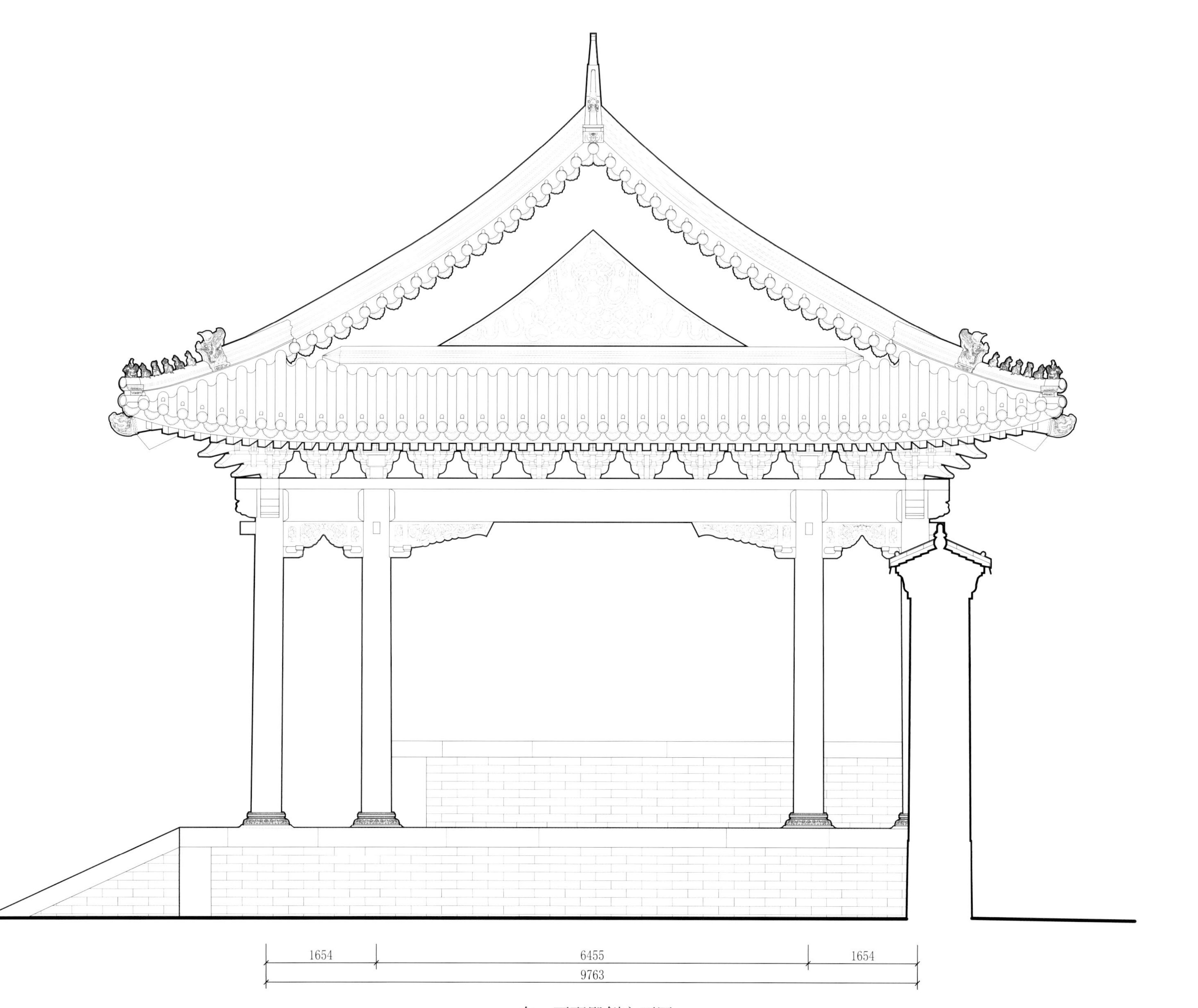

东、西配殿侧立面图

Side elevation of East or West Side Hall

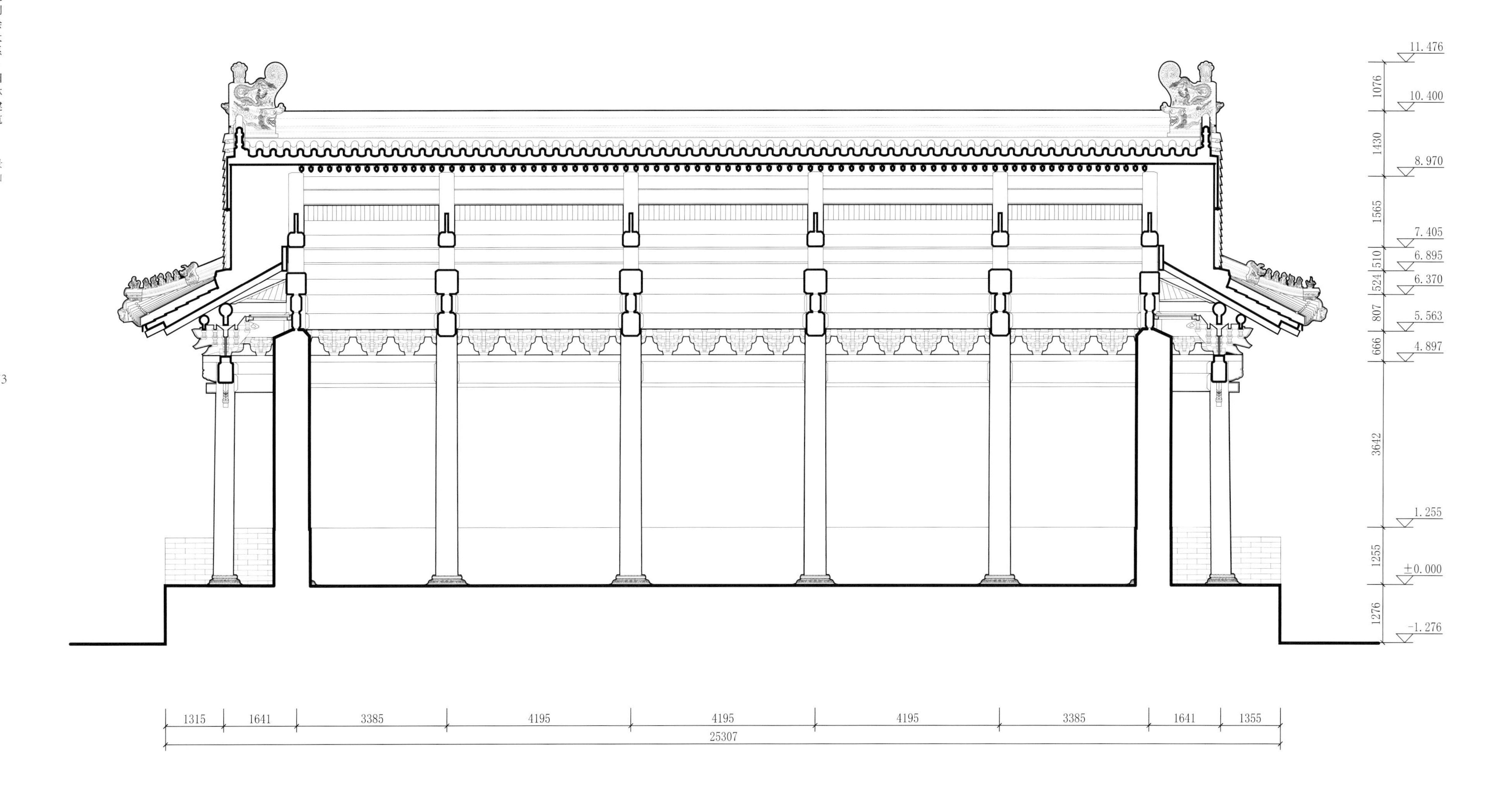

东、西配殿纵剖面图

Longitudinal section of East or West Side Hall

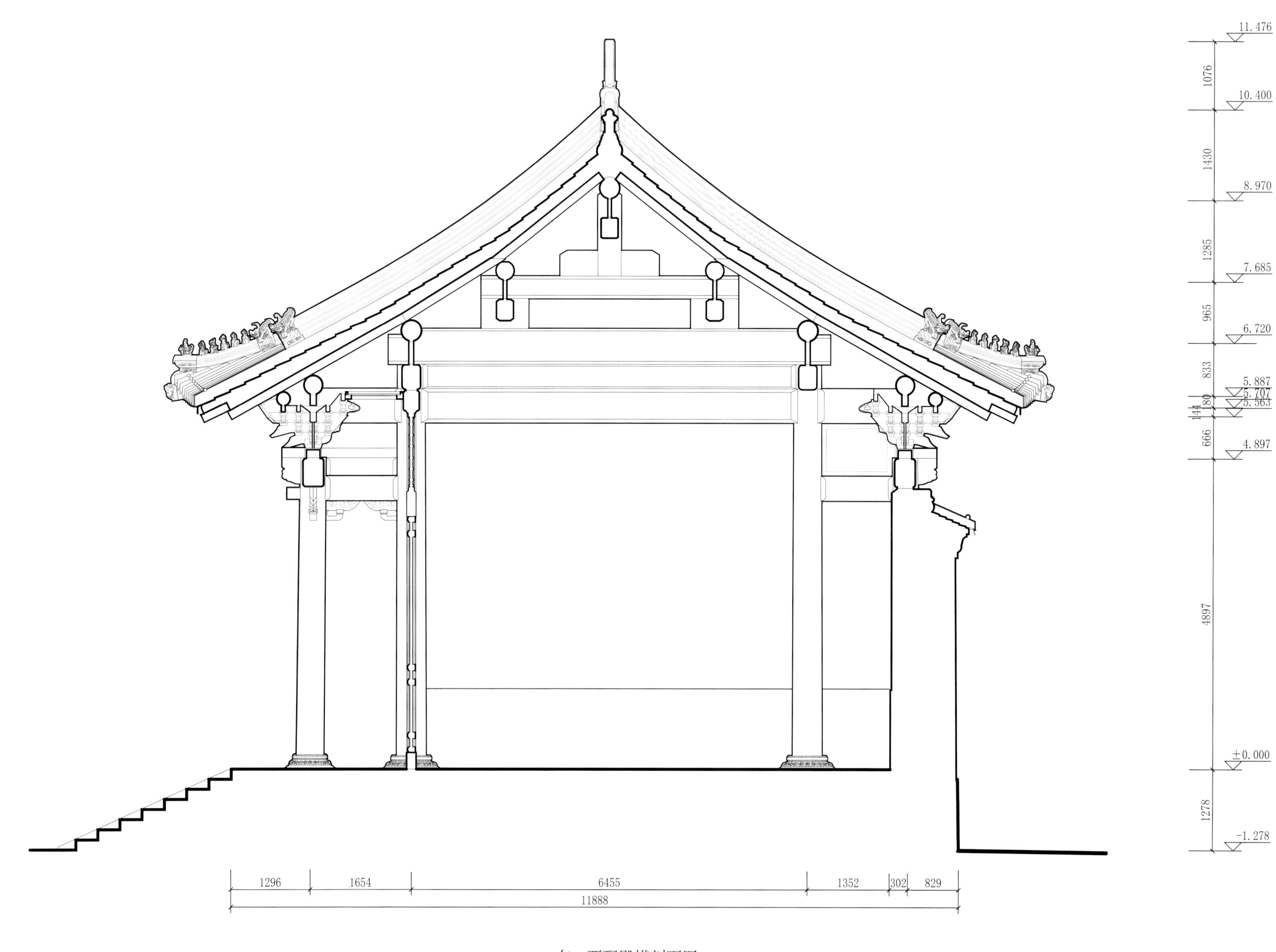

东、西配殿横剖面图
Transverse section of East or West Side Hall

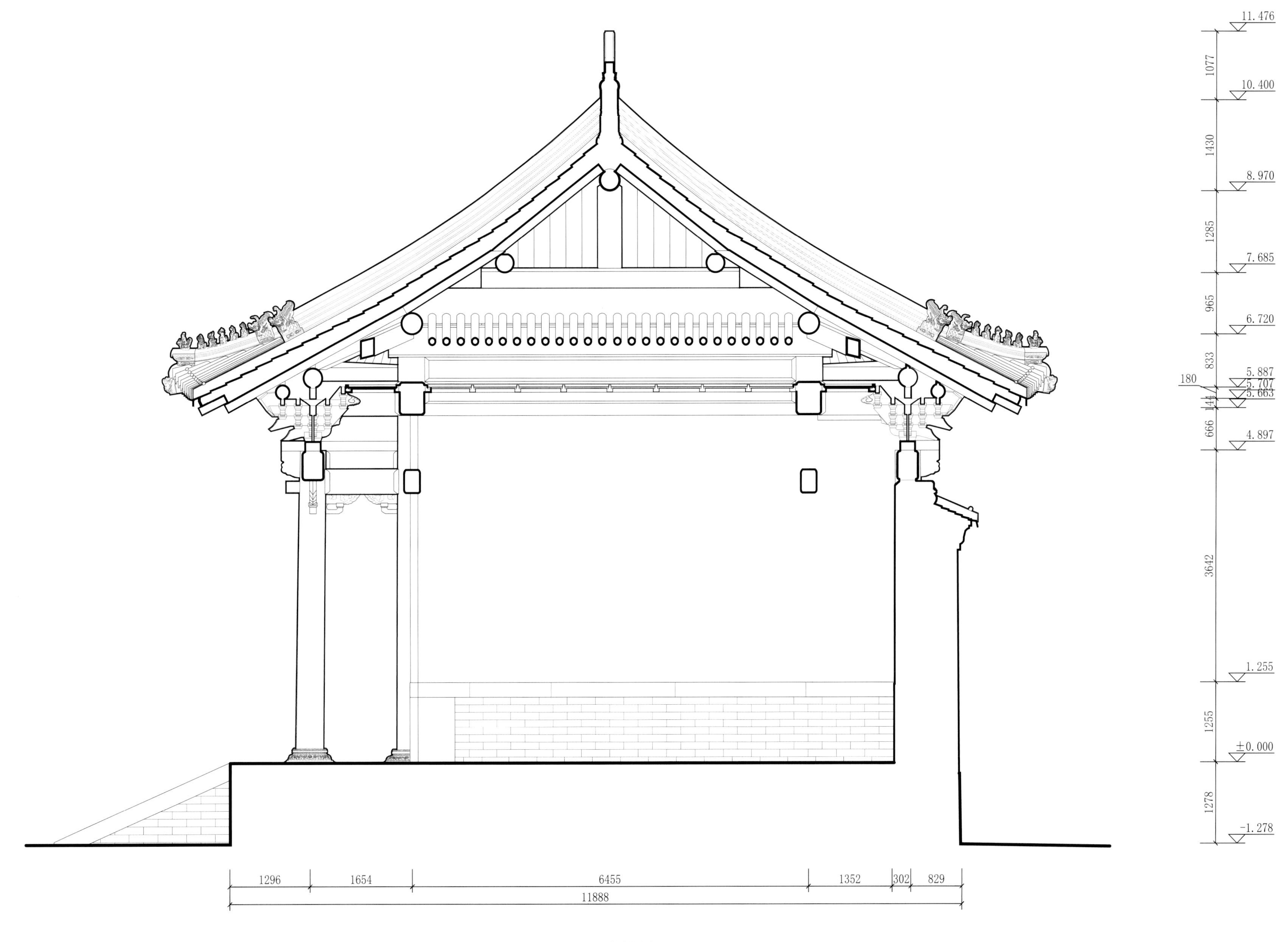

东、西配殿外廊横剖面图

Transverse section of East or West Side Hall

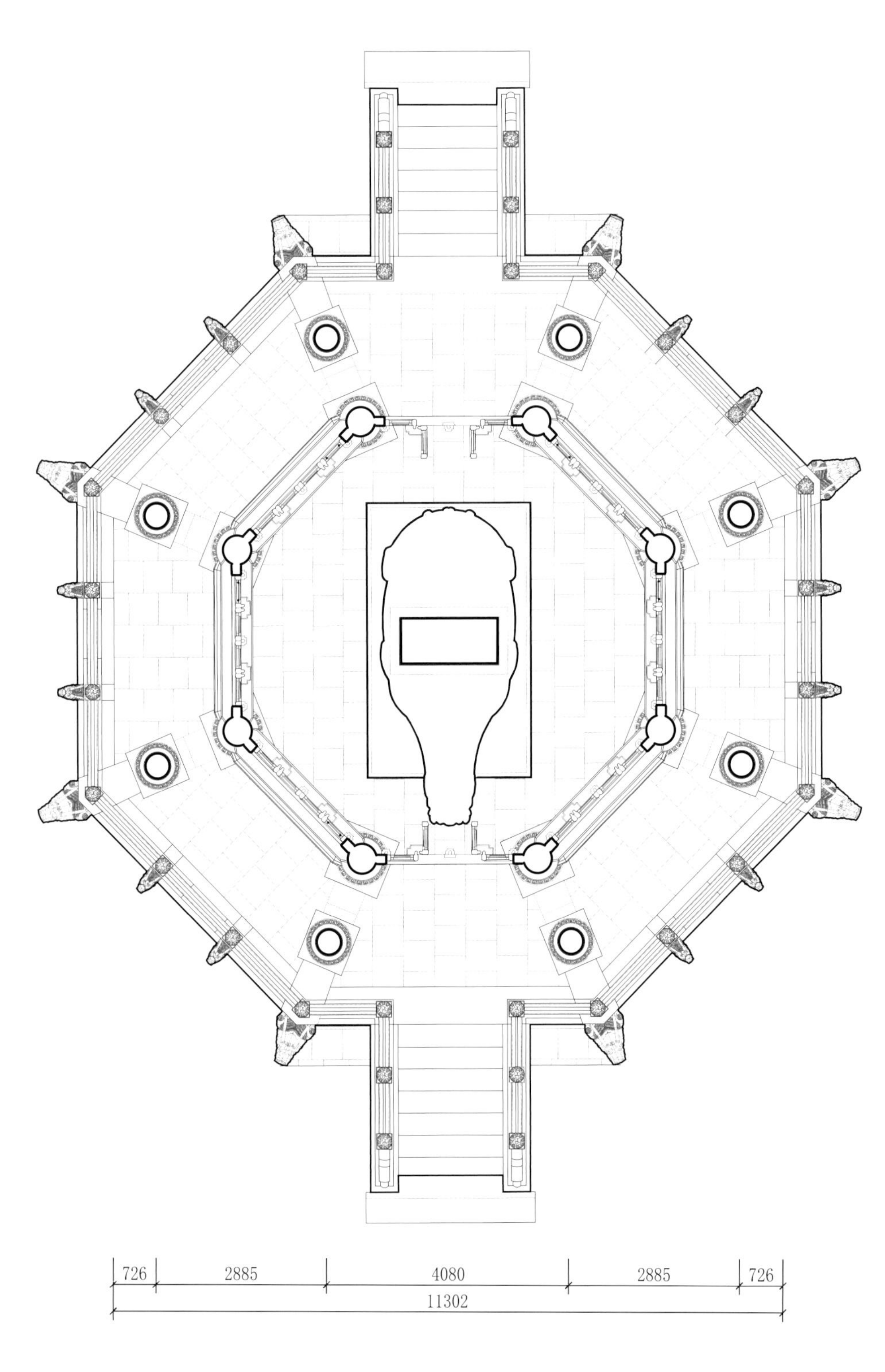

东、西碑亭平面图
Plan of East or West Tablet Pavilion

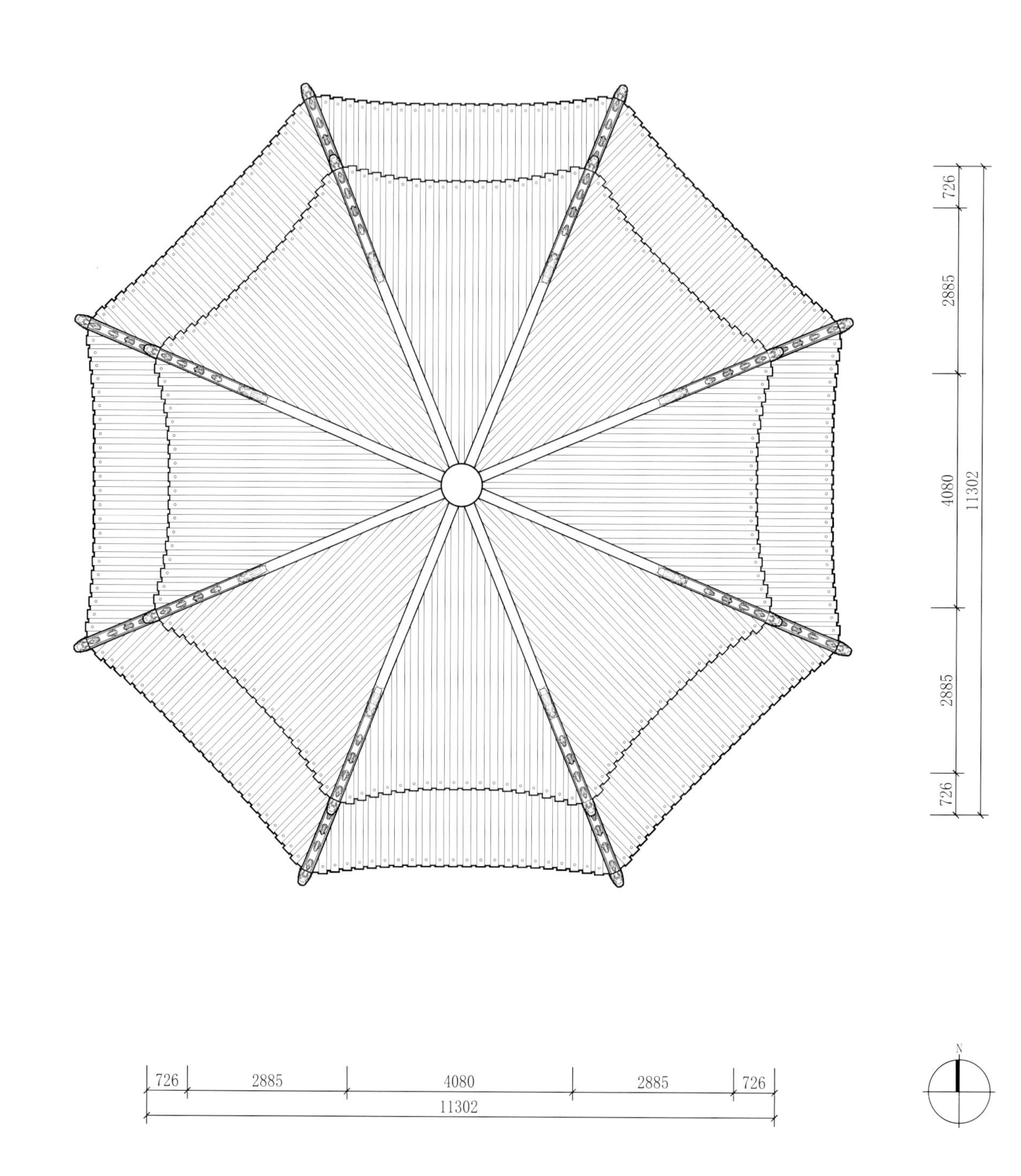

东、西碑亭屋顶平面图
Roof Plan of East or West Tablet Pavilion

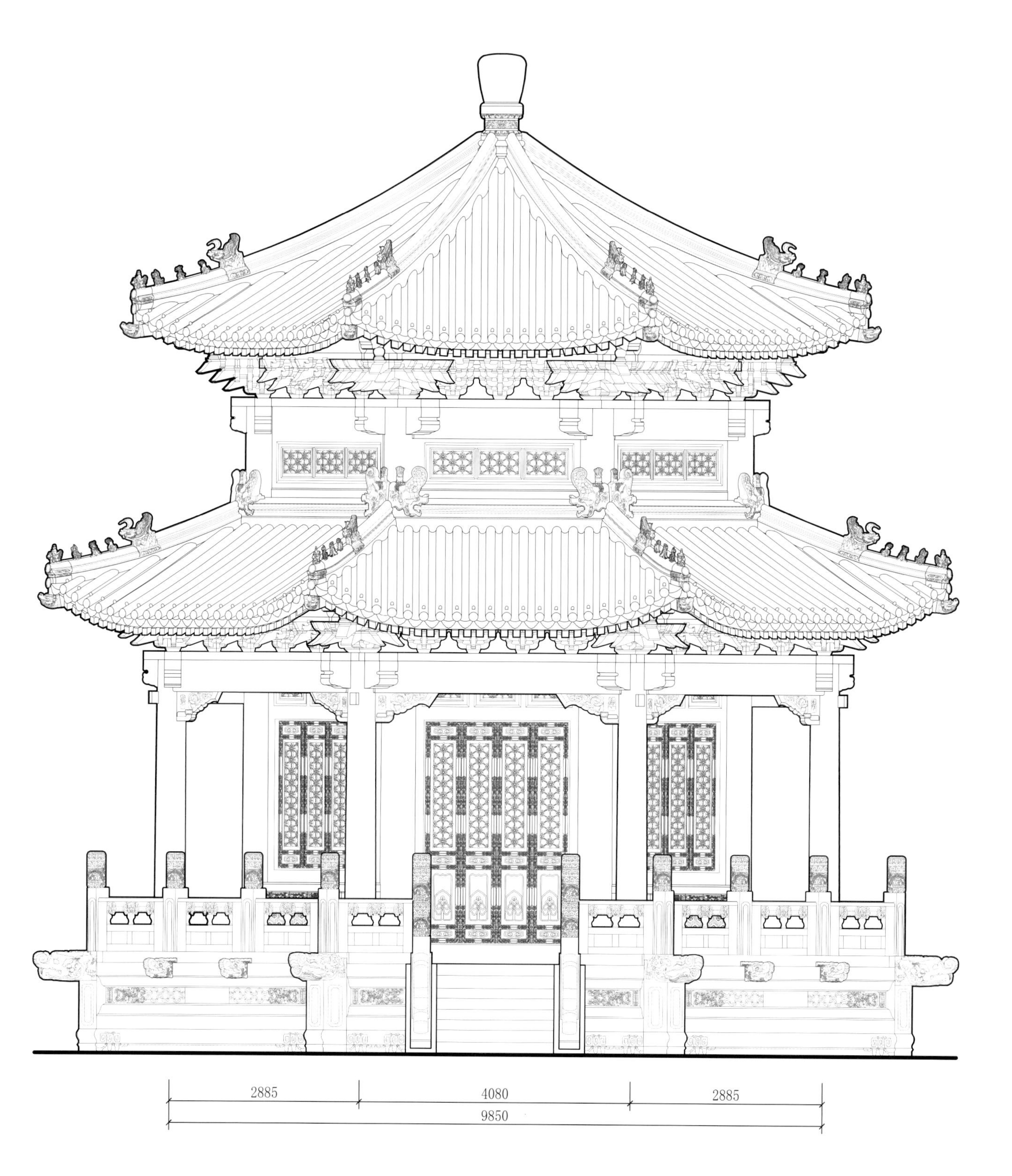

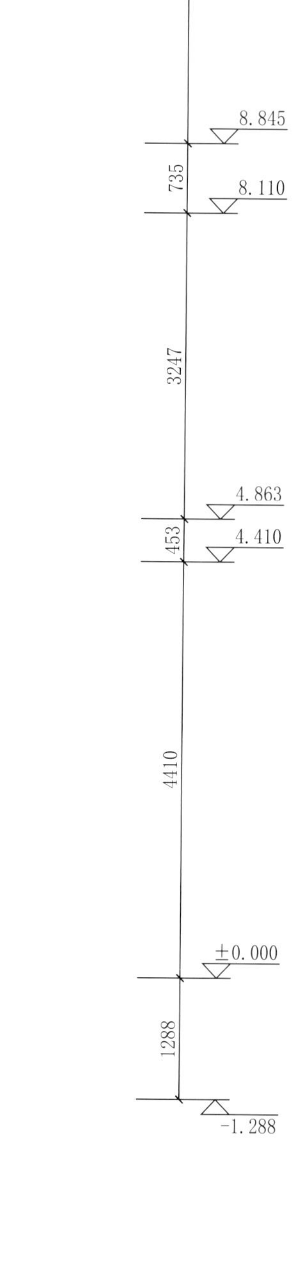

东、西碑亭正立面图

Front elevation of East or West Tablet Pavilion

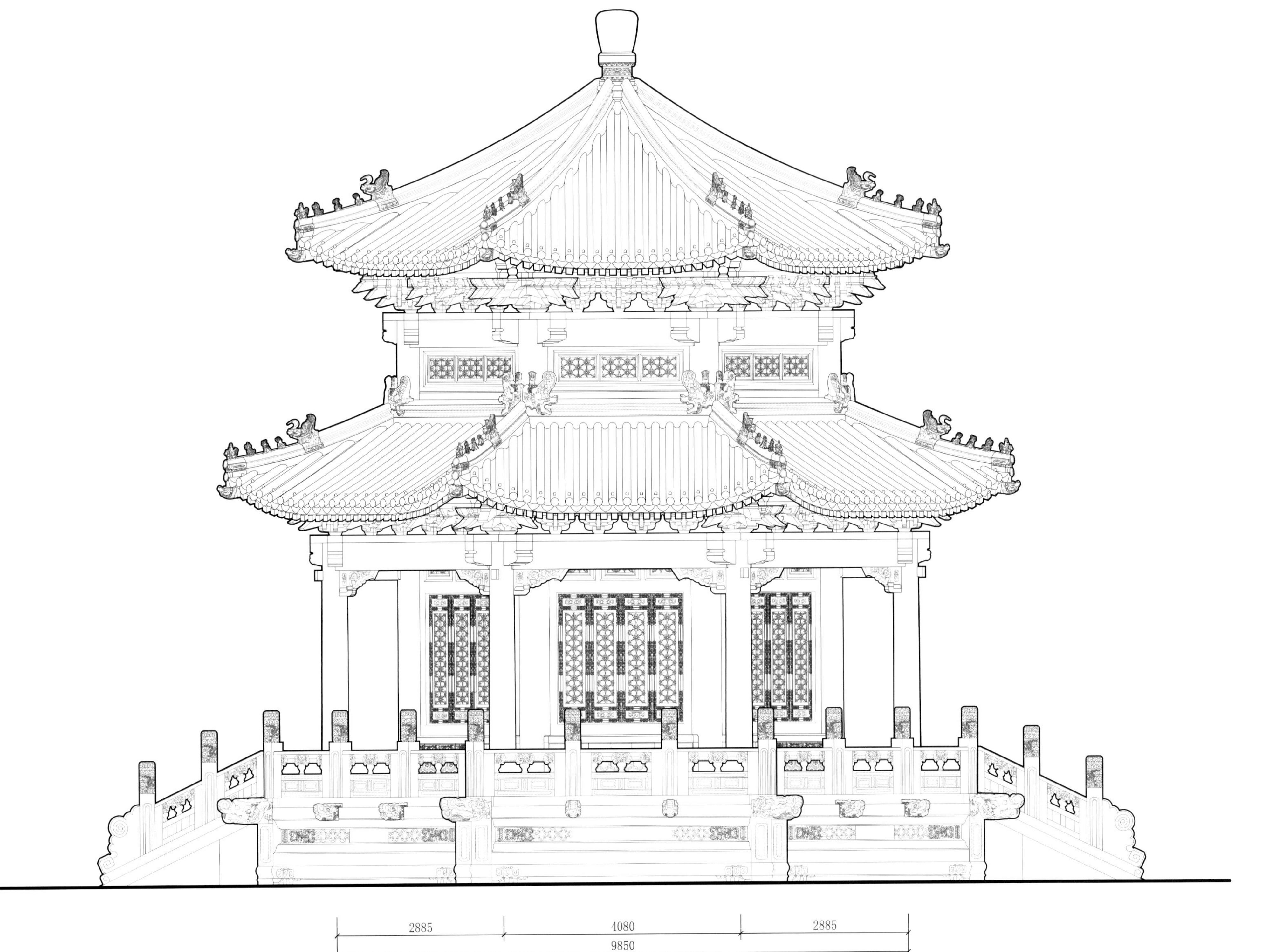

东、西碑亭侧立面图
Side elevation of East or West Tablet Pavilion

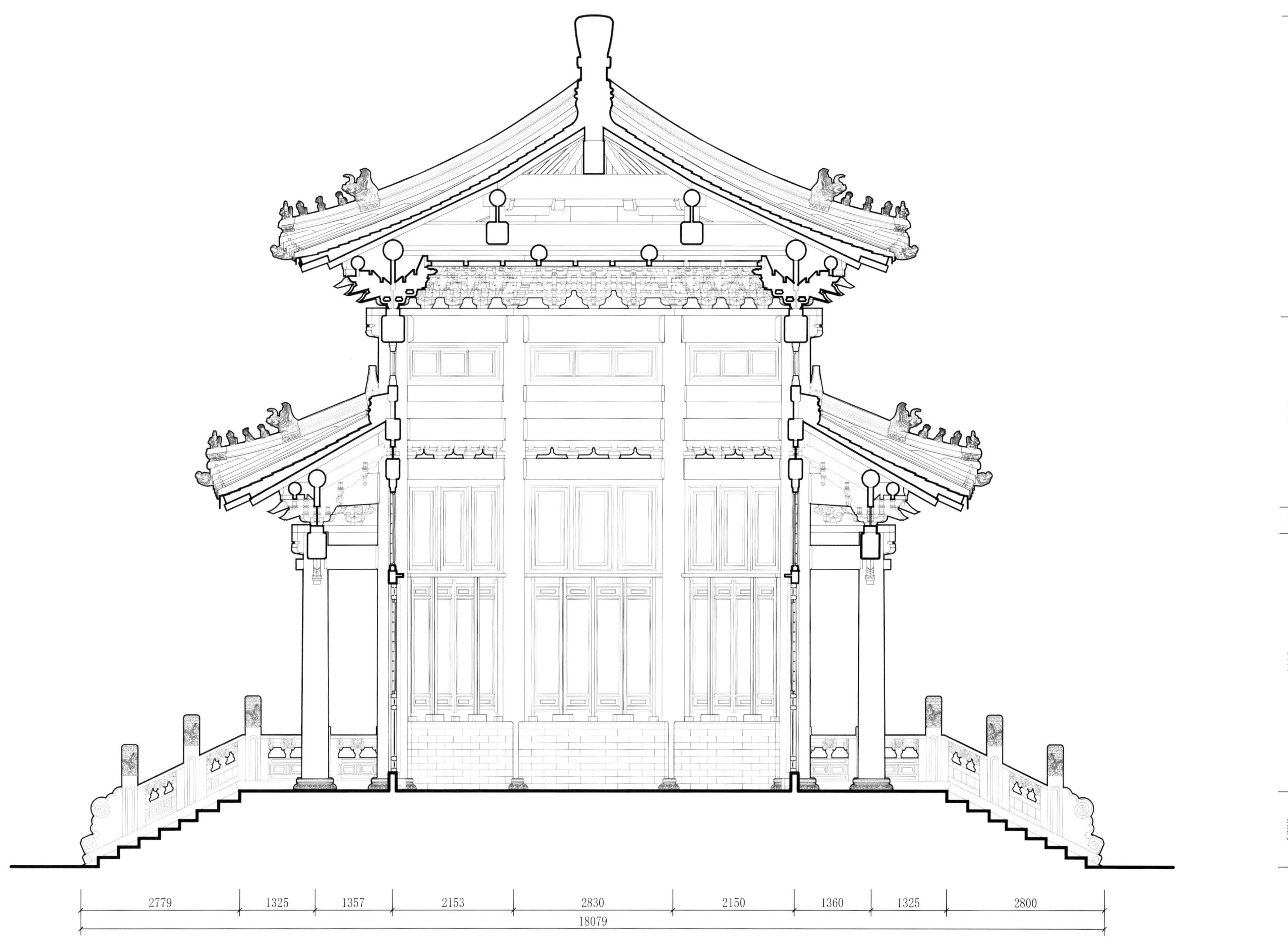

东、西碑亭横剖面图

Transverse section of East or West Tablet Pavilion

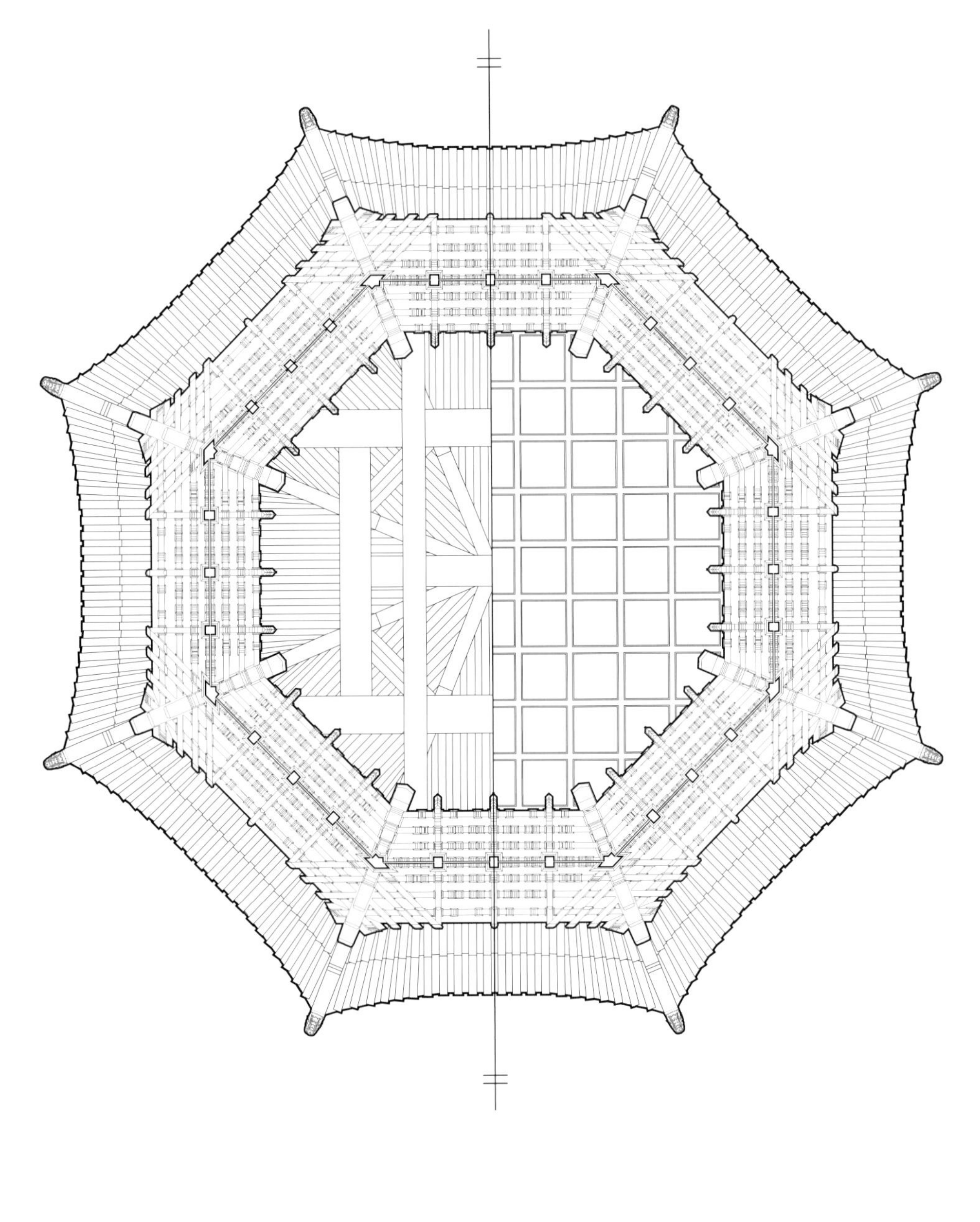

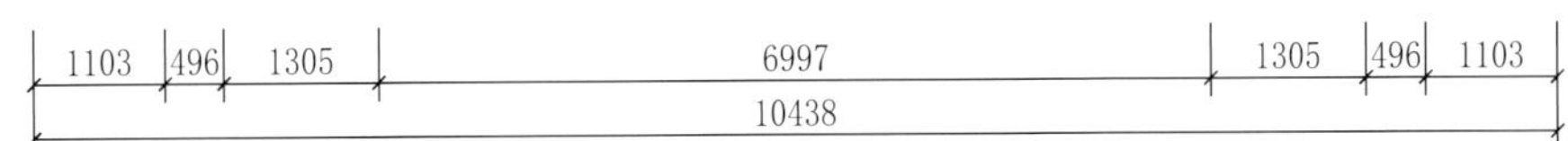

东、西碑亭上檐梁架仰视图
Bottom view of beam frame of the upper cornice of East or West Tablet Pavilion

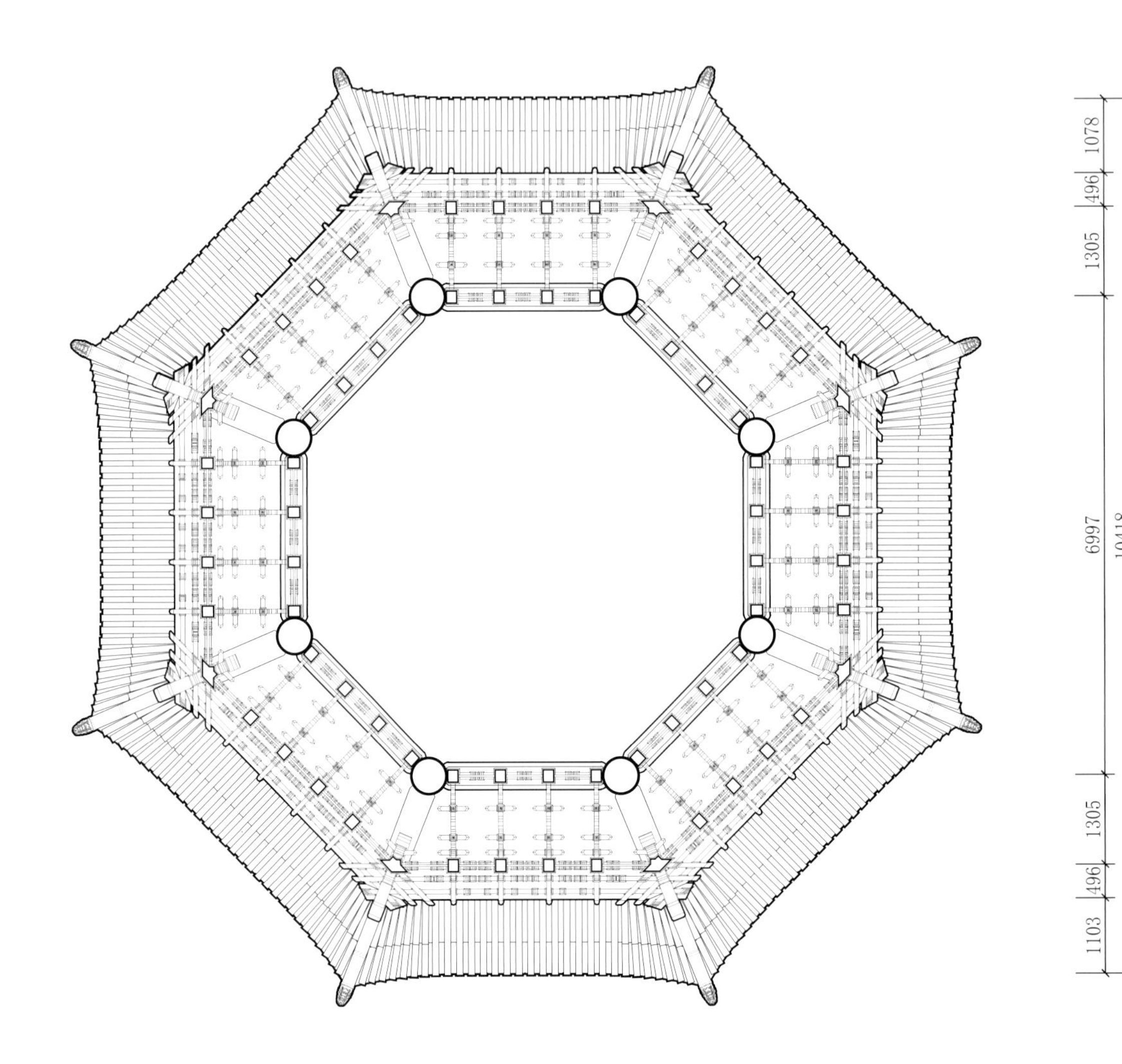

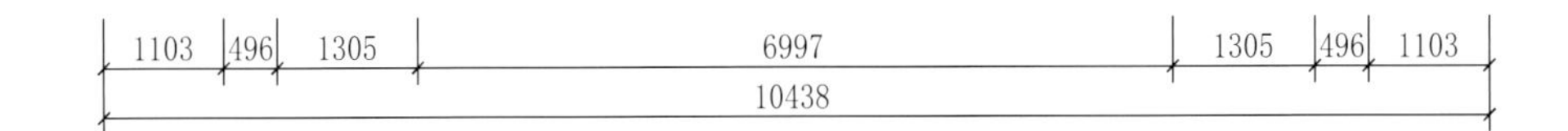

东、西碑亭下檐梁架仰视图
Bottom view of beam frame of the under cornice of East or West Tablet Pavilion

东、西碑亭槛墙琉璃纹大样图
Detail drawing of the texture of East or West Tablet Pavilion

0 0.15 0.45m

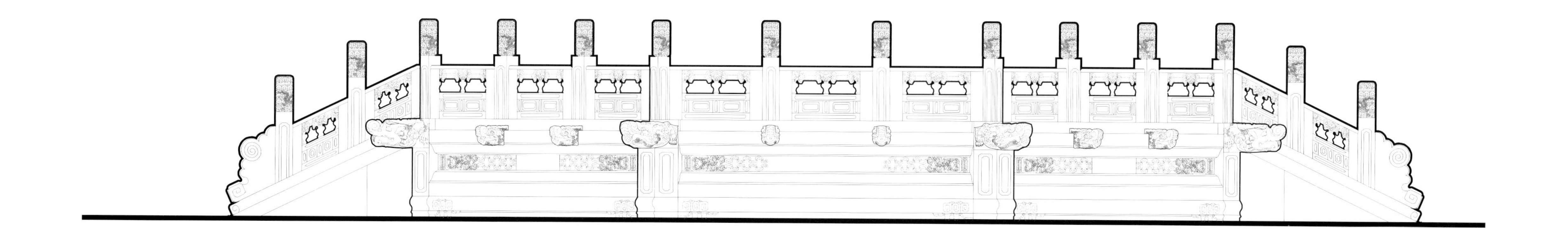

东、西碑亭须弥座大样图
Detail drawing of stone Xumizuo stylobate of East or West Tablet Pavilion

0 1.0 3.0m

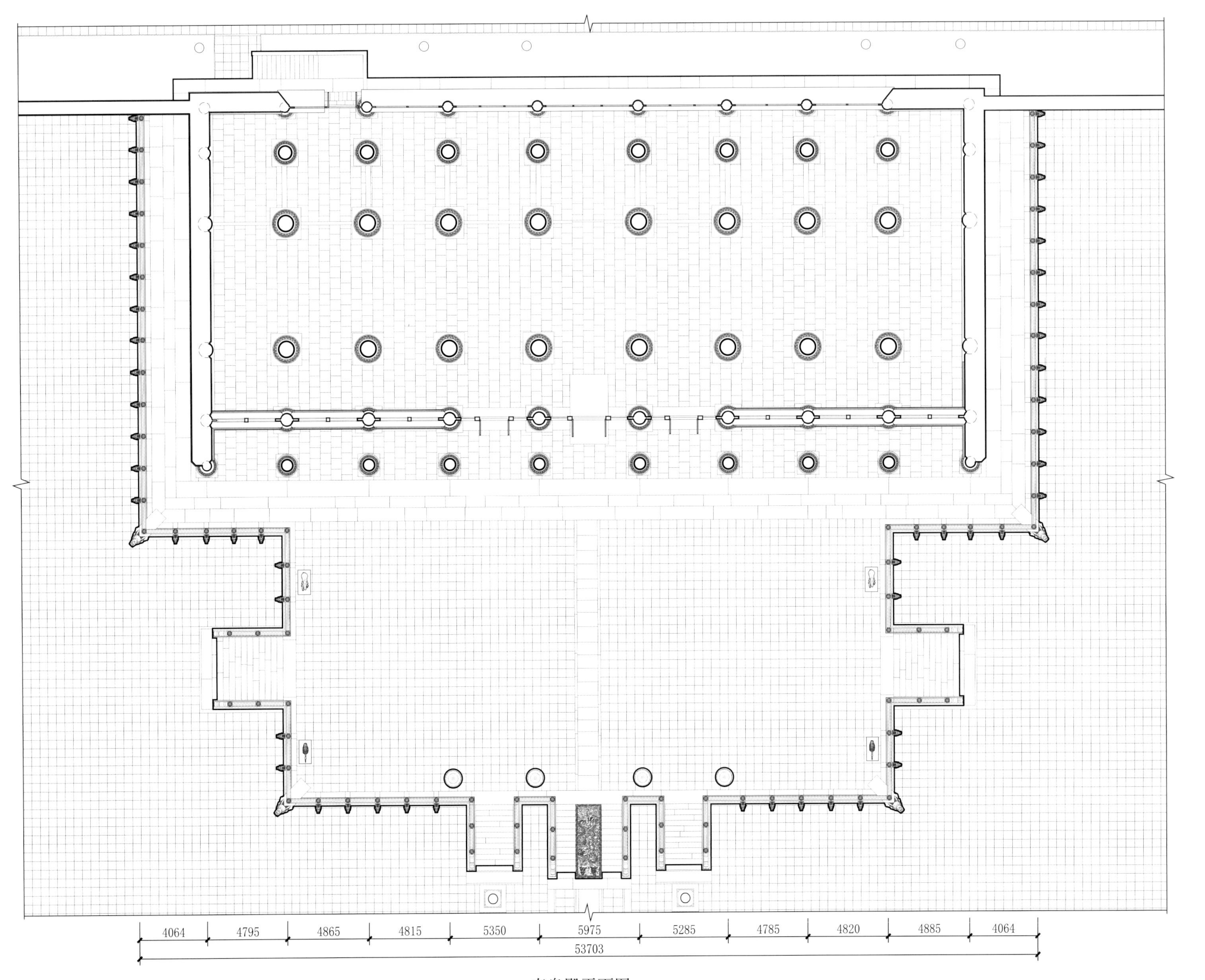

寿皇殿平面图
Plan of Shouhuang Hall

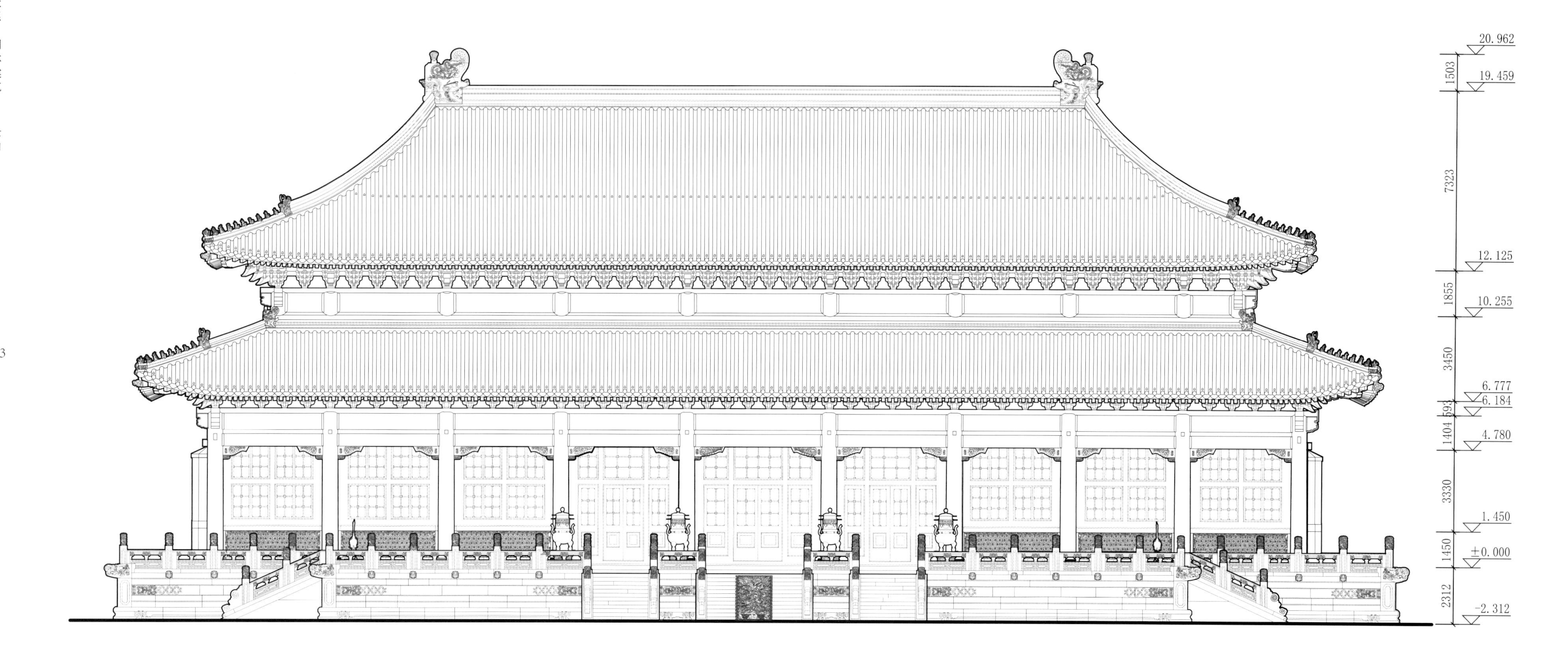

寿皇殿正立面图
Front elevation of Shouhuang Hall

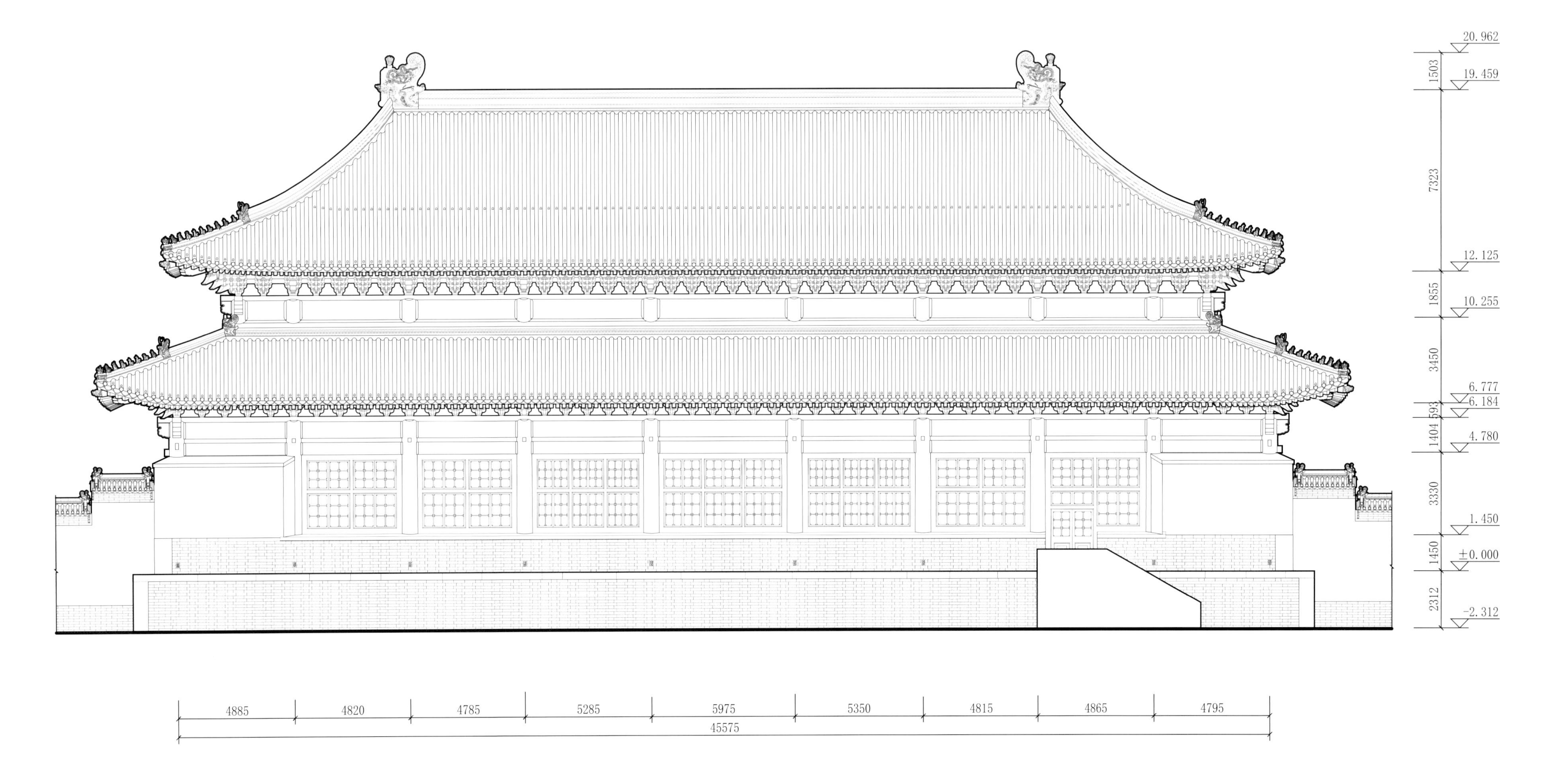

寿皇殿背立面图

Back elevation of Shouhuang Hall

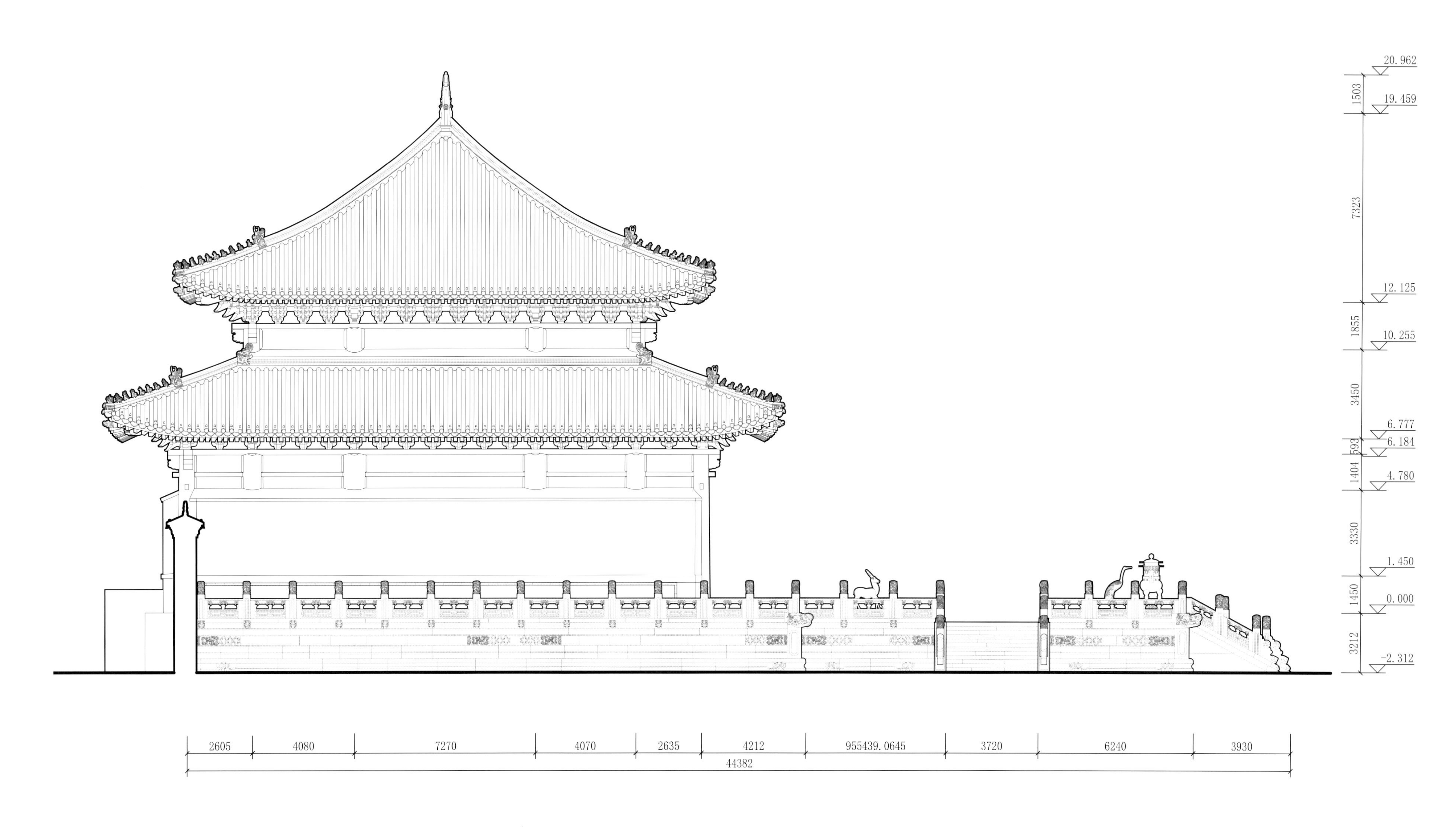

寿皇殿侧立面图

Side elevation of Shouhuang Hall

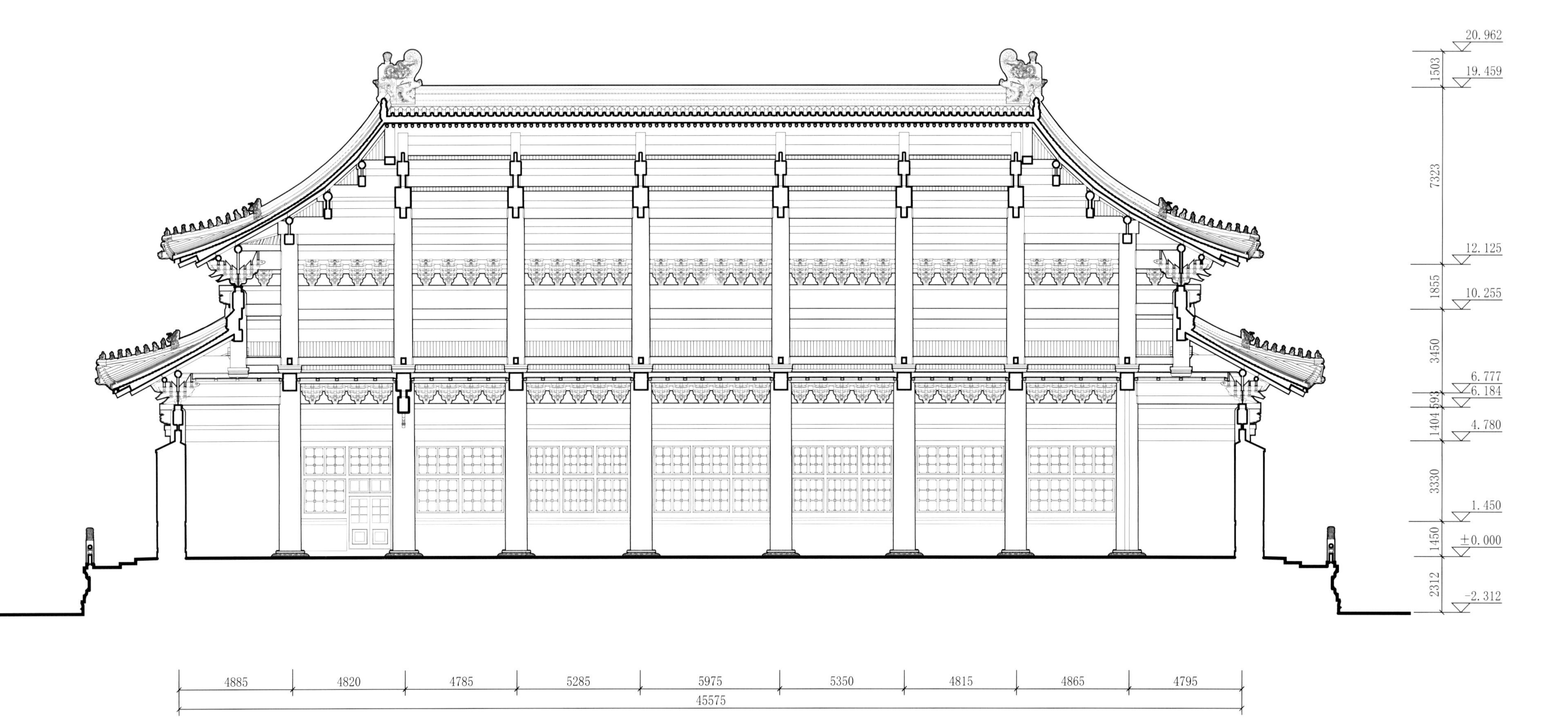

寿皇殿纵剖面图

Longitudinal section of Shouhuang Hall

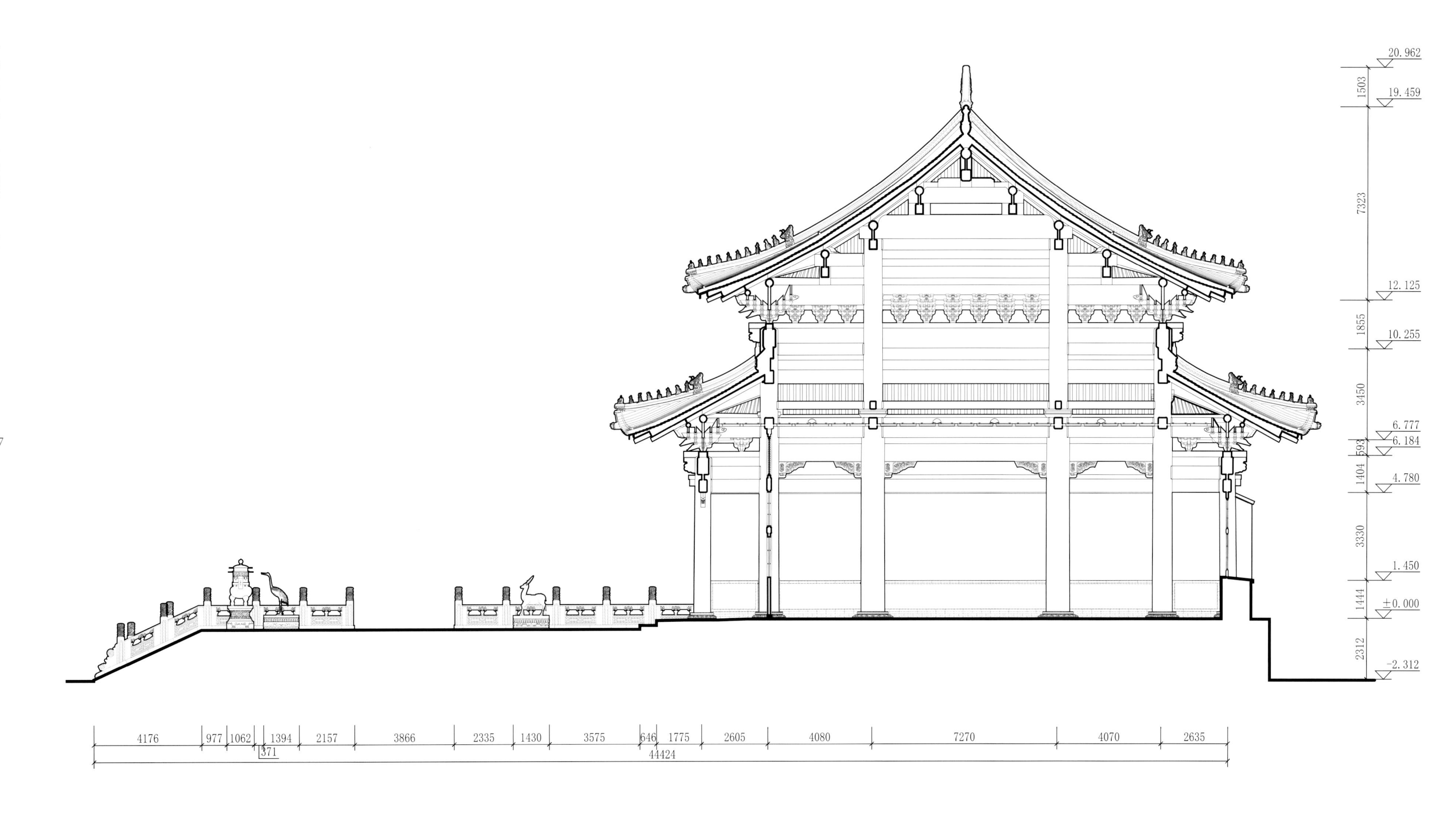

寿皇殿明间横剖面图

Transverse section of Shouhuang Hall

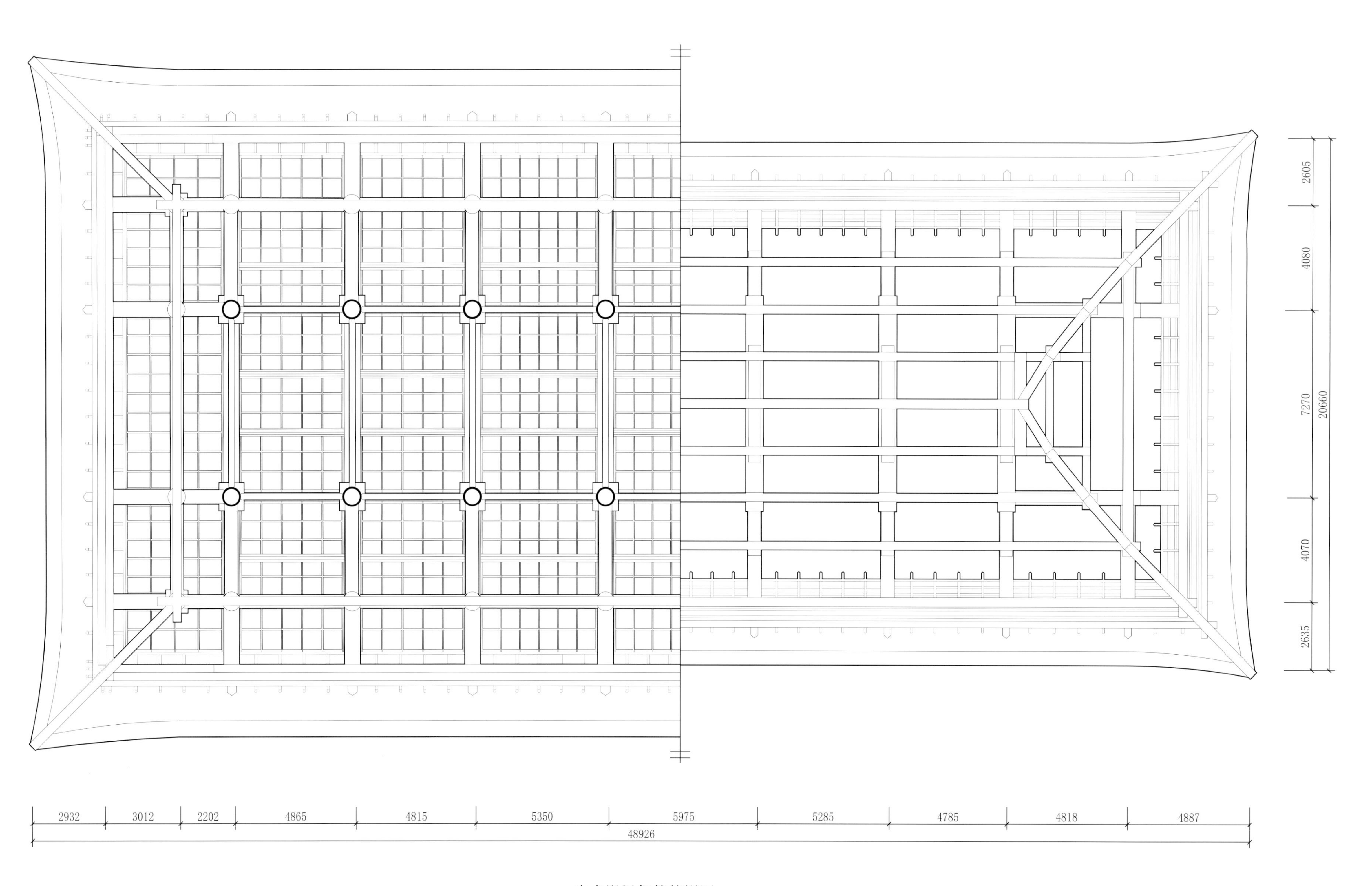

寿皇殿梁架仰俯视图

Bottom and Vertical view of beam frame of Shouhuang Hall

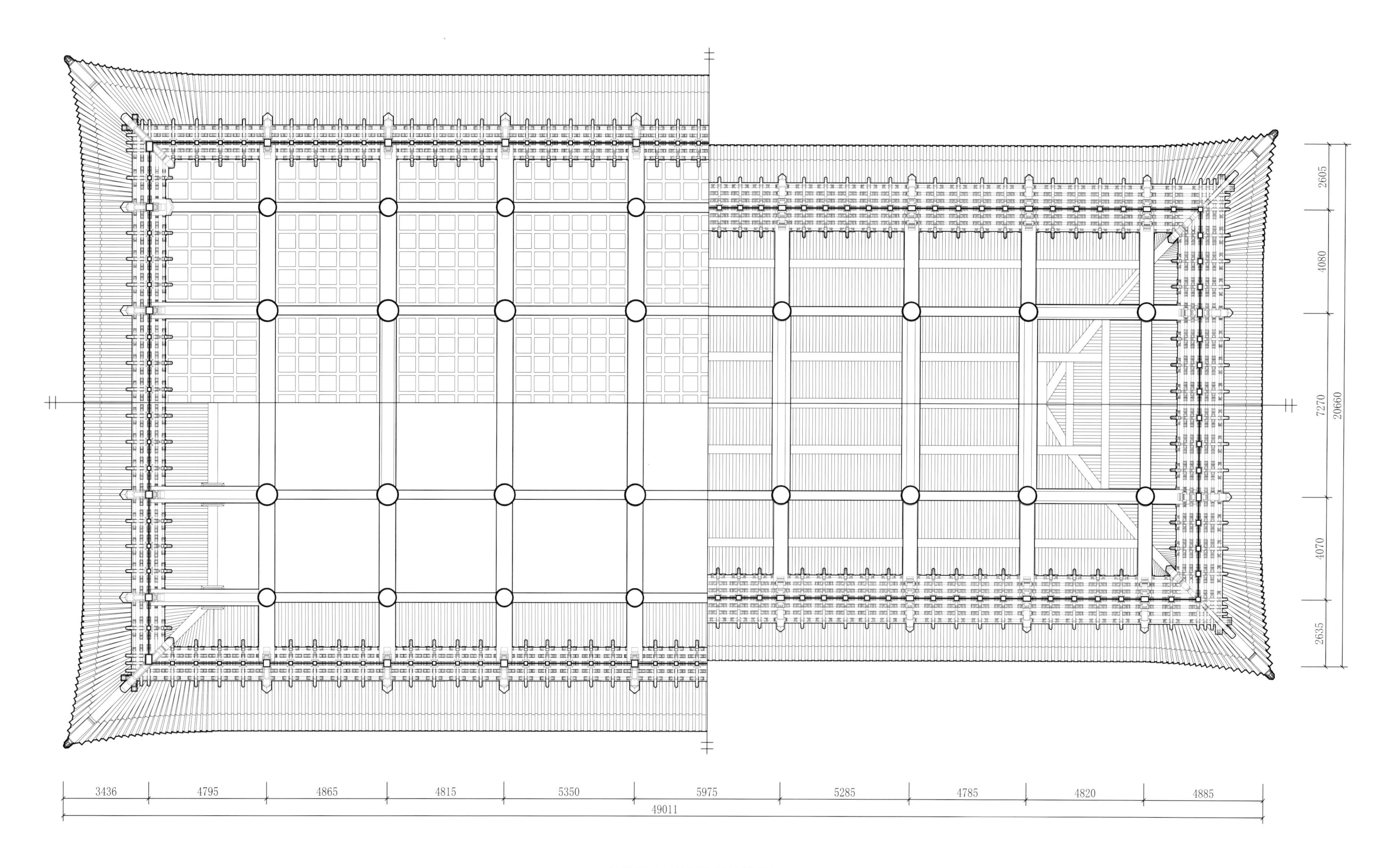

寿皇殿上下檐梁架仰视图

Bottom view of beam frame of the upper and under cornice of Shouhuang Hall

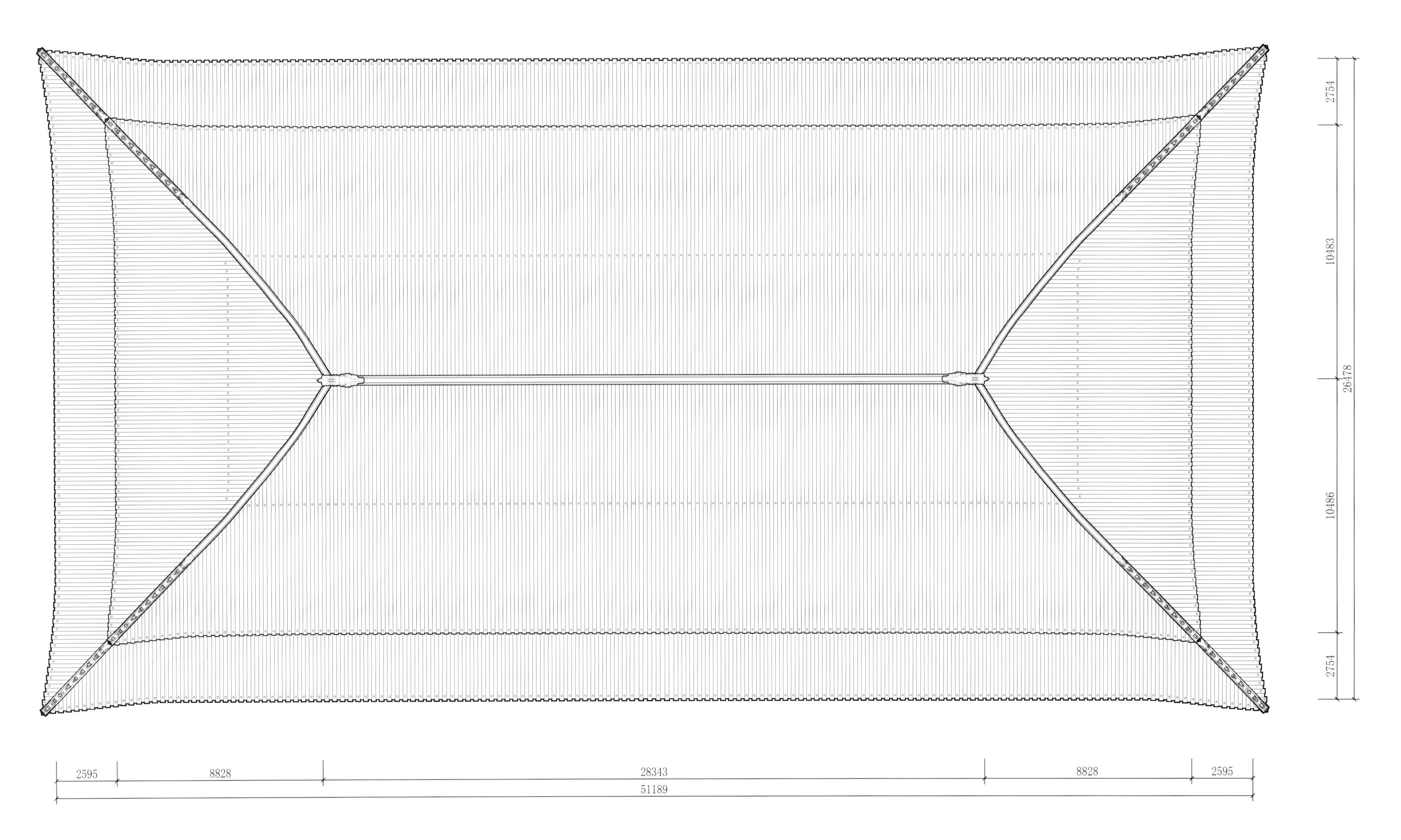

寿皇殿屋顶平面图
Roof plan of Shouhuang Hall

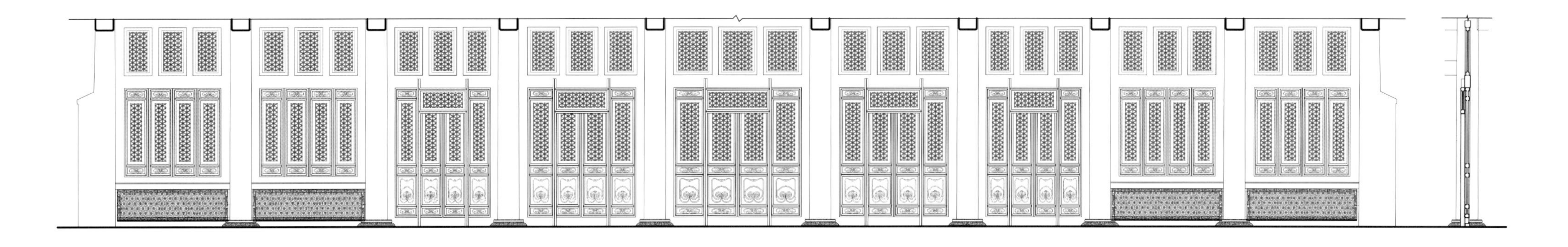

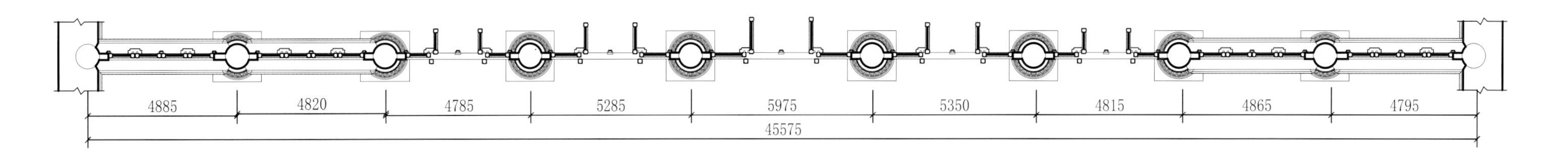

寿皇殿门窗复原大样图

Detail drawing of Shouhuang Hall

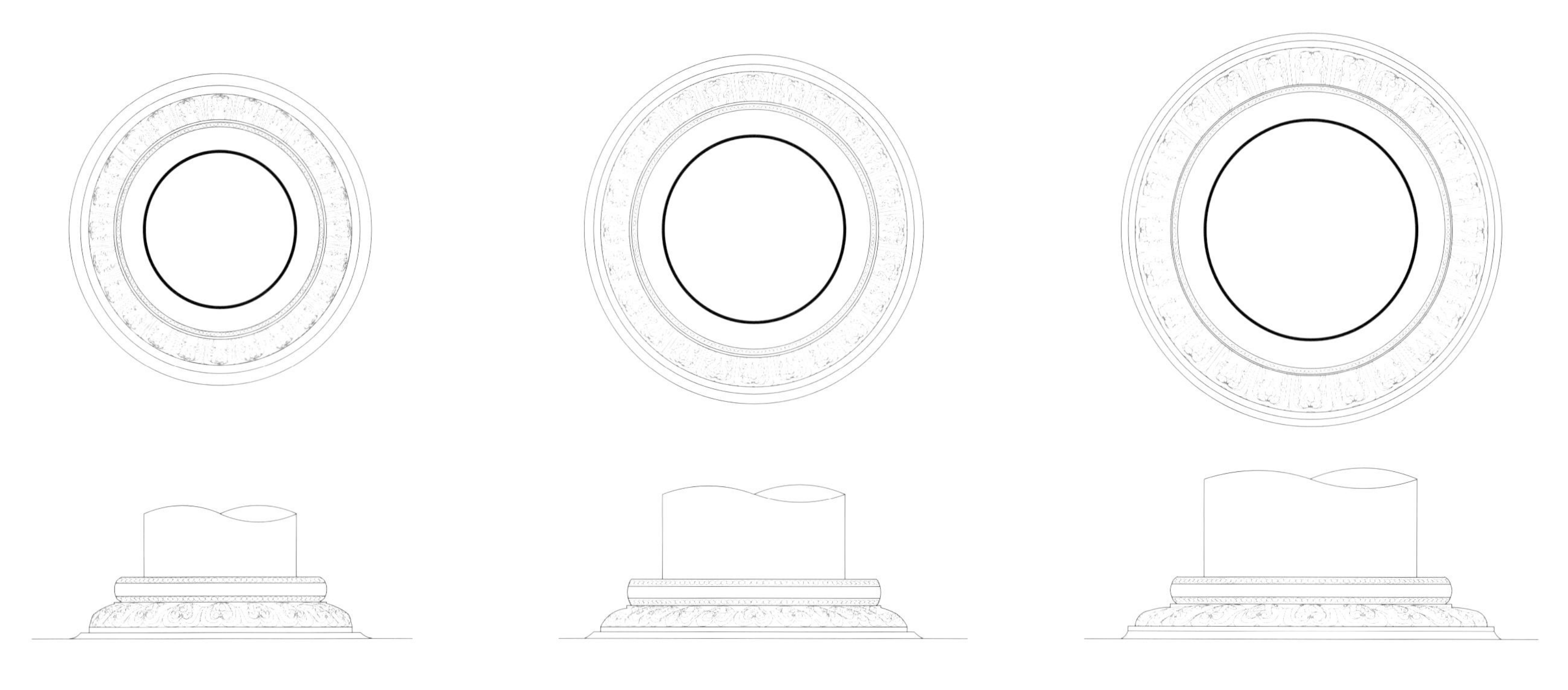

寿皇殿柱础大样图
Detail drawing of the plinth of Shouhuang Hall

寿皇殿槛墙琉璃纹大样图
Detail drawing of the texture of Shouhuang Hall

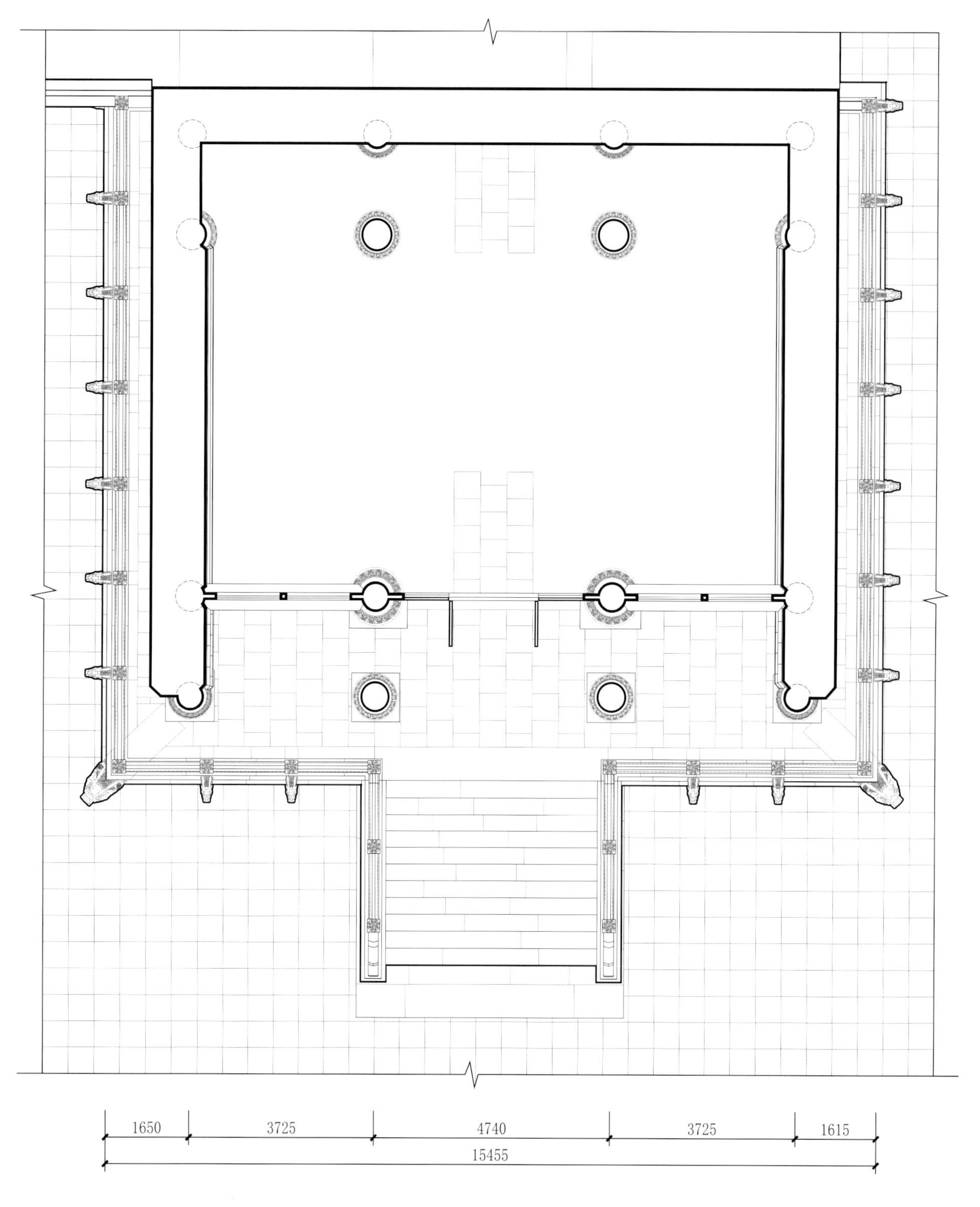

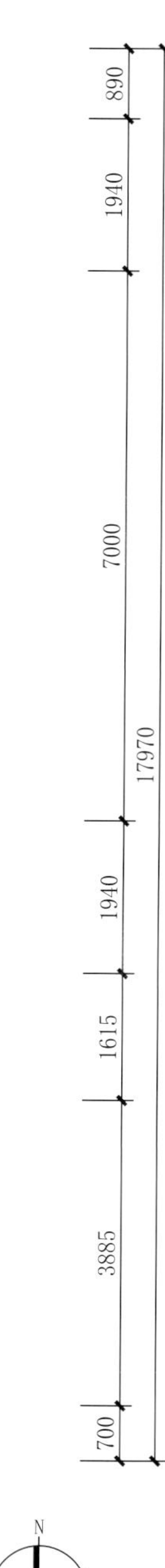

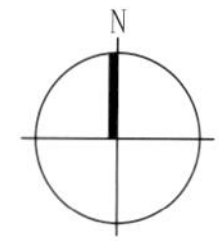

衍庆殿、绵禧殿平面图
Plan of Yanqing or Mianxi Hall

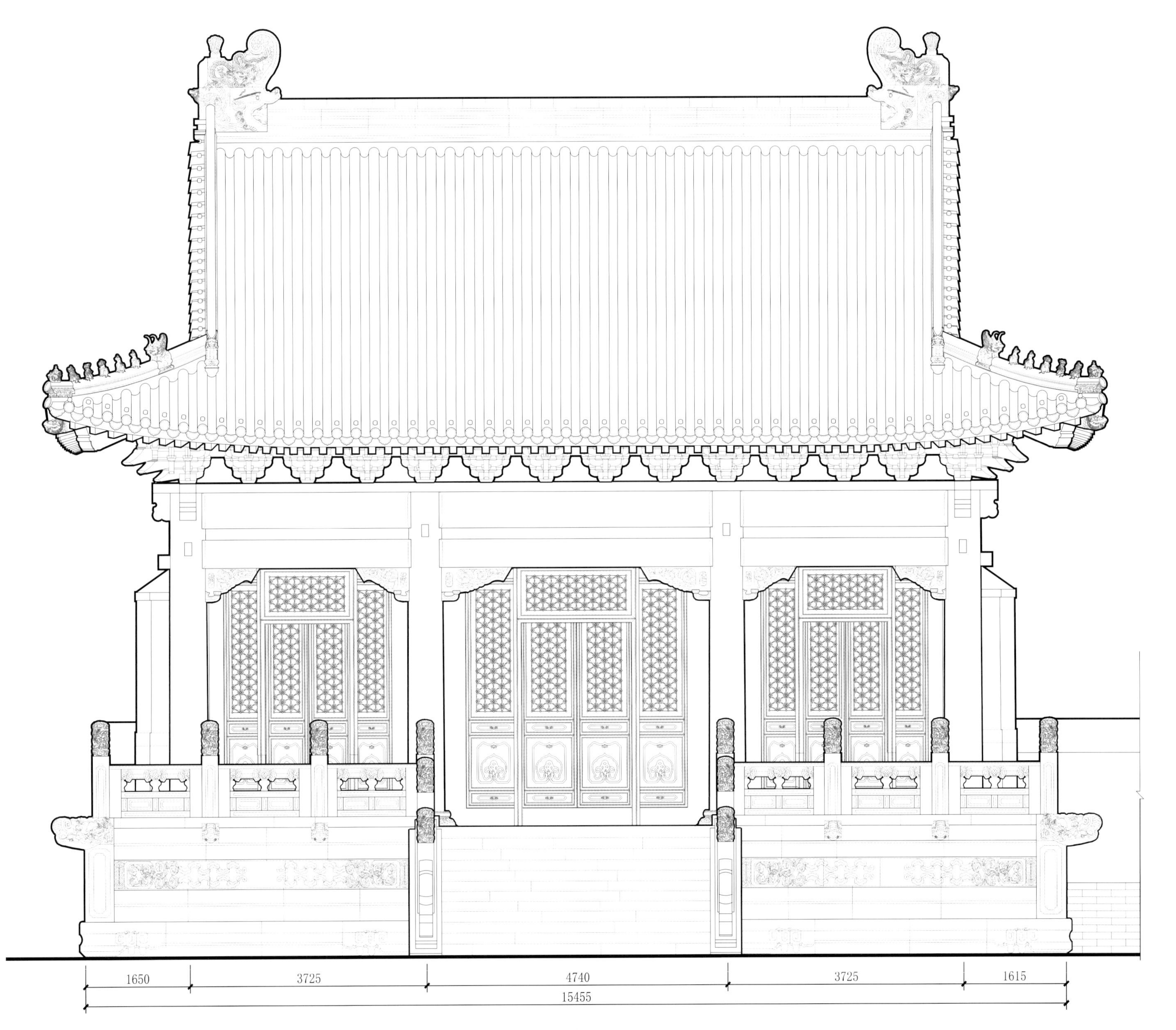

衍庆殿、绵禧殿正立面图
Front elevation of Yanqing or Mianxi Hall

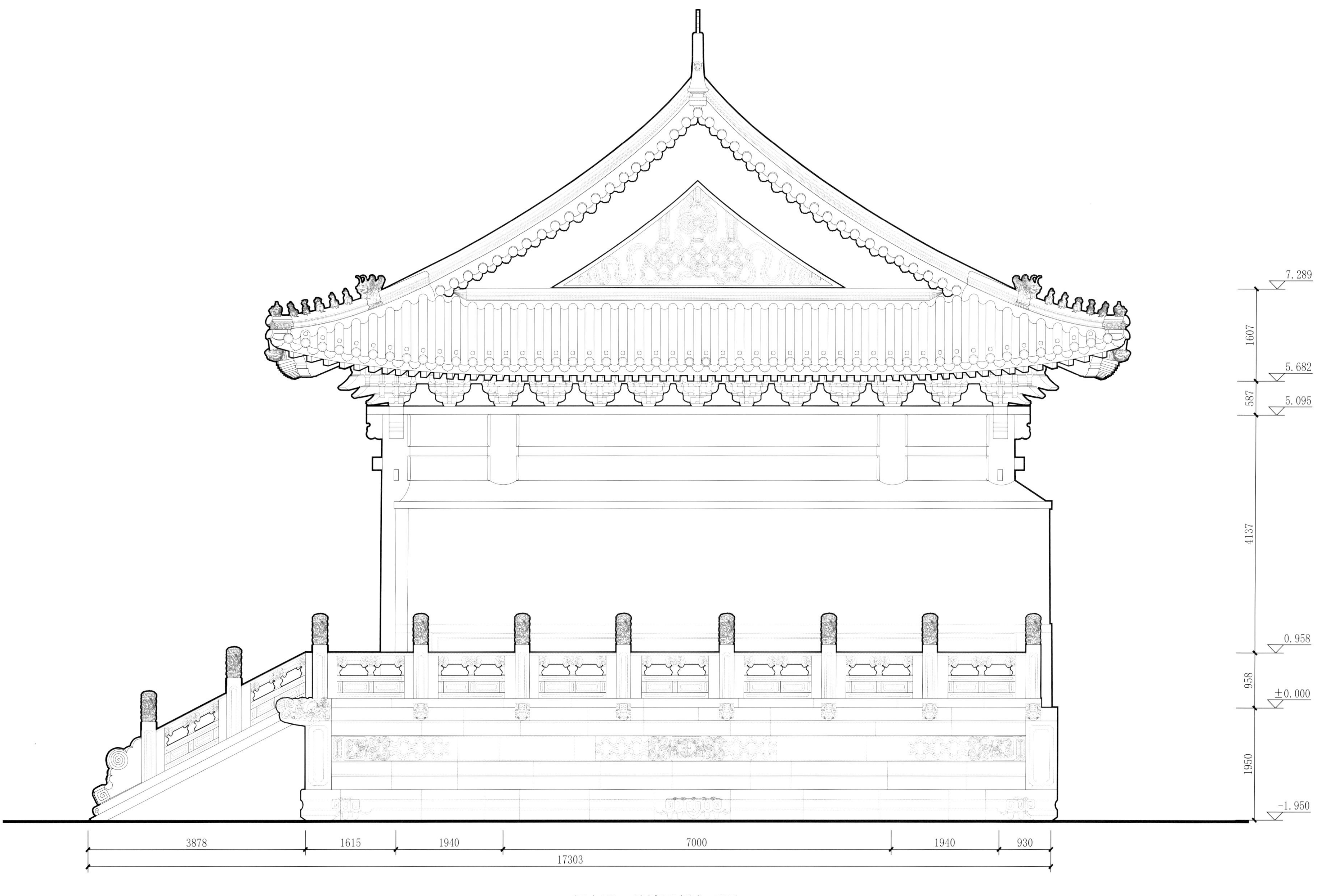

衍庆殿、绵禧殿侧立面图
Side elevation of Yanqing or Mianxi Hall

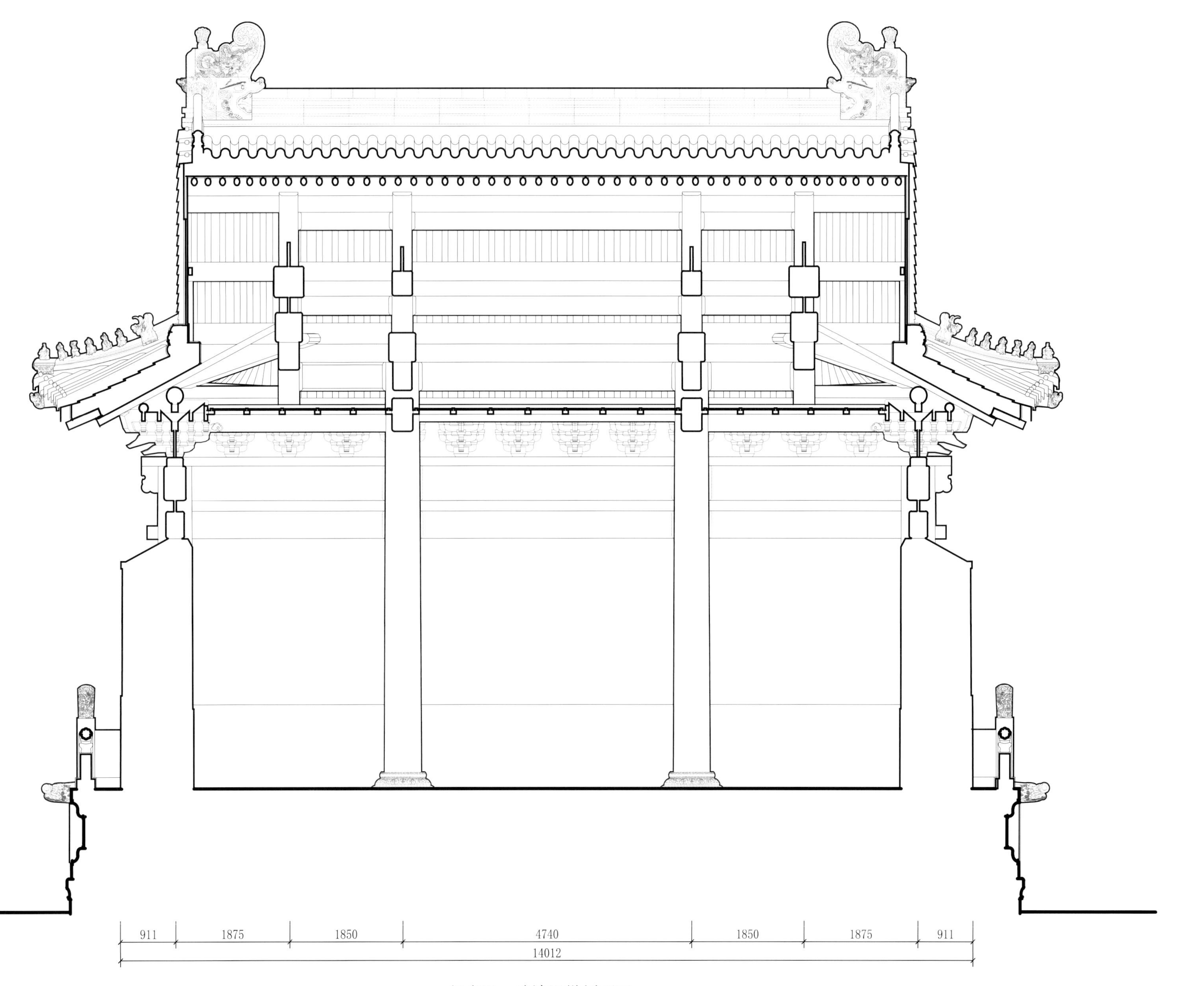

衍庆殿、绵禧殿纵剖面图
Longitudinal section of Yanqing or Mianxi Hall

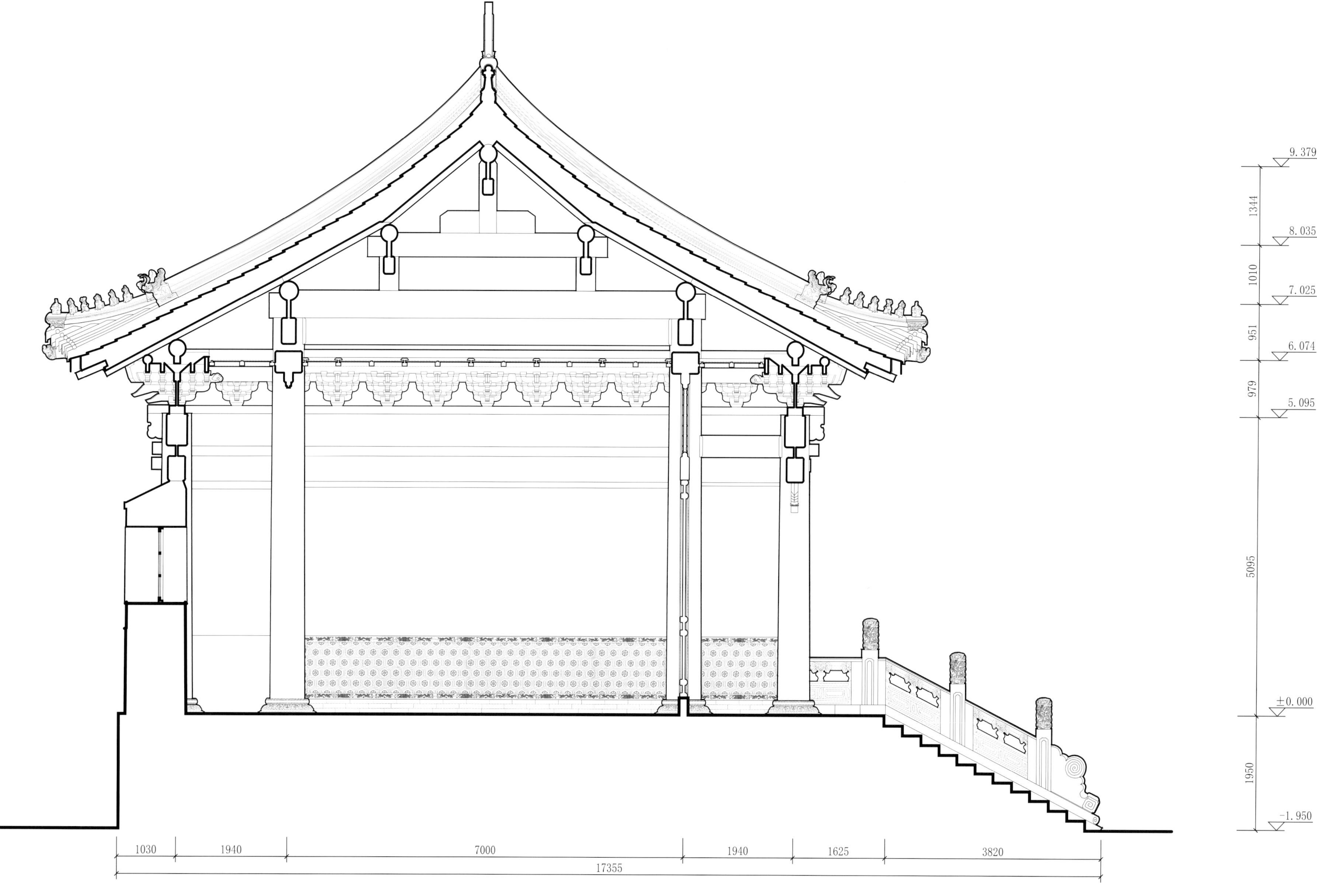

衍庆殿、绵禧殿横剖面图

Transverse section of Yanqing or Mianxi Hall

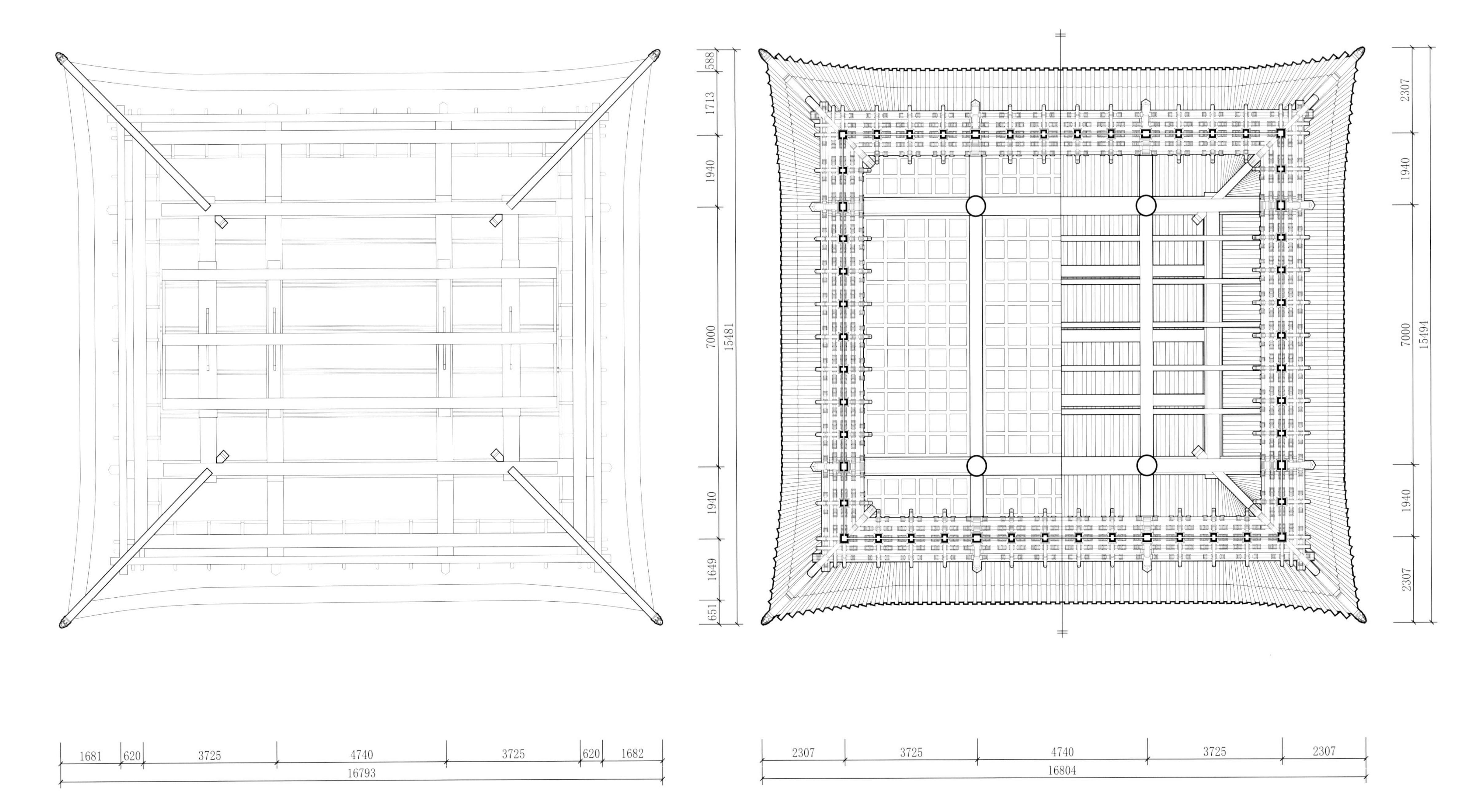

衍庆殿、绵禧殿梁架俯视图
Vertical view of beam frame of Yanqing or Mianxi Hall

衍庆殿、绵禧殿梁架仰视图
Bottom view of beam frame of Yanqing or Mianxi Hall

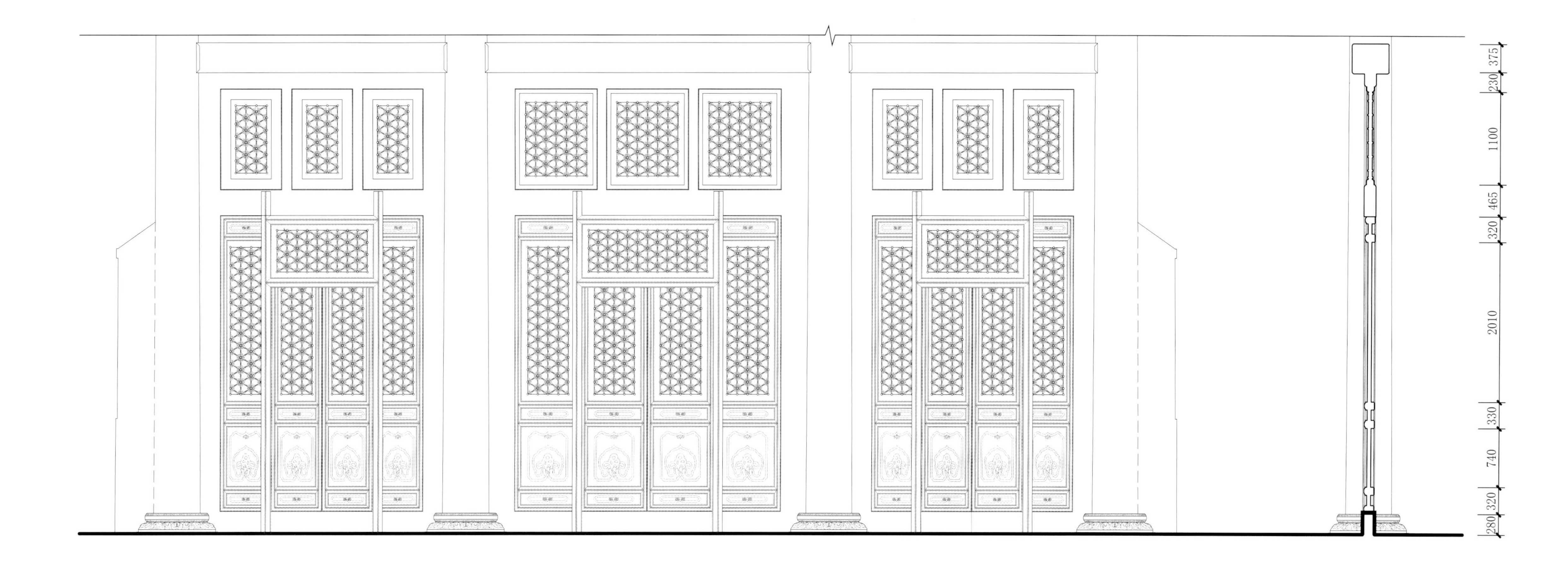

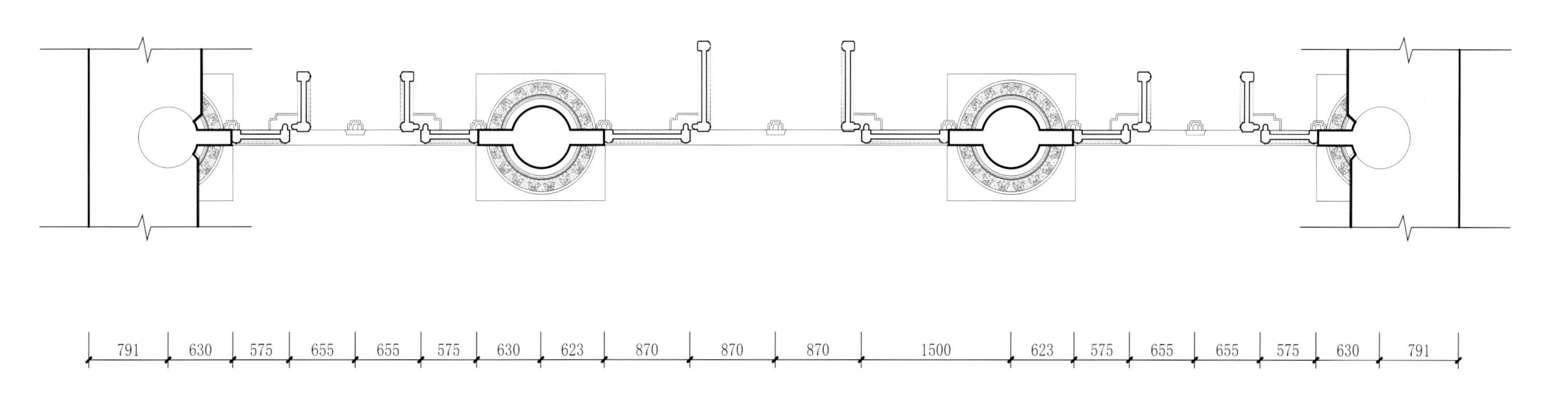

衍庆殿、绵禧殿门窗复原大样图
Detail drawing of Yanqing or Mianxi Hall

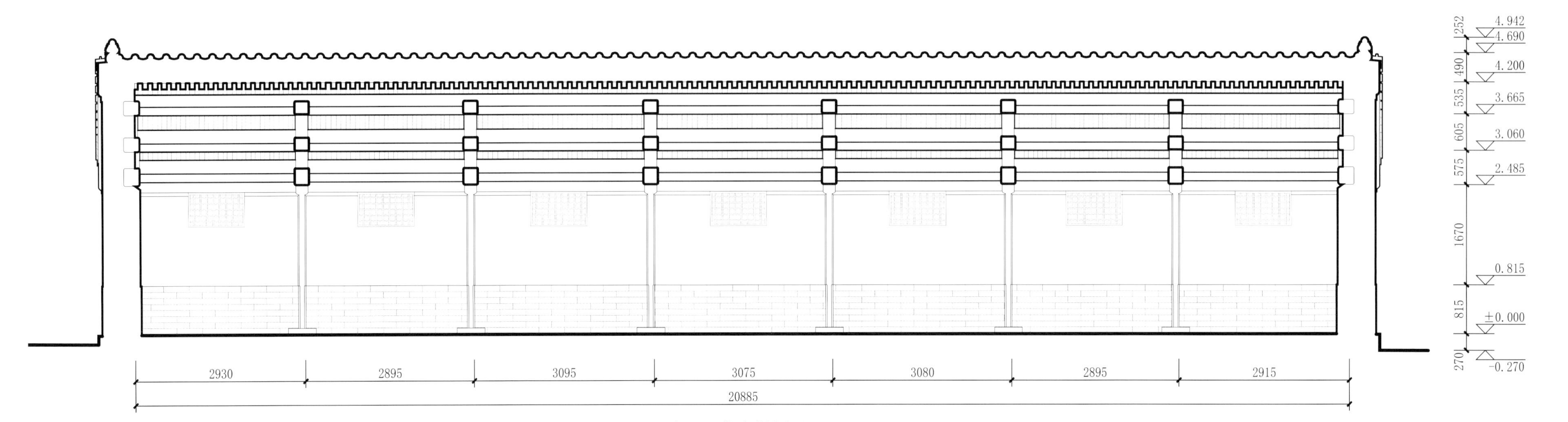

东、西值房纵剖面图
Longitudinal section of East or West Duty Room

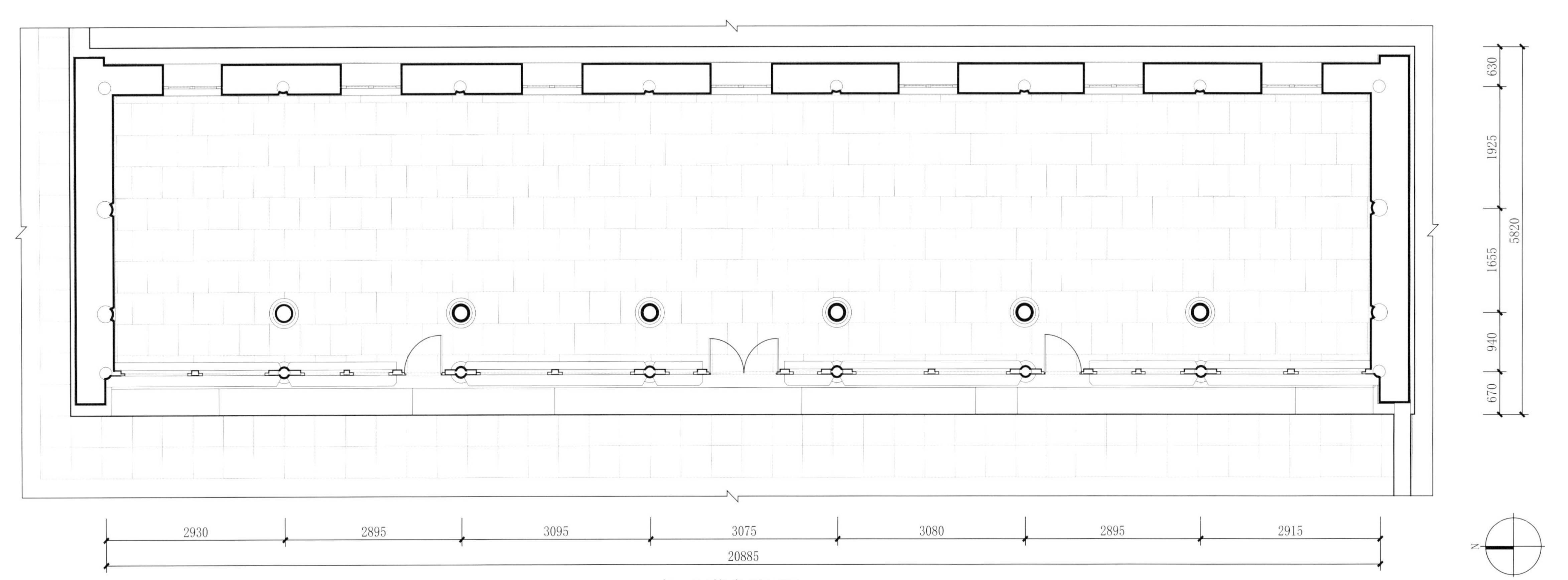

东、西值房平面图
Plan of East or West Duty Room

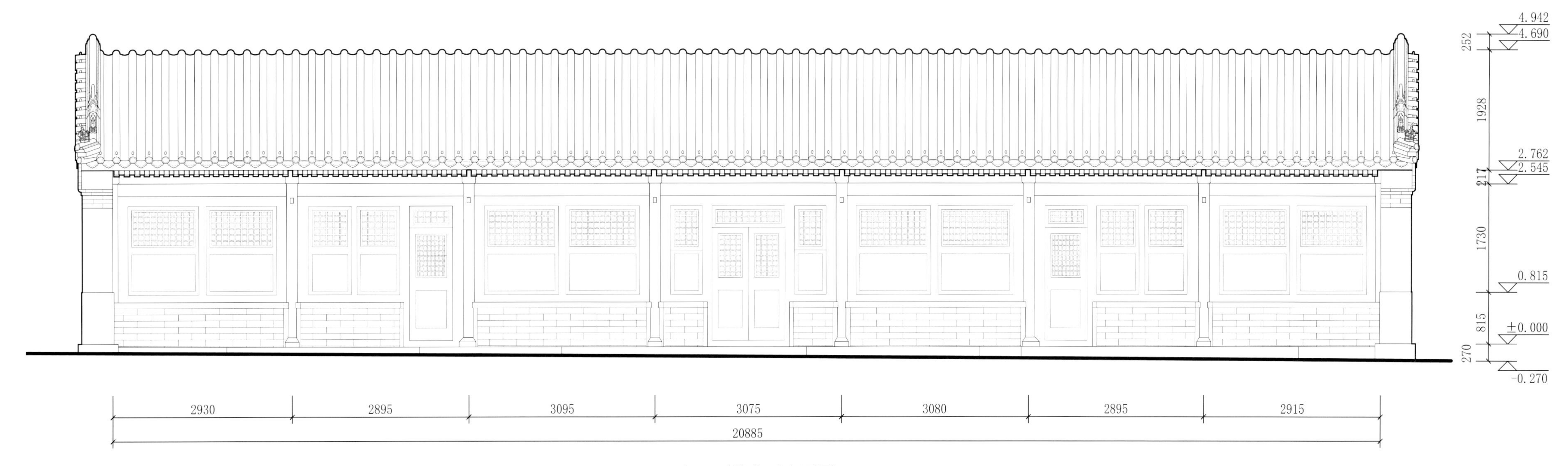

东、西值房正立面图
Front elevation of East or West Duty Room

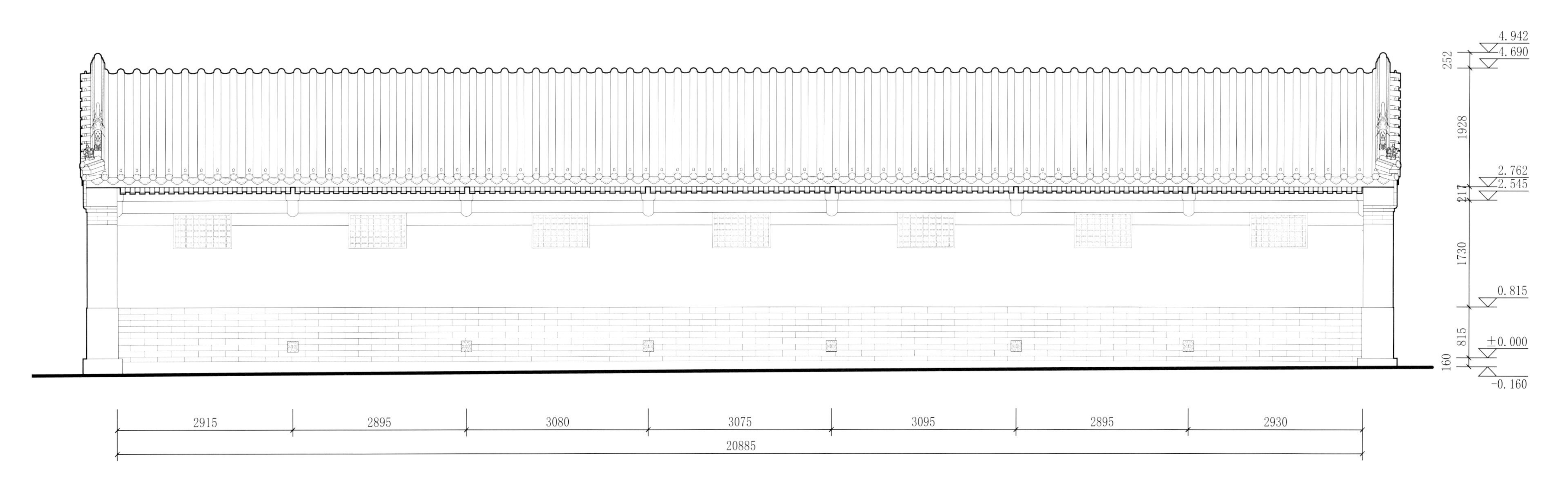

东、西值房背立面图
Back elevation of East or West Duty Room

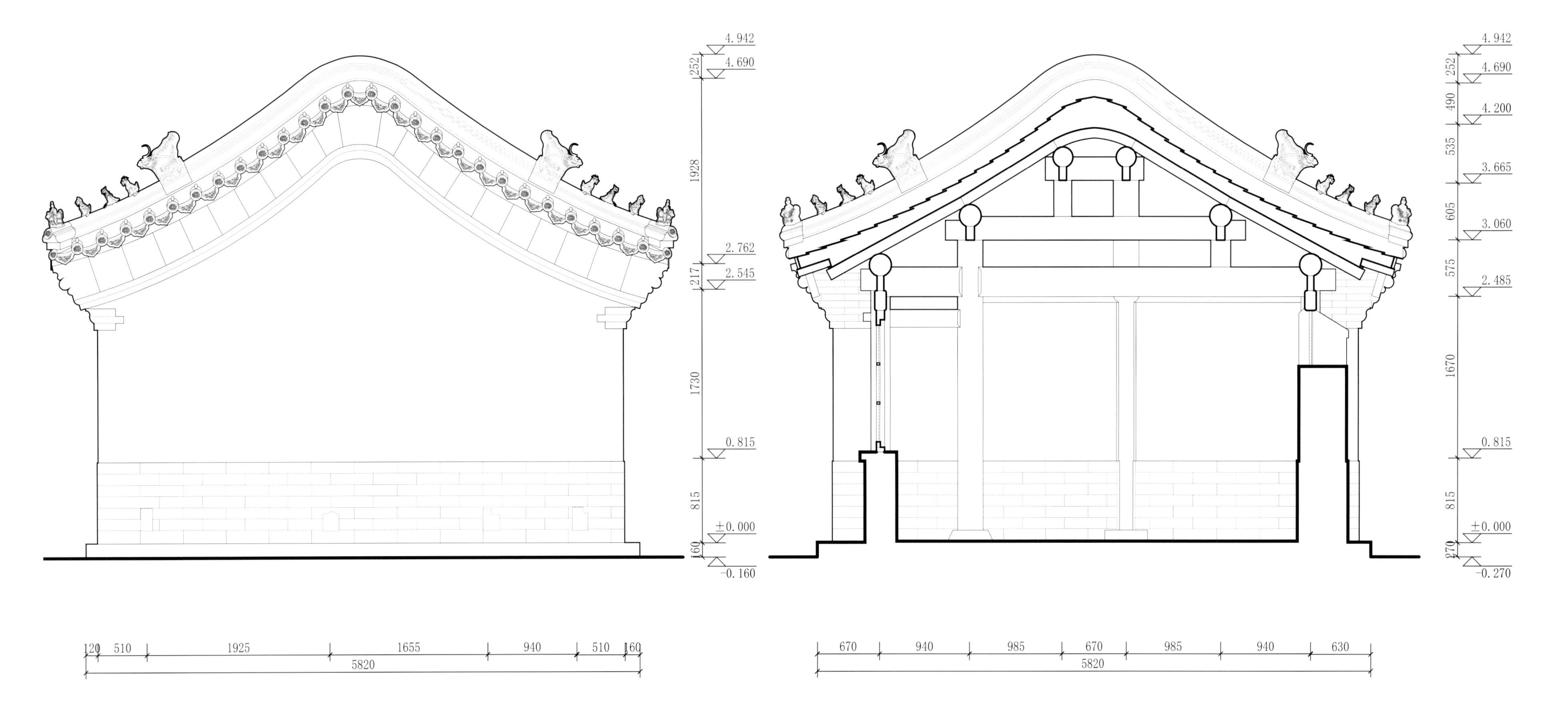

东、西值房侧立面图
Side elevation of East or West Duty Room

东、西值房明间横剖面图
Transverse section of East or West Duty Room

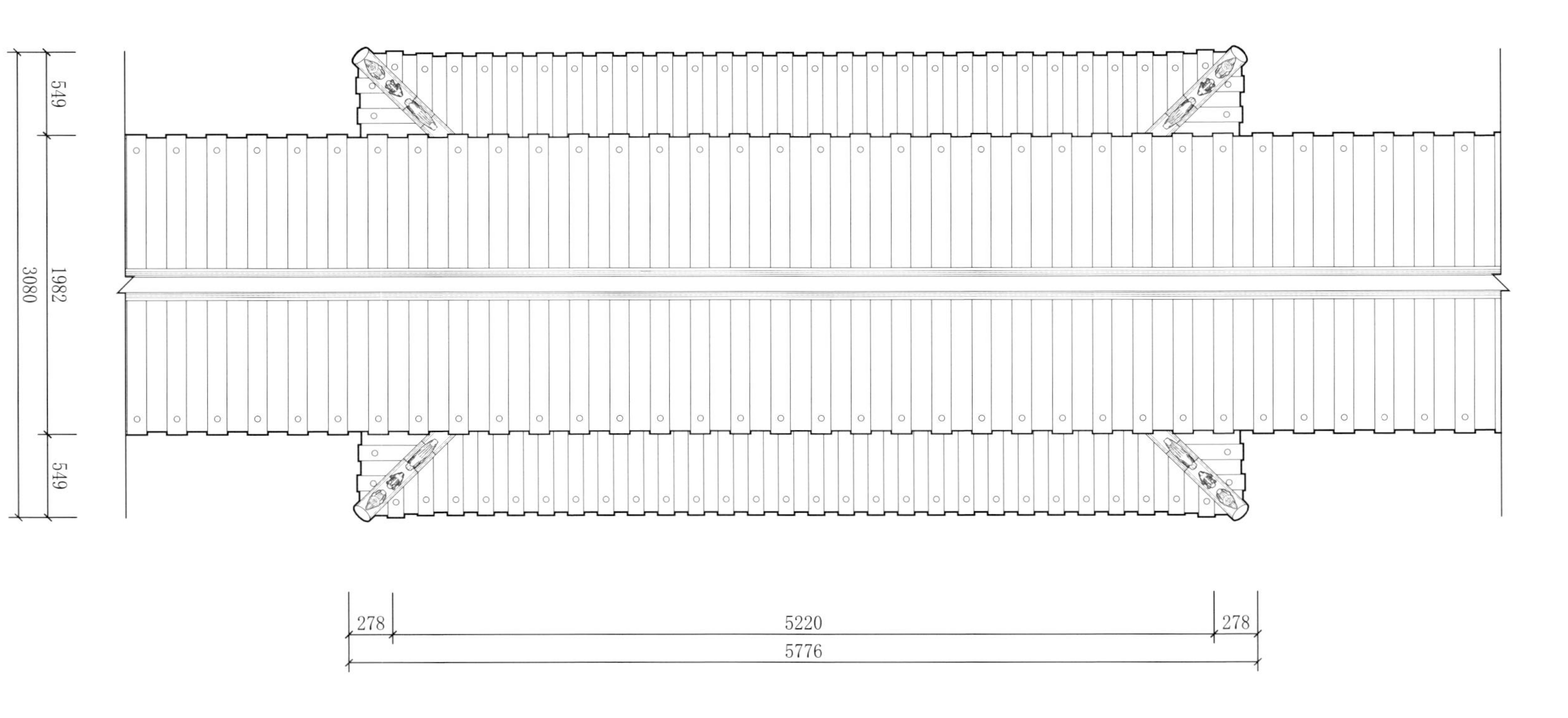

东、西砖城门屋顶平面图
Roof plan of East or West Bricked Gate

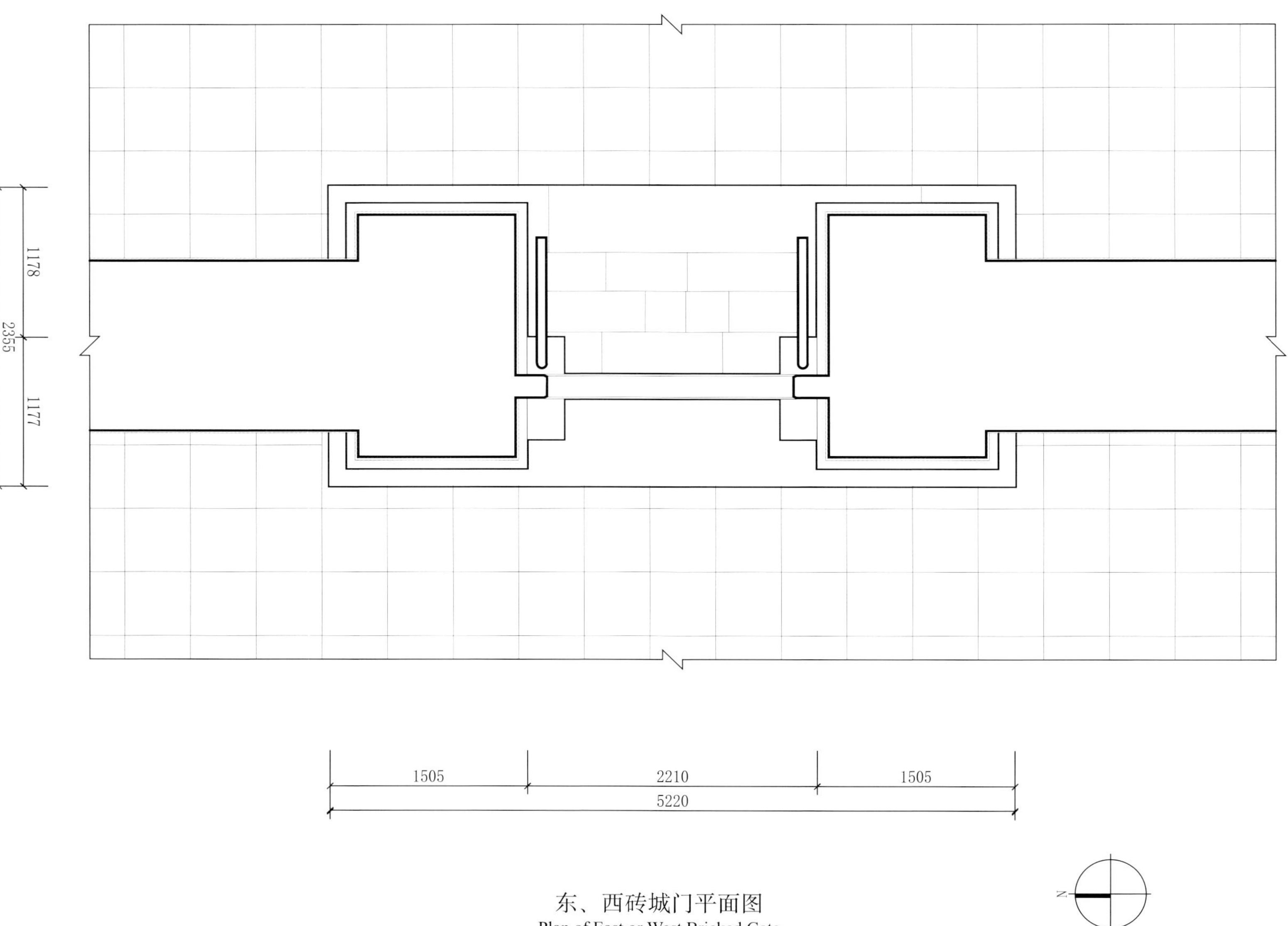

东、西砖城门平面图
Plan of East or West Bricked Gate

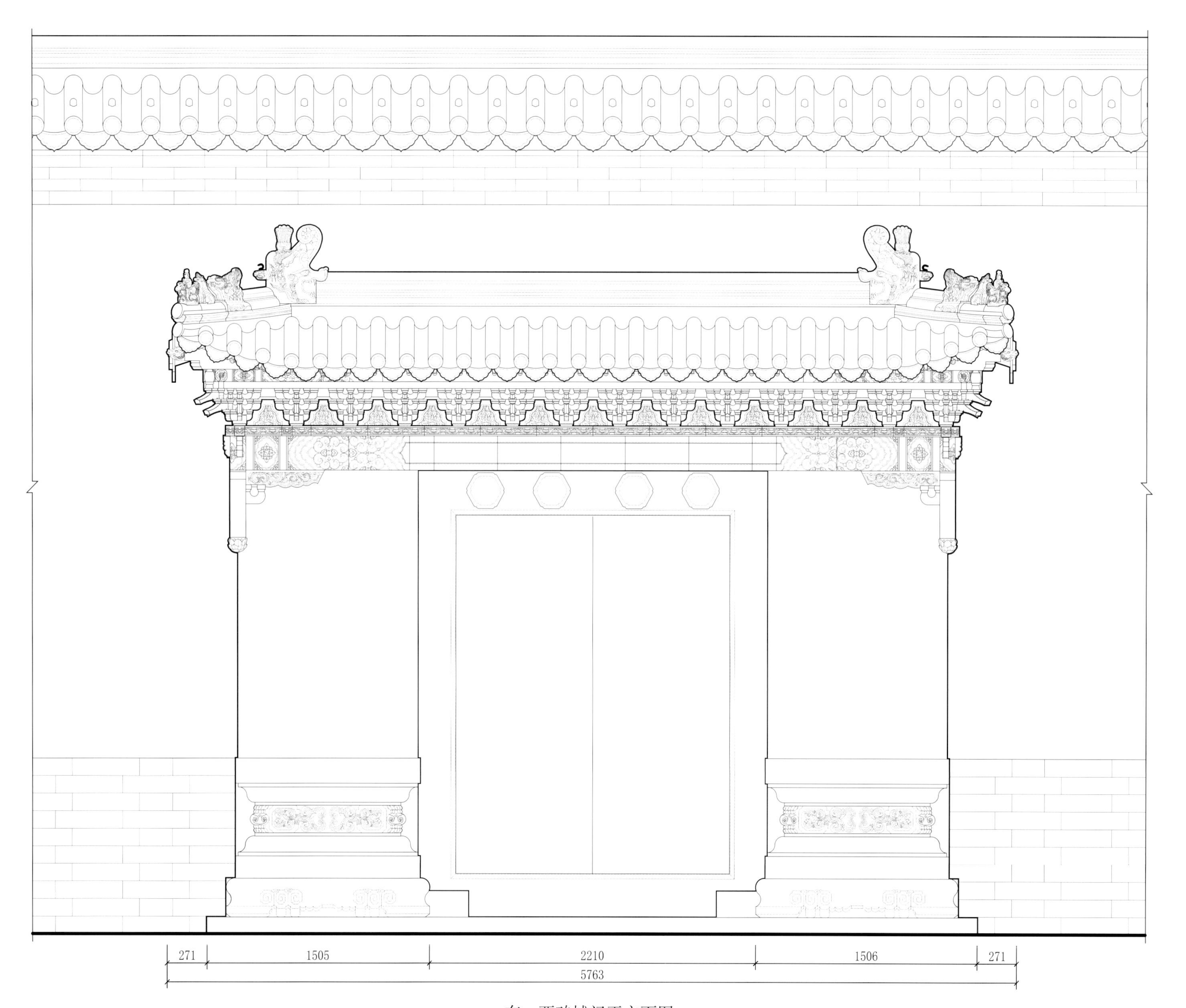

东、西砖城门正立面图
Front elevation of East or West Bricked Gate

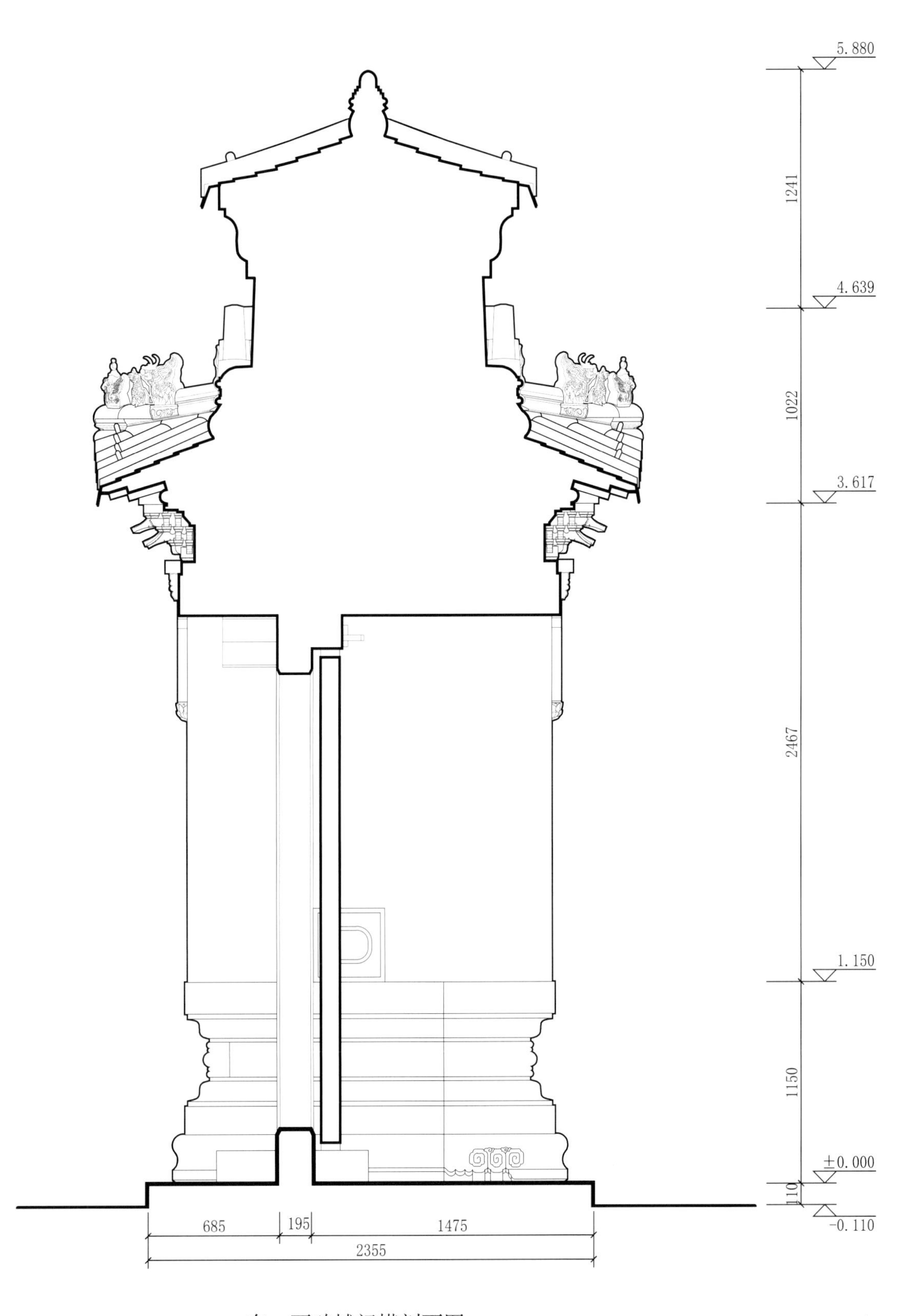

东、西砖城门横剖面图
Transverse section of East or West Bricked Gate

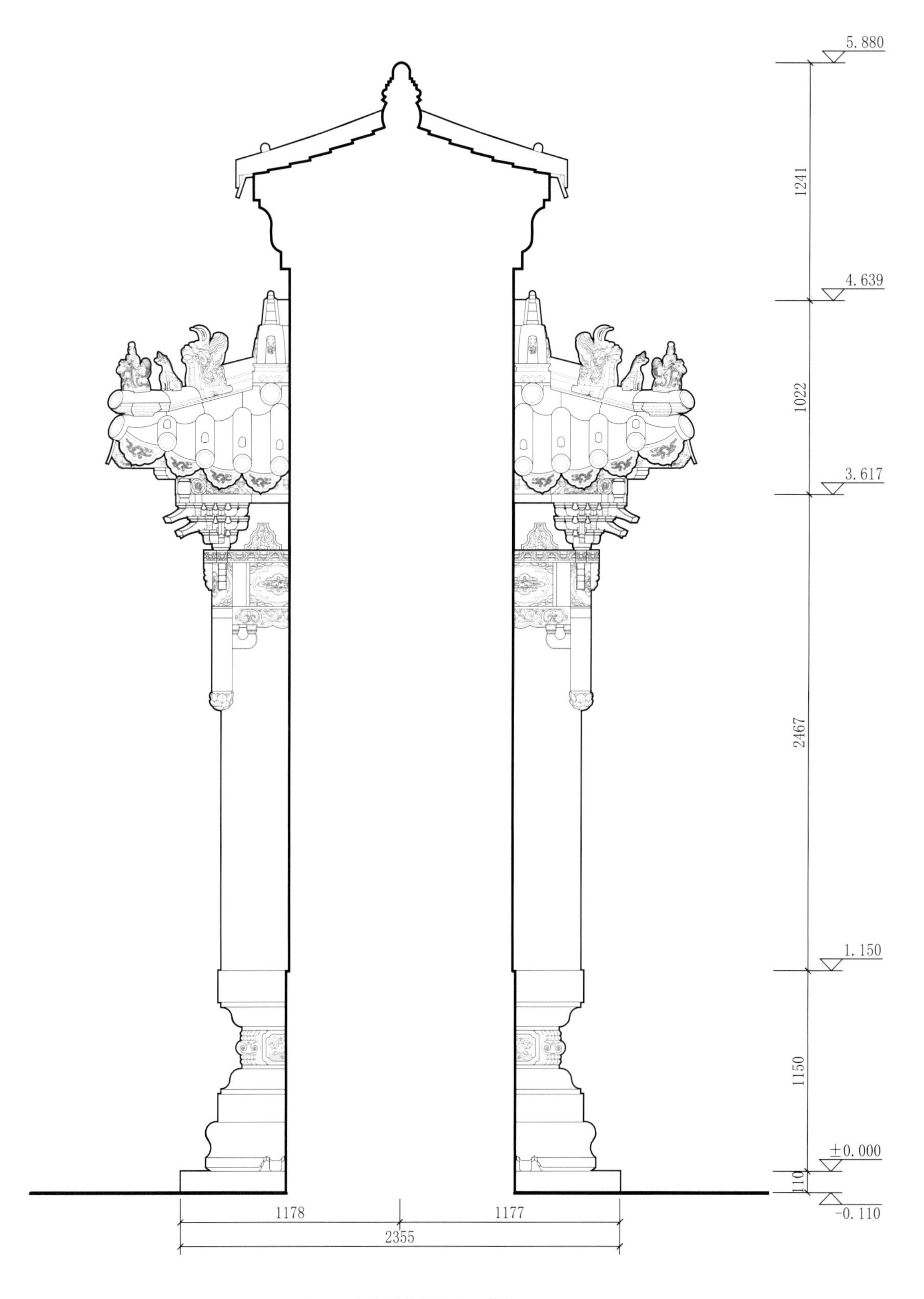

东、西砖城门侧立面图
Side elevation of East or West Bricked Gate

观德殿建筑群

The Groups of Guande Hall

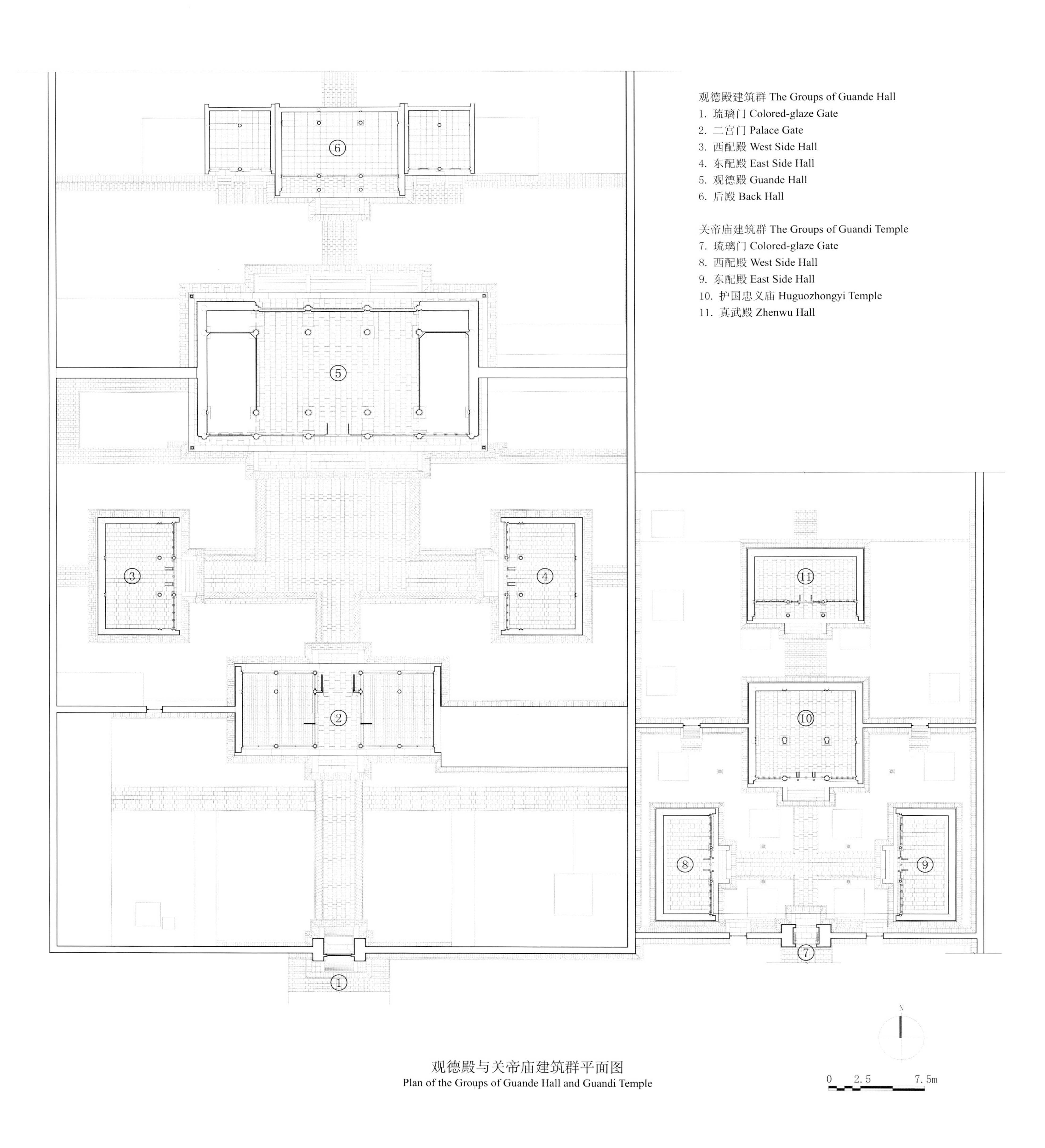

观德殿与关帝庙建筑群平面图
Plan of the Groups of Guande Hall and Guandi Temple

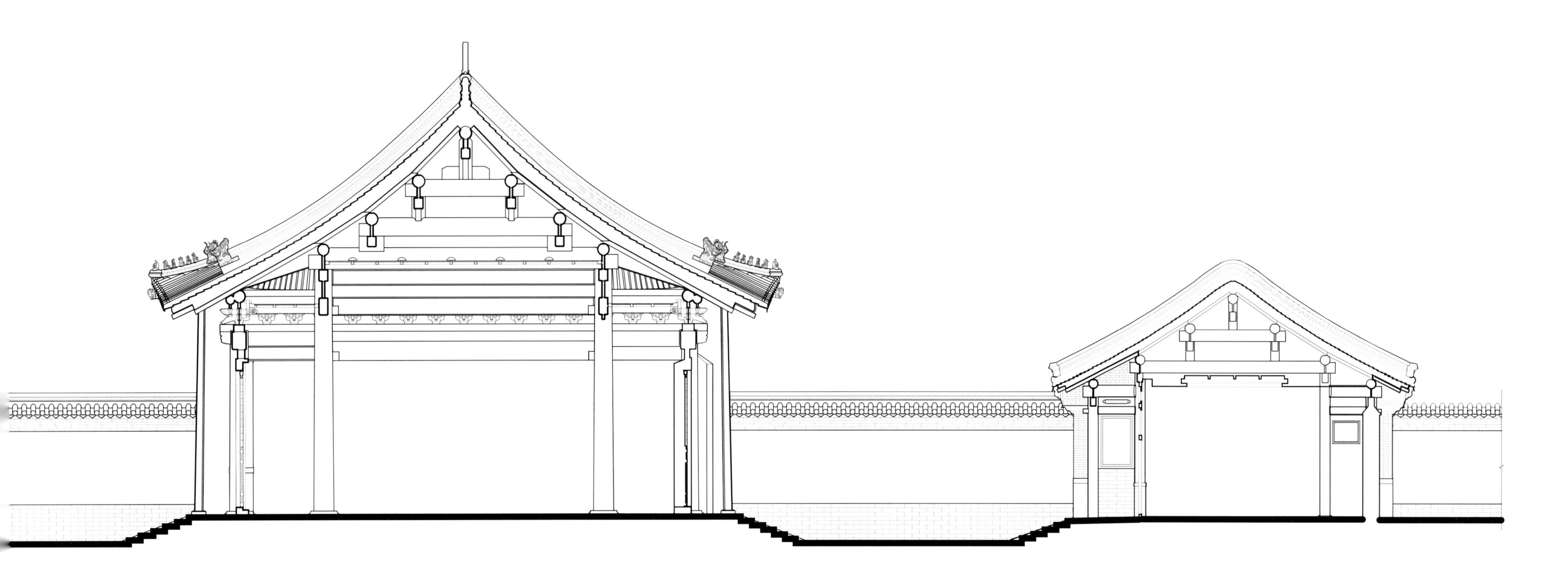

0
1.2
3.6m

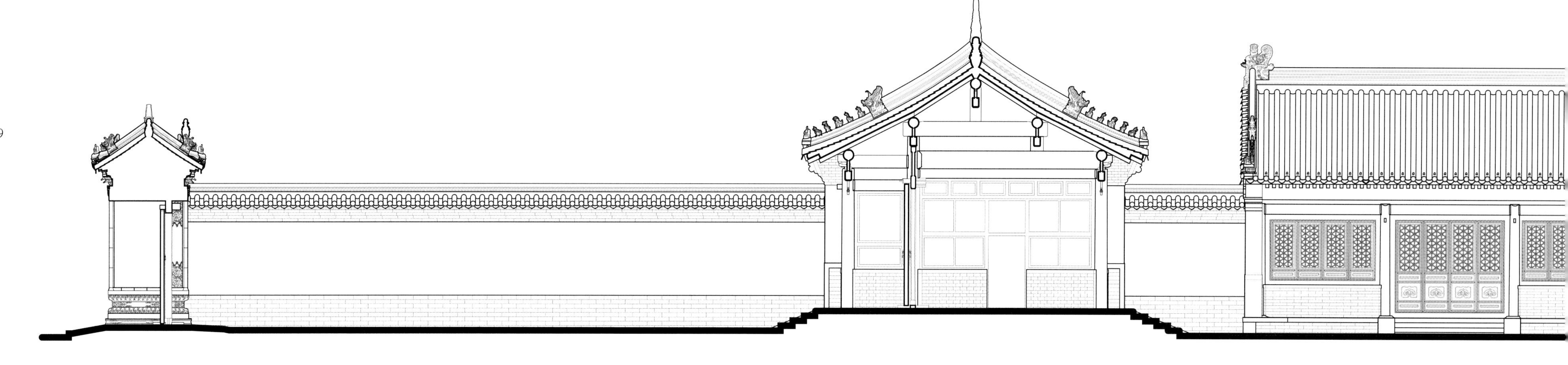

观德殿建筑群剖面图

Section of the Groups of the Guande Hall

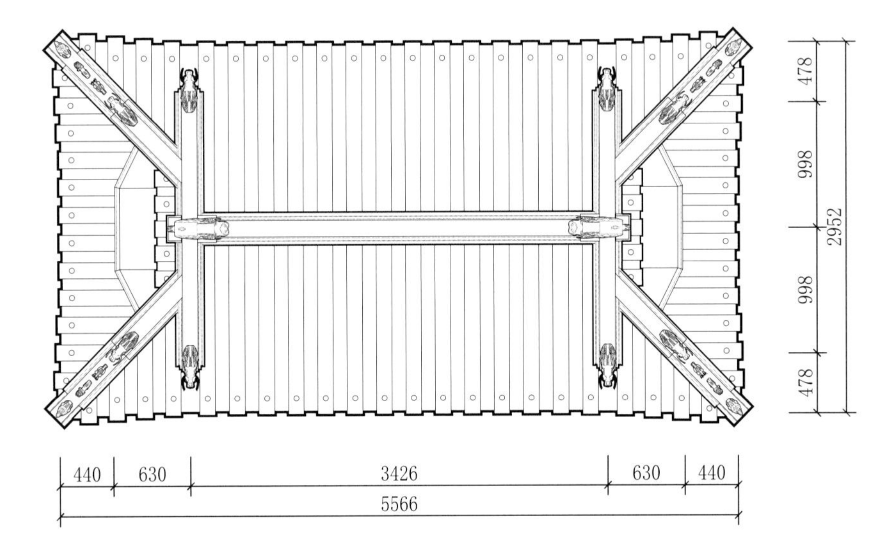

观德殿琉璃门屋顶平面图

Roof plan of Colored-glaze Gate of Guande Hall

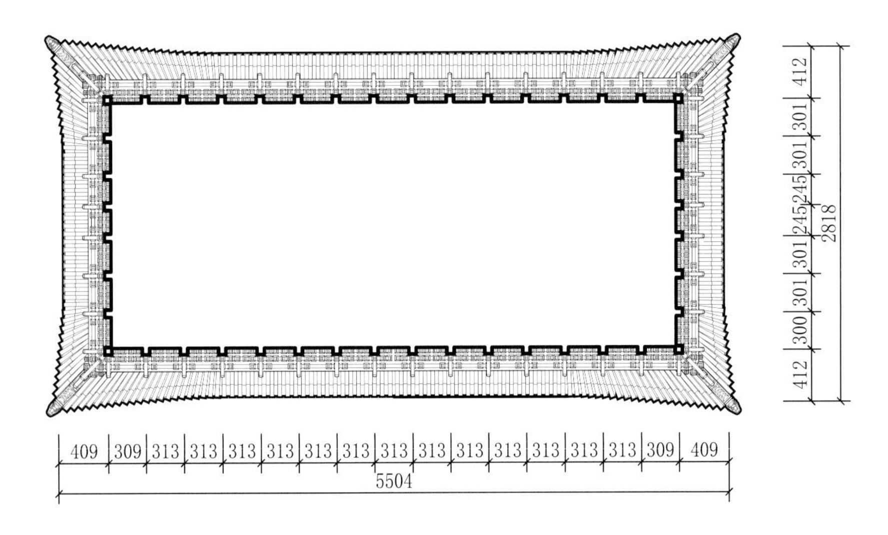

观德殿琉璃门梁架仰视图

Bottom view of beam frame of Colored-glaze Gate of Guande Hall

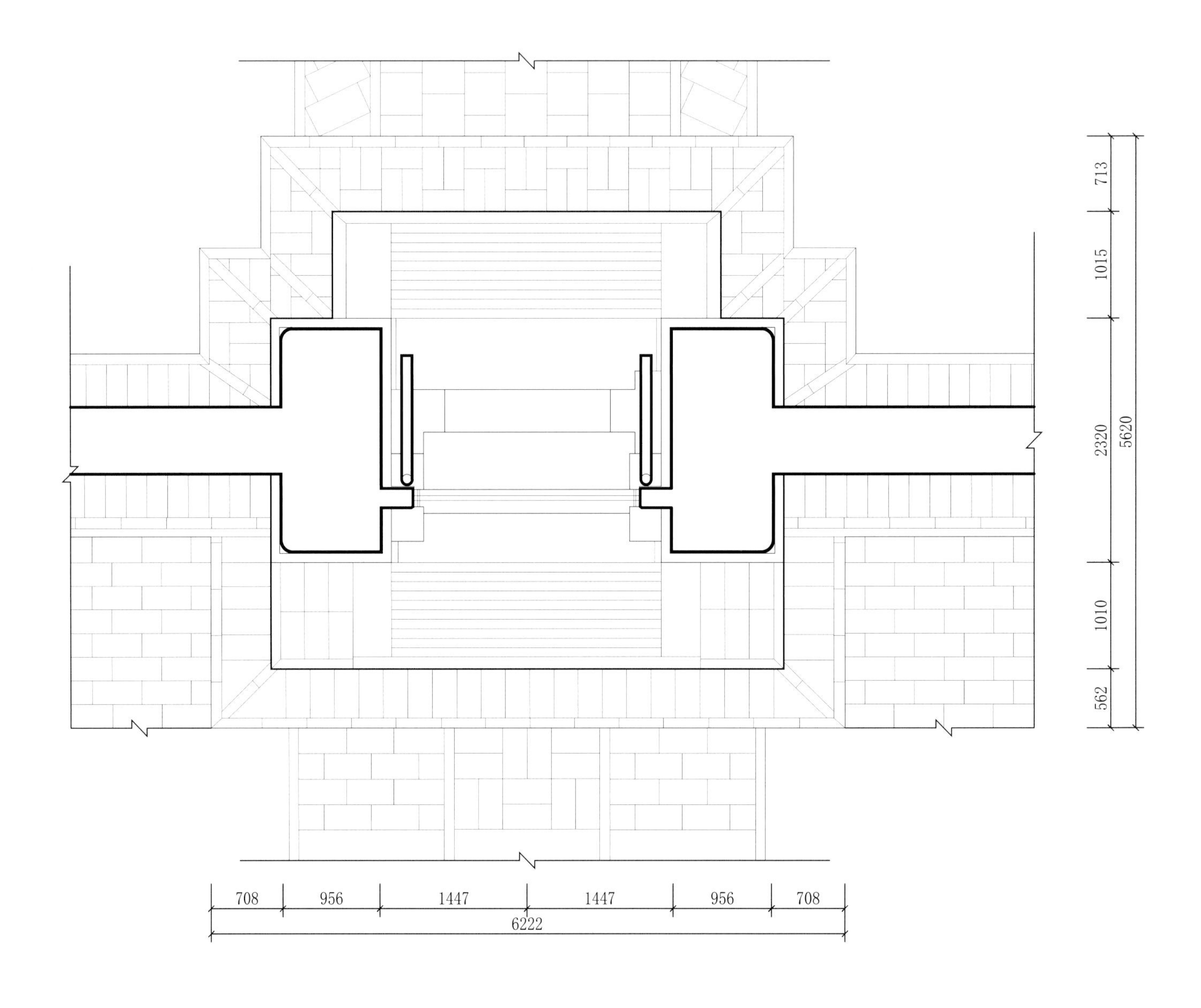

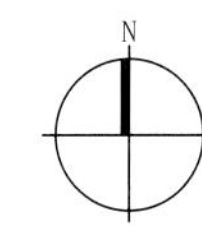

观德殿琉璃门平面图

Plan of Colored-glaze Gate of Guande Hall

观德殿琉璃门正立面图

Front elevation of Colored-glaze Gate of Guande Hall

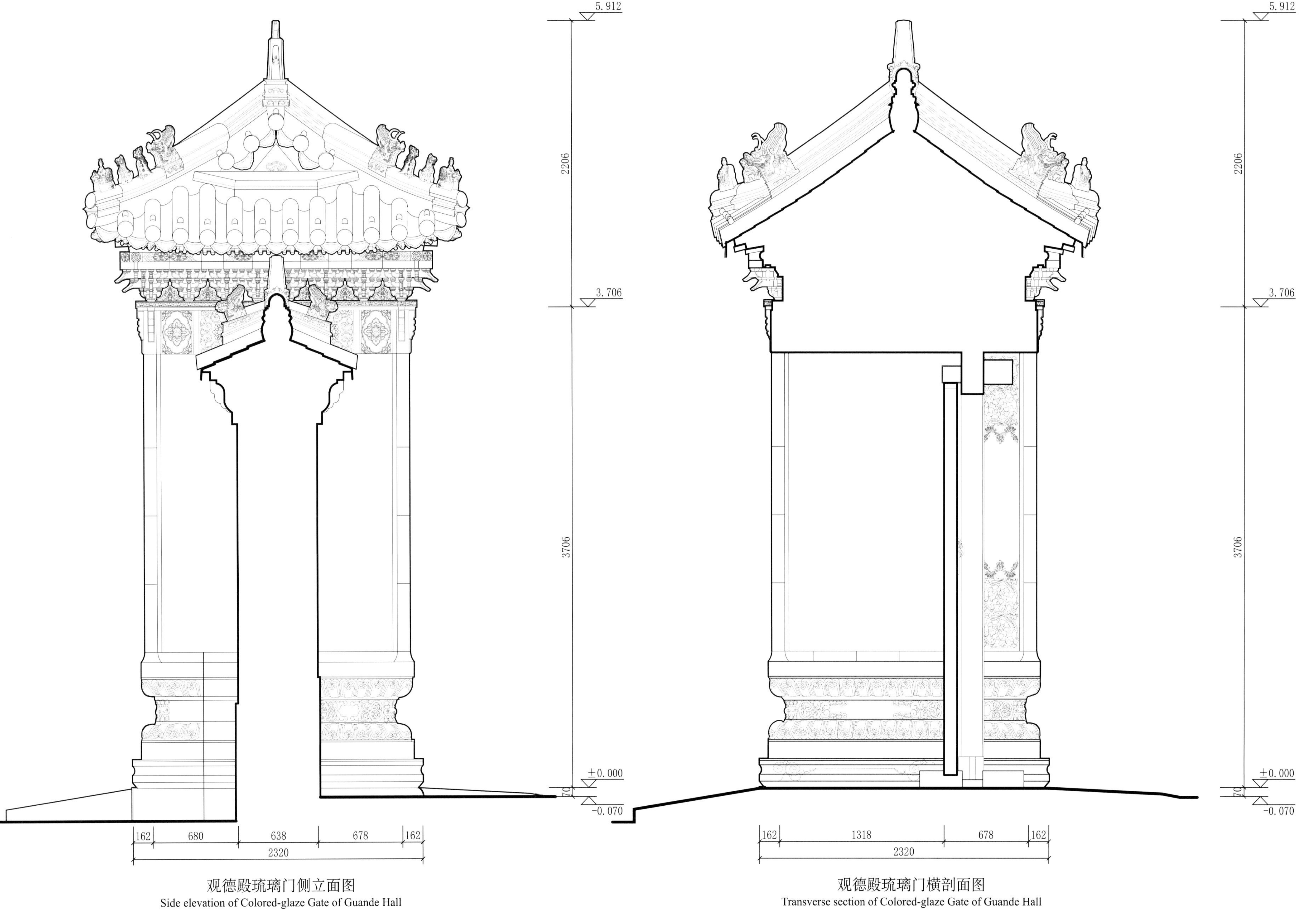

观德殿琉璃门侧立面图
Side elevation of Colored-glaze Gate of Guande Hall

观德殿琉璃门横剖面图
Transverse section of Colored-glaze Gate of Guande Hall

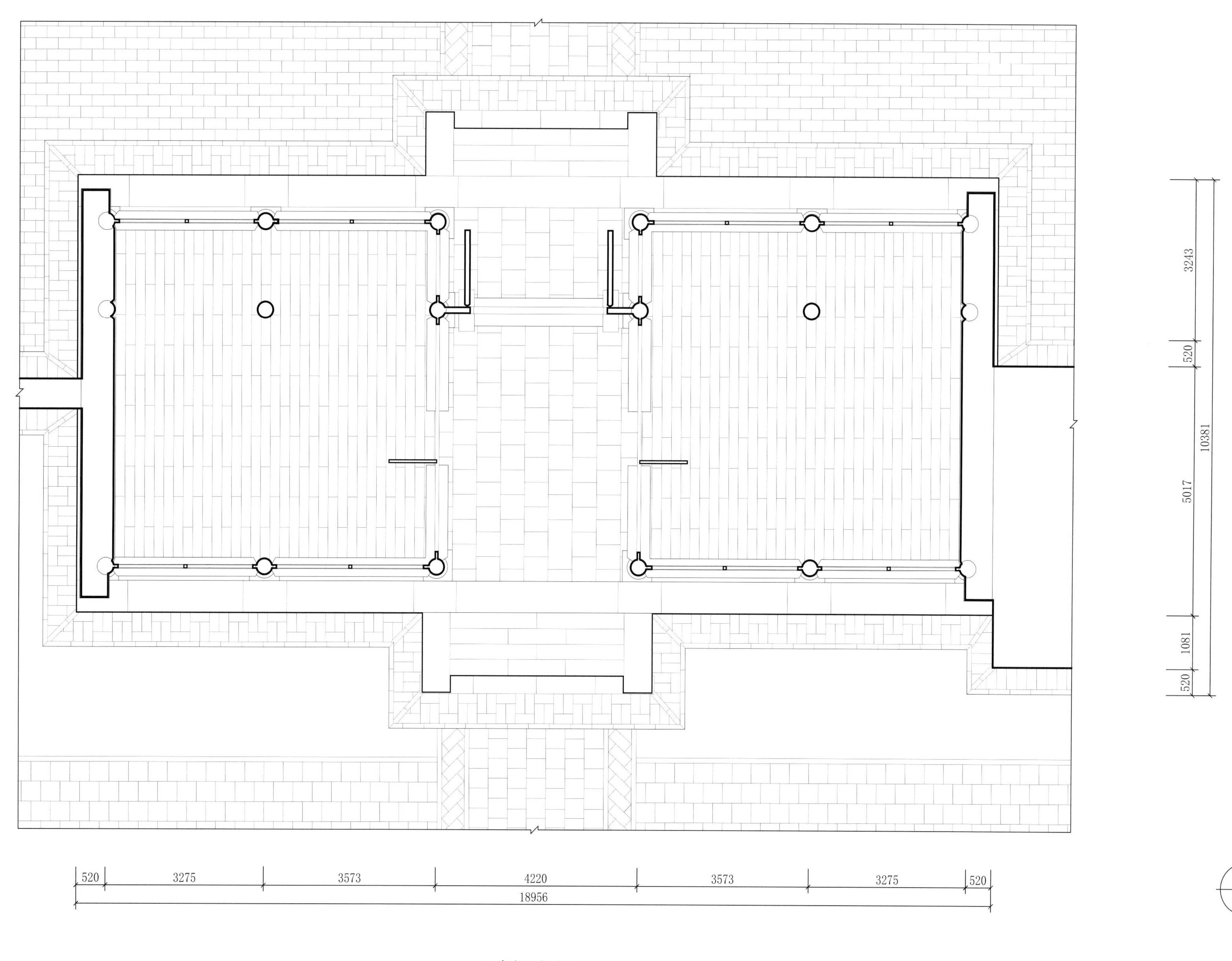

二宫门平面图
Plan of Palace Gate

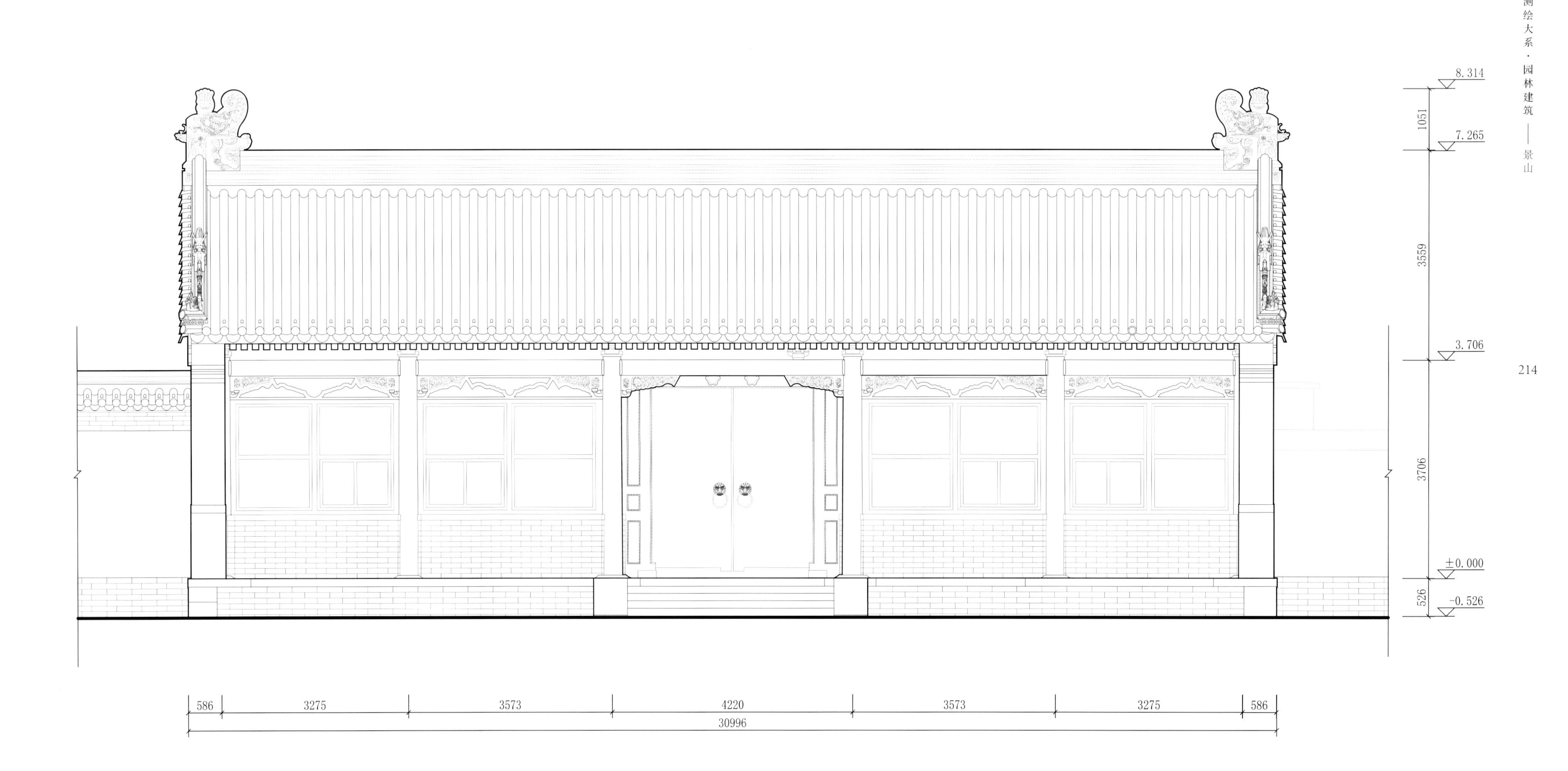

二宫门正立面图
Front elevation of Palace Gate

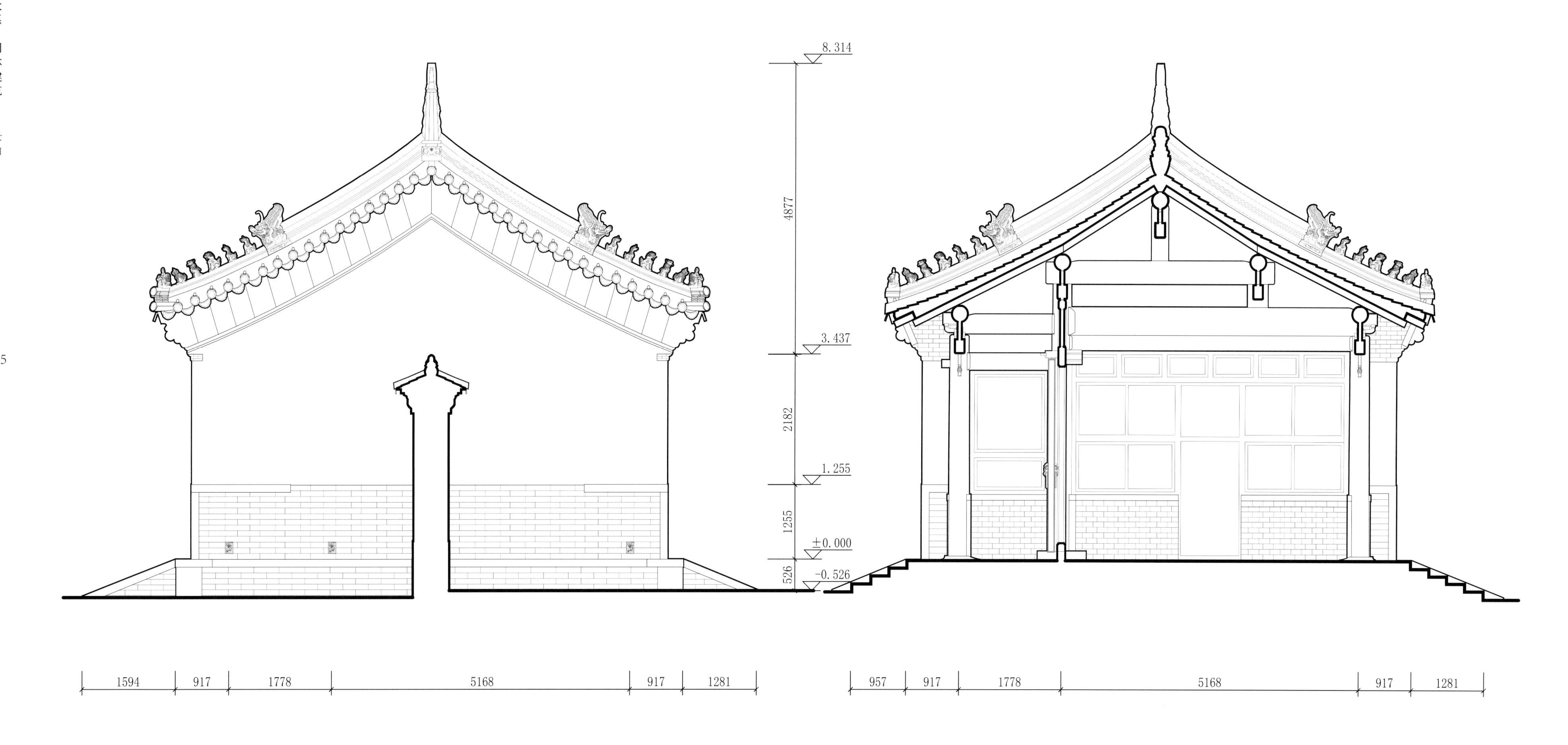

二宫门侧立面图
Side elevation of Palace Gate

二宫门横剖面图
Transverse section of Palace Gate

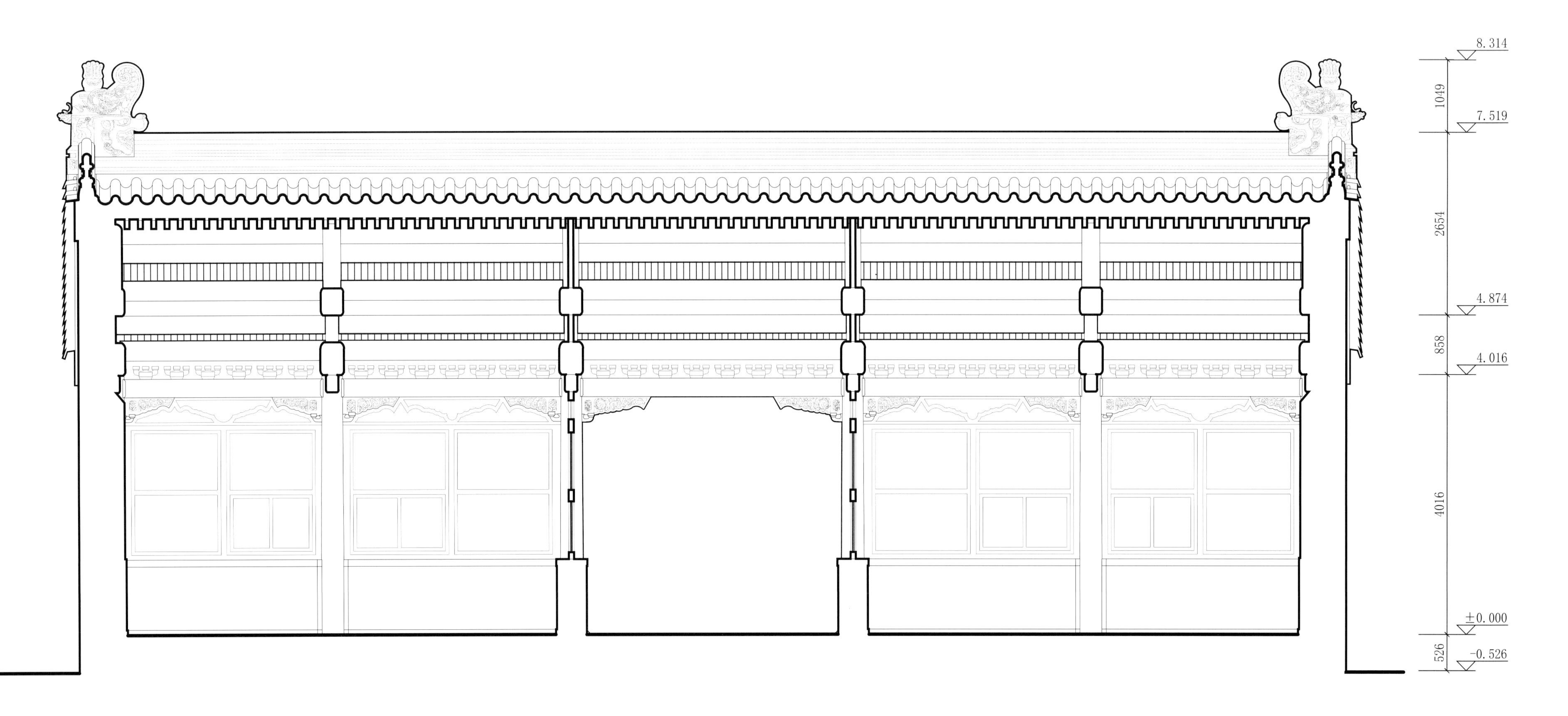

二宫门纵剖面图
Longitudinal section of Palace Gate

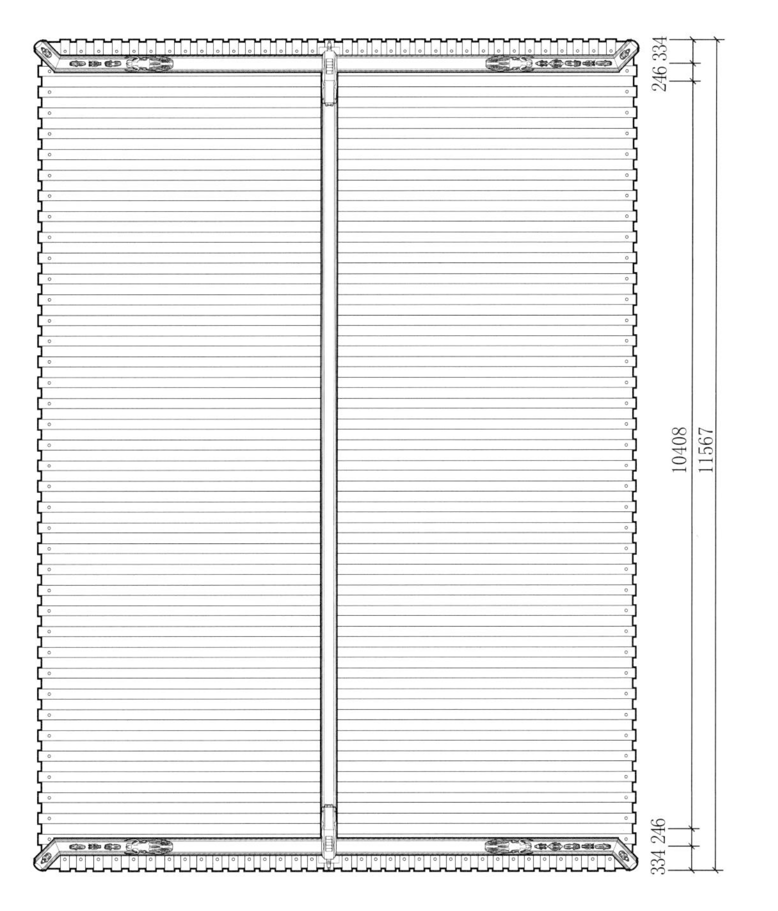

东、西配殿屋顶平面图
Roof plan of East or West Side Hall

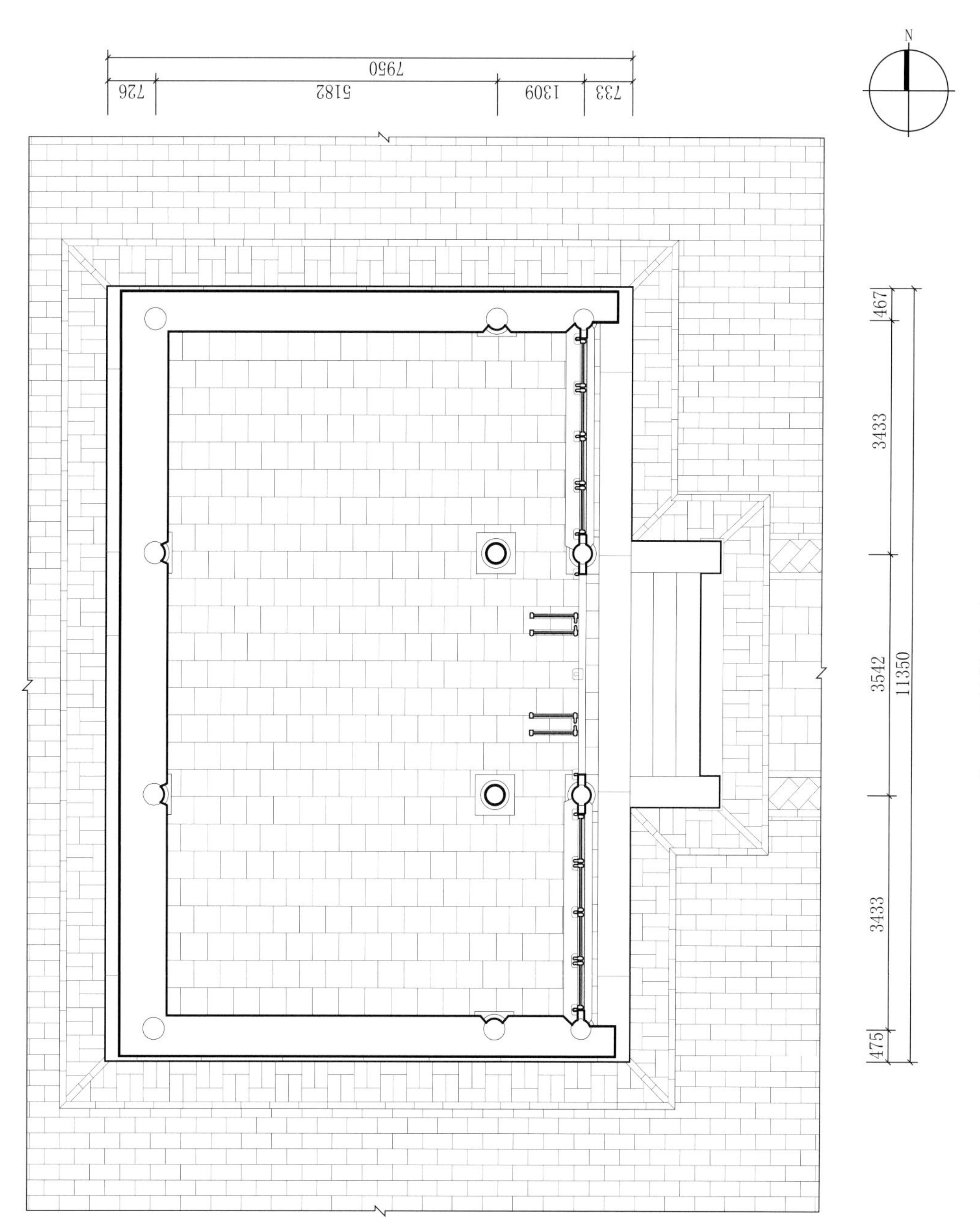

东、西配殿平面图
Plan of East or West Side Hall

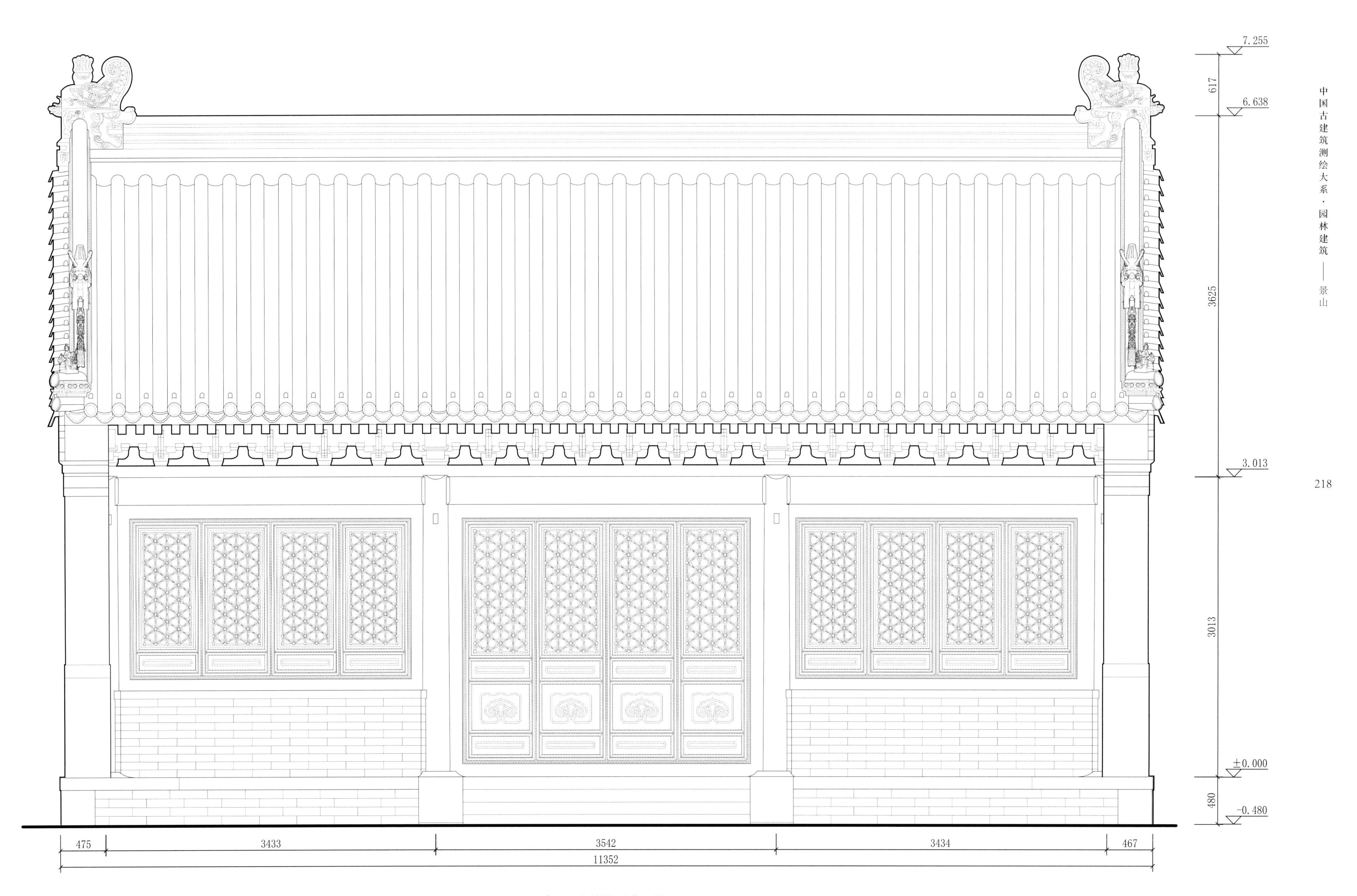

东、西配殿正立面图
Front elevation of East or West Side Hall

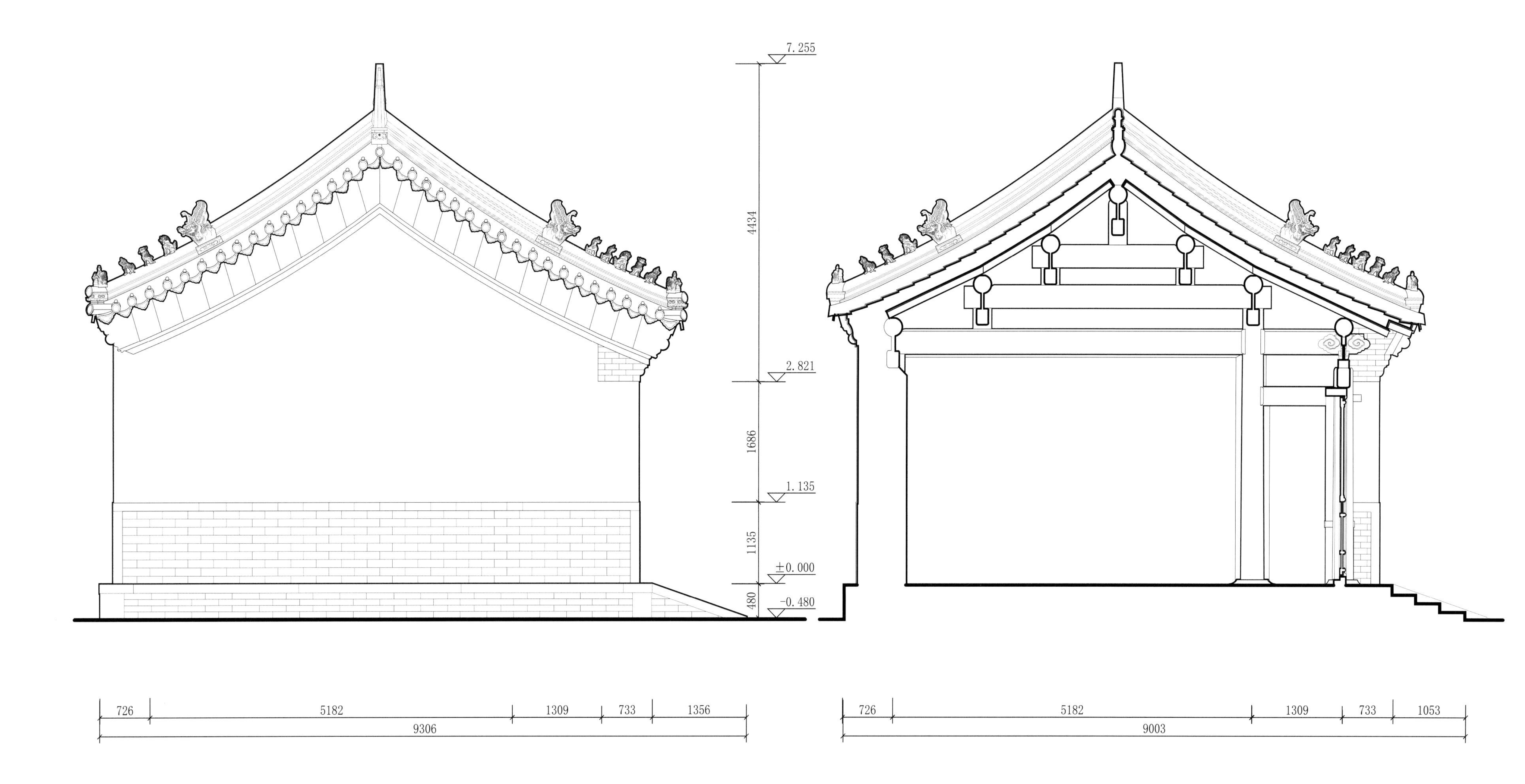

东、西配殿侧立面图
Side elevation of East or West Side Hall

东、西配殿明间横剖面图
Transverse section of East or West Side Hall

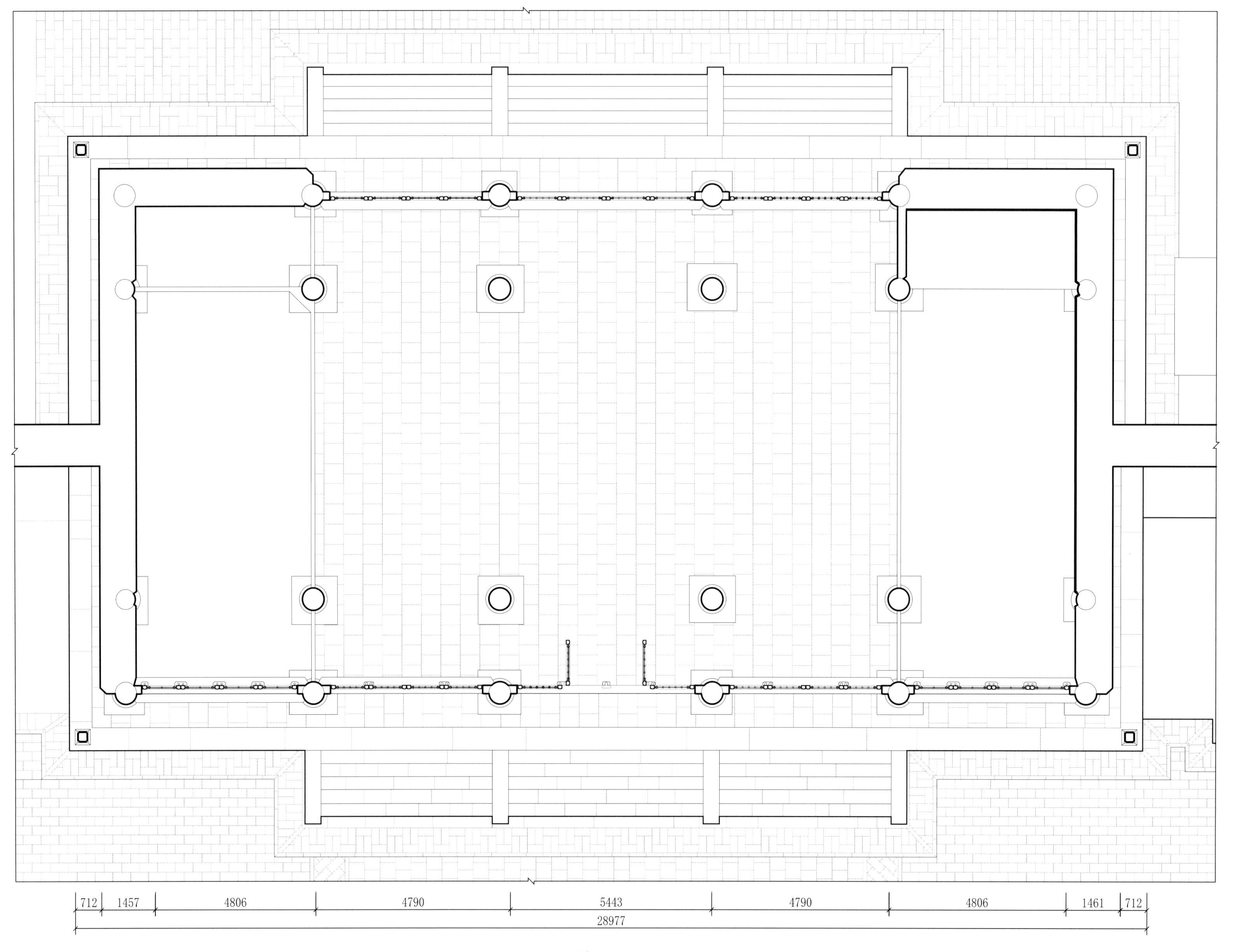

观德殿平面图
Plan of Guande Hall

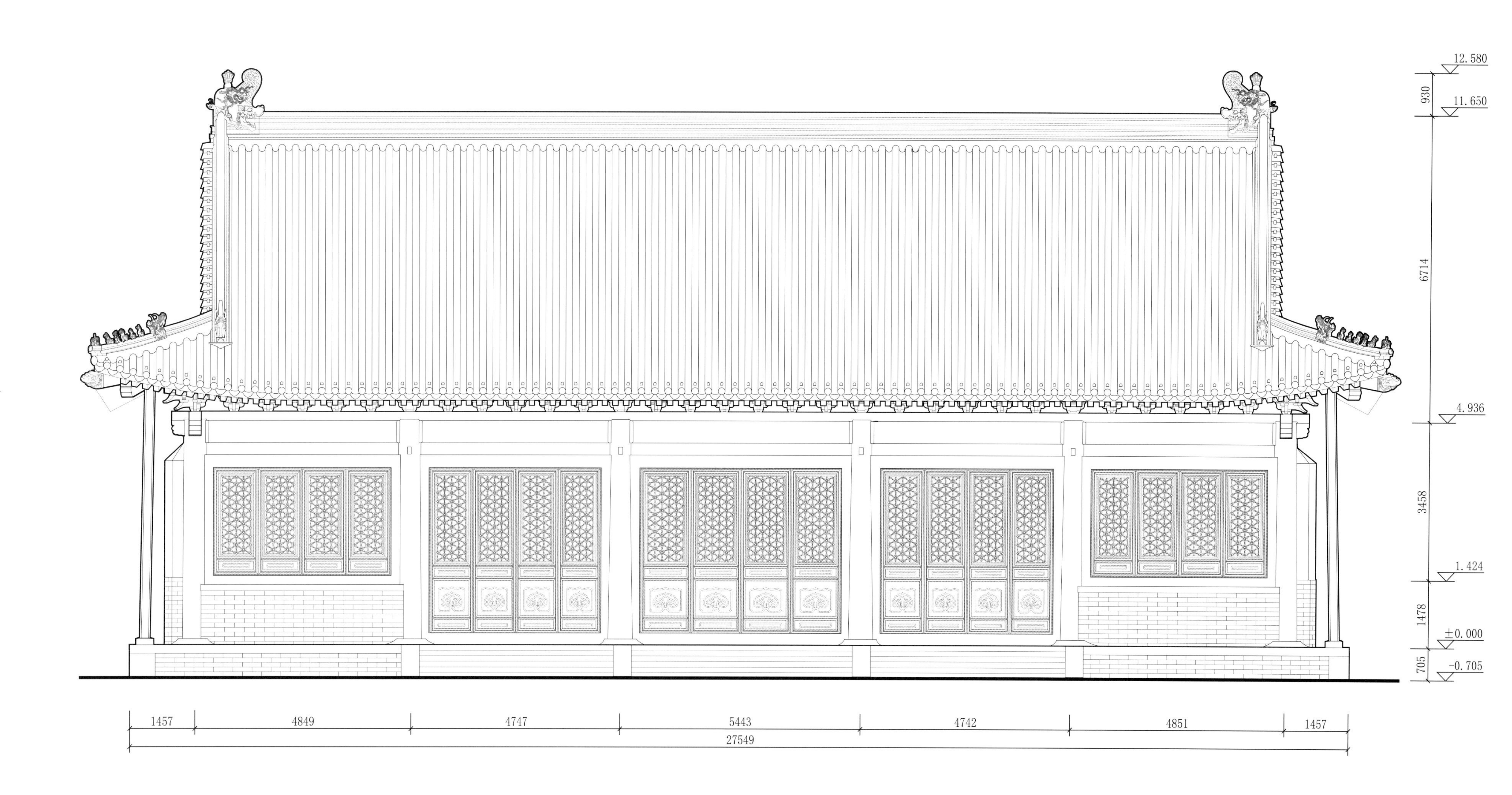

观德殿正立面图

Front elevation of Guande Hall

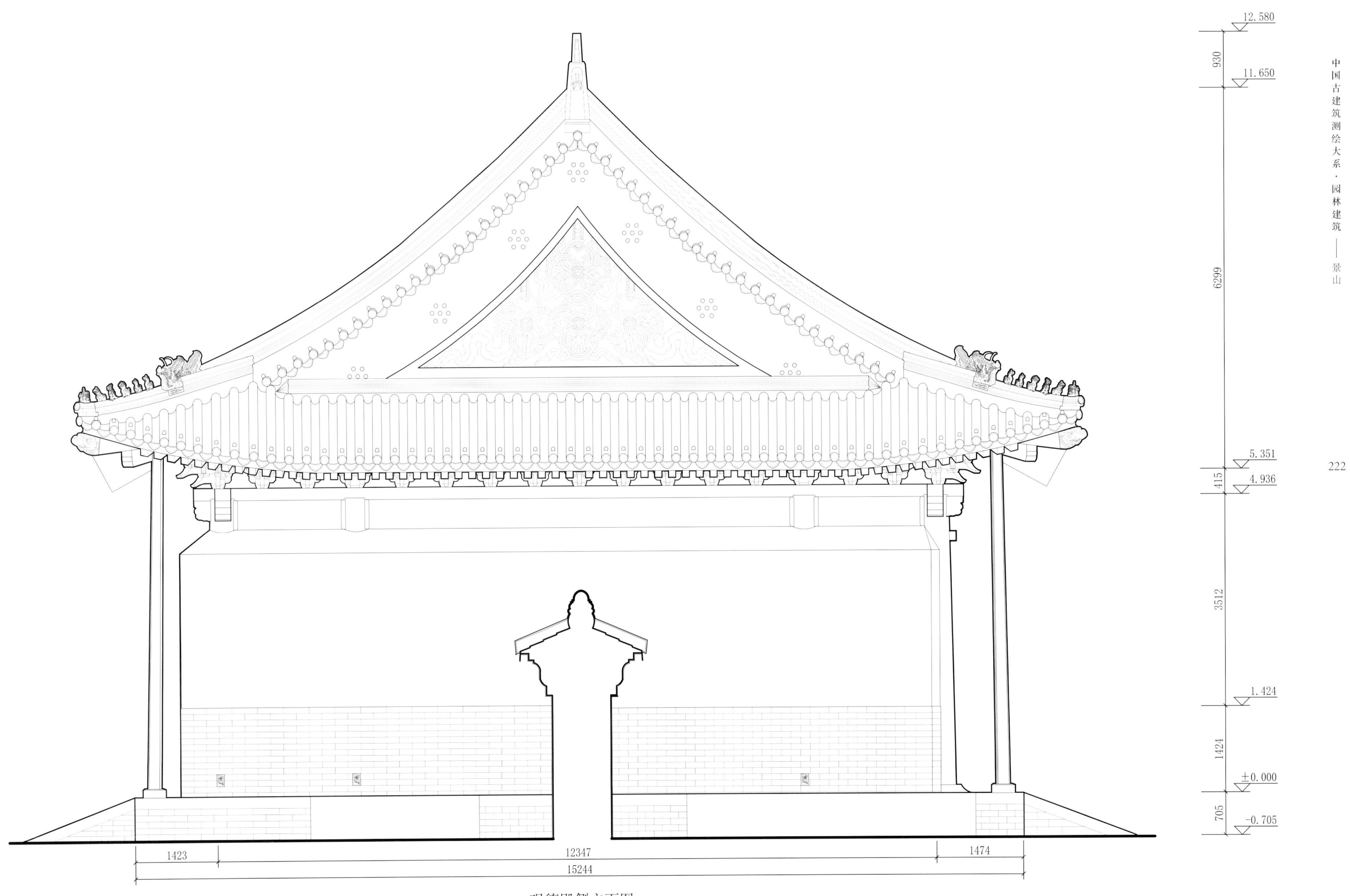

观德殿侧立面图
Side elevation of Guande Hall

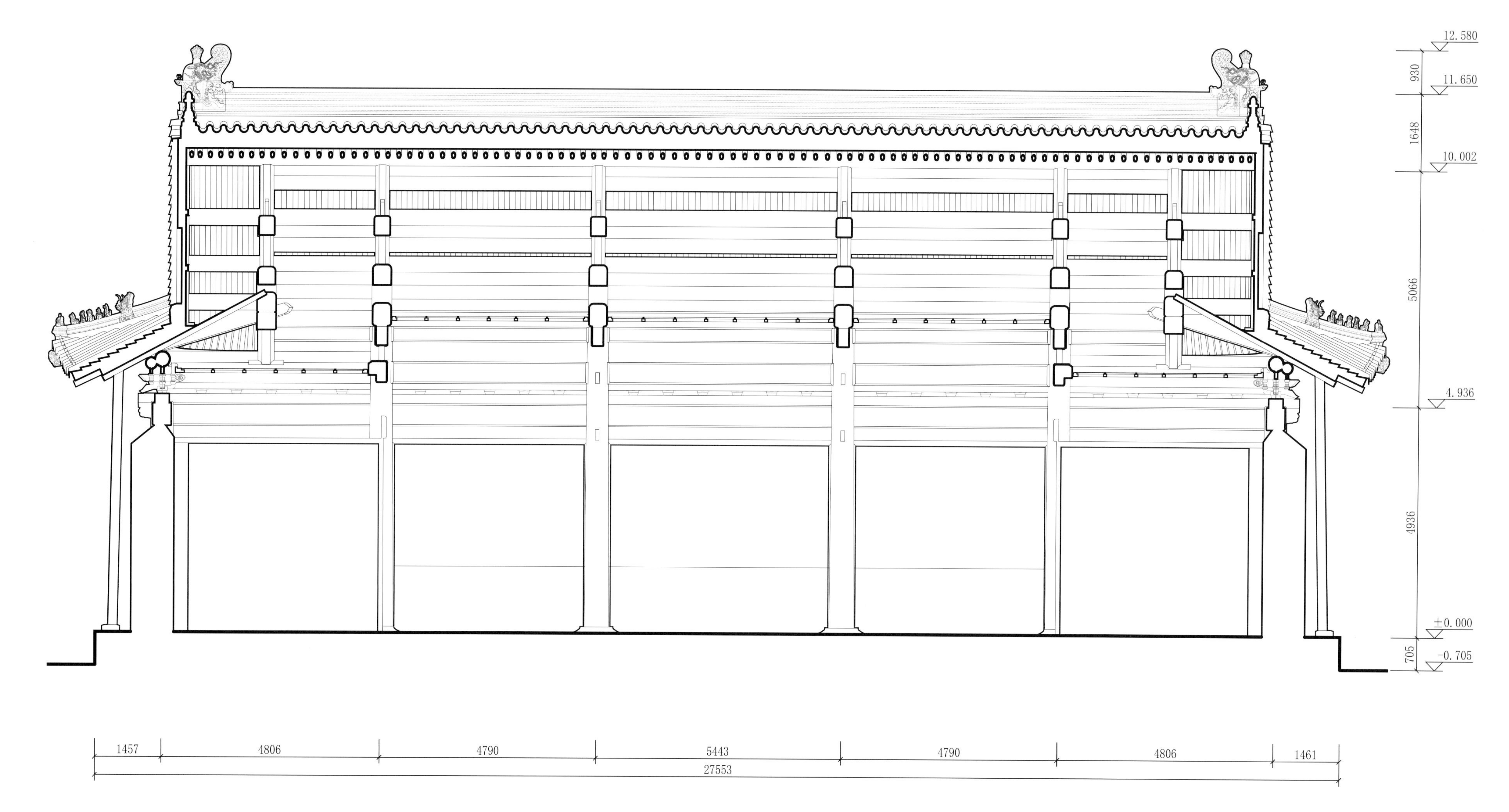

观德殿纵剖面图

Longitudinal section of Guande Hall

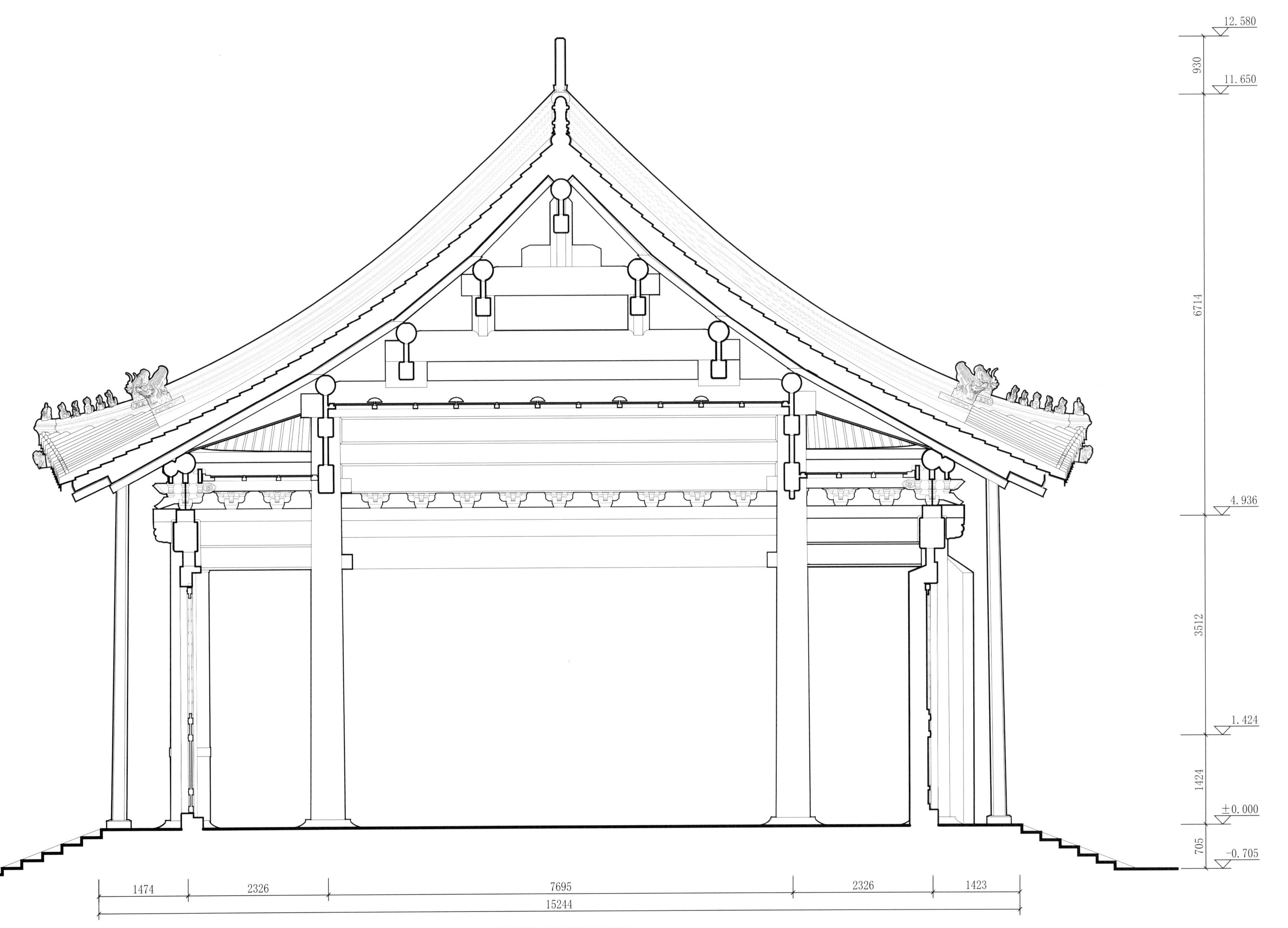

观德殿明间横剖面图
Central bay section of Guande Hall

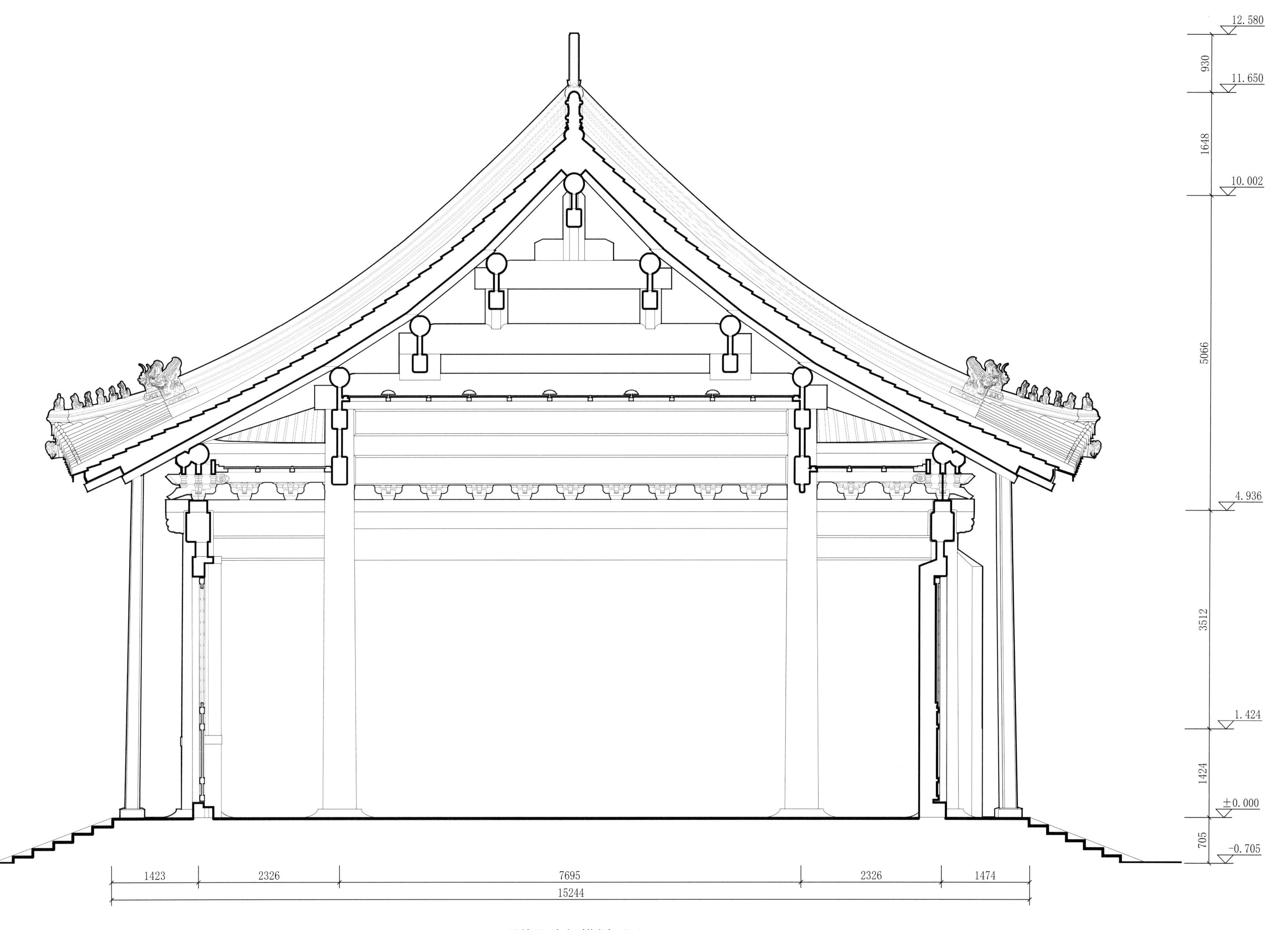

观德殿次间横剖面图

Lateral bay section of Guande Hall

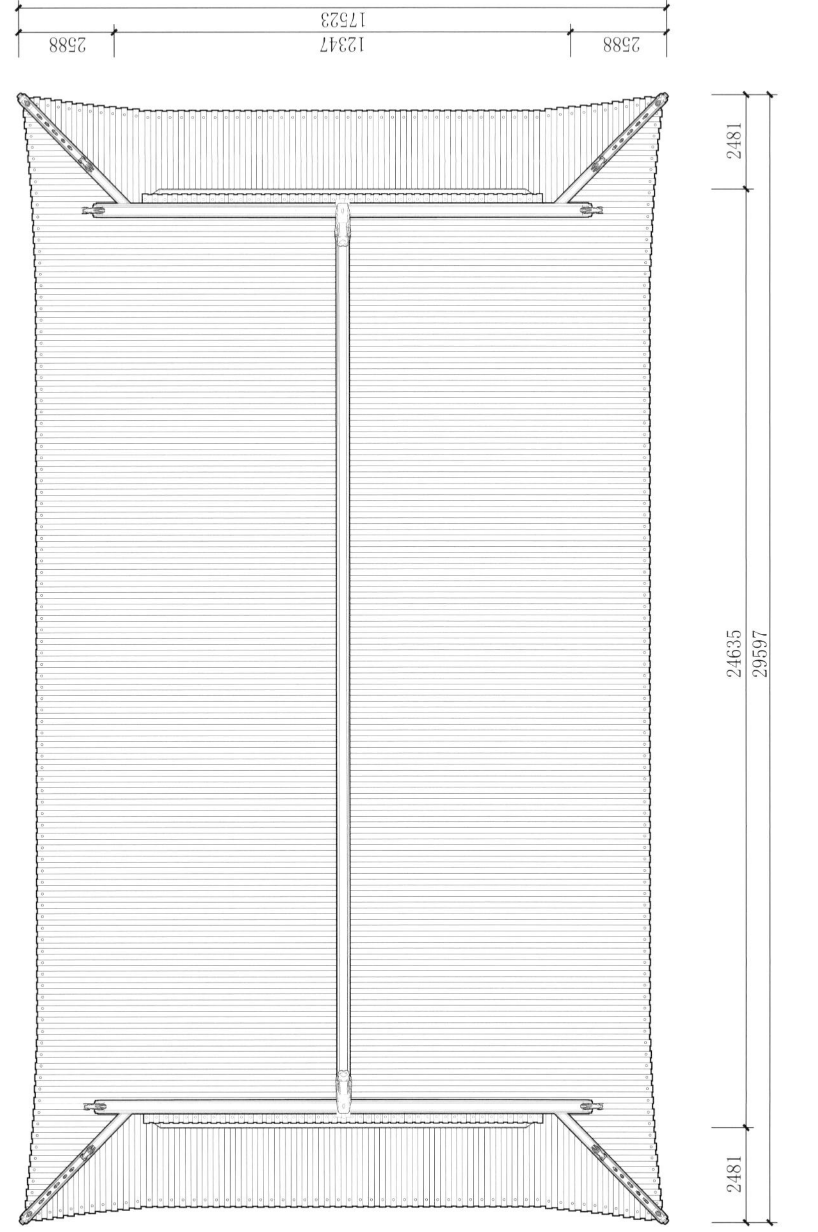

观德殿屋顶平面图
Roof plan of Guande Hall

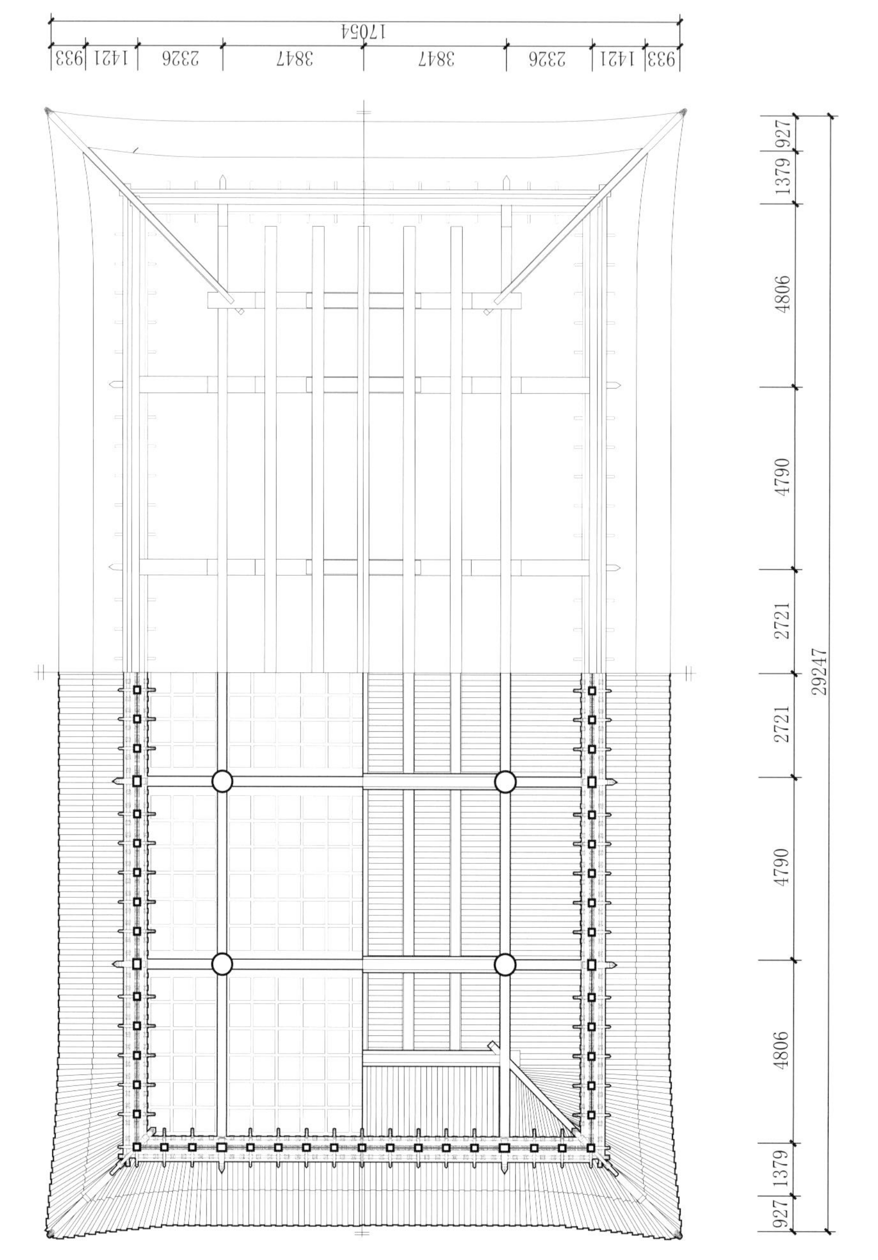

观德殿梁架仰俯视图
Vertical and bottom view of beam frame of Guande Hall

关帝庙建筑群

The Groups of Guandi Temple

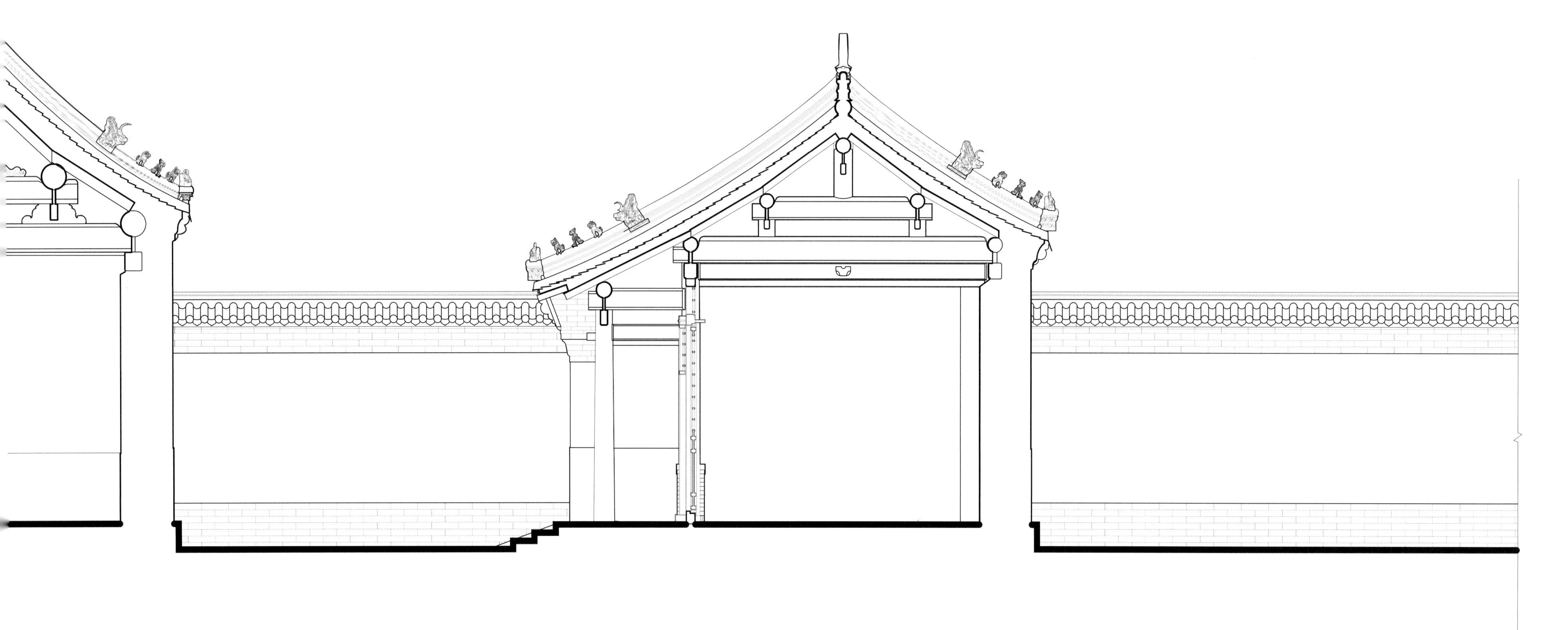

0
0.75
2.25m

关帝庙建筑群剖面图
Section of the Groups of Guandi Temple

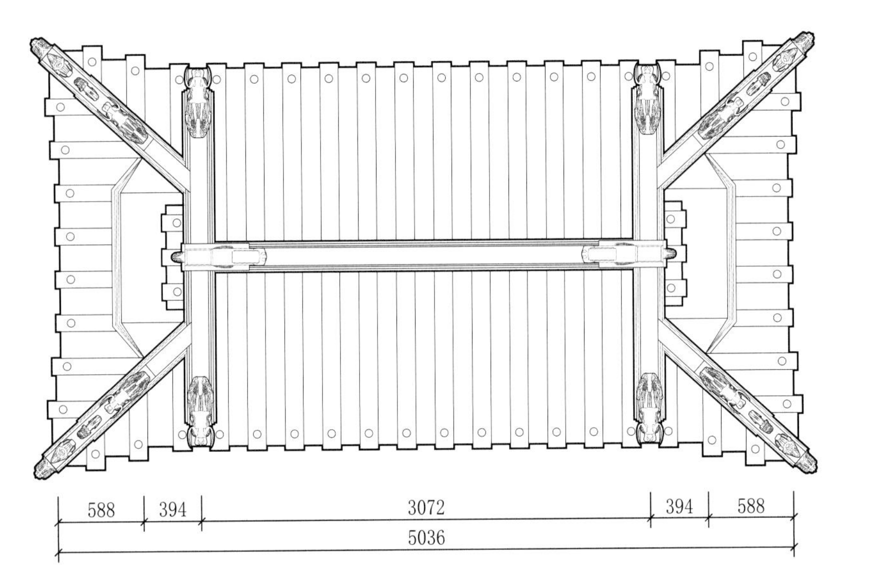

关帝庙琉璃门屋顶平面图

Roof plan of Colored-glaze Gate of Guandi Temple

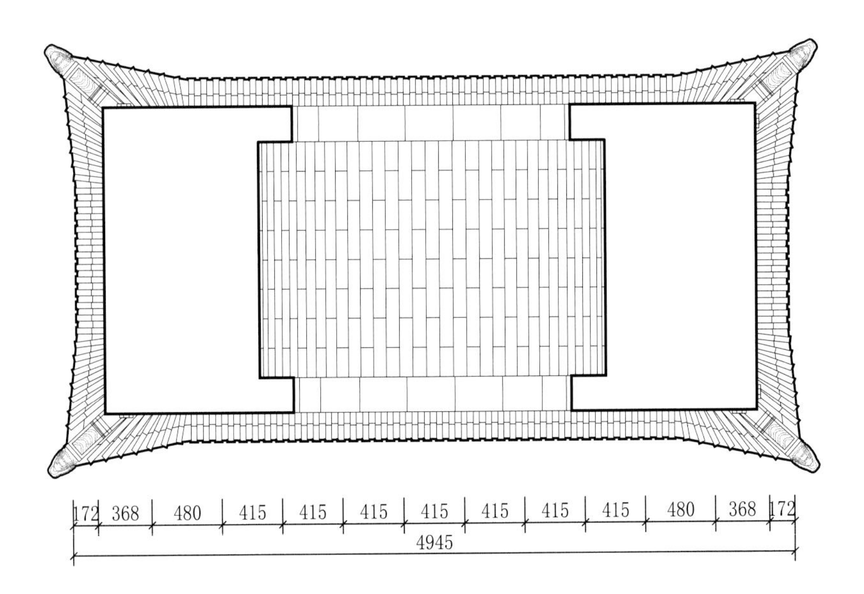

关帝庙琉璃门梁架仰视图

Bottom view of beam frame of Colored-glaze Gate of Guandi Temple

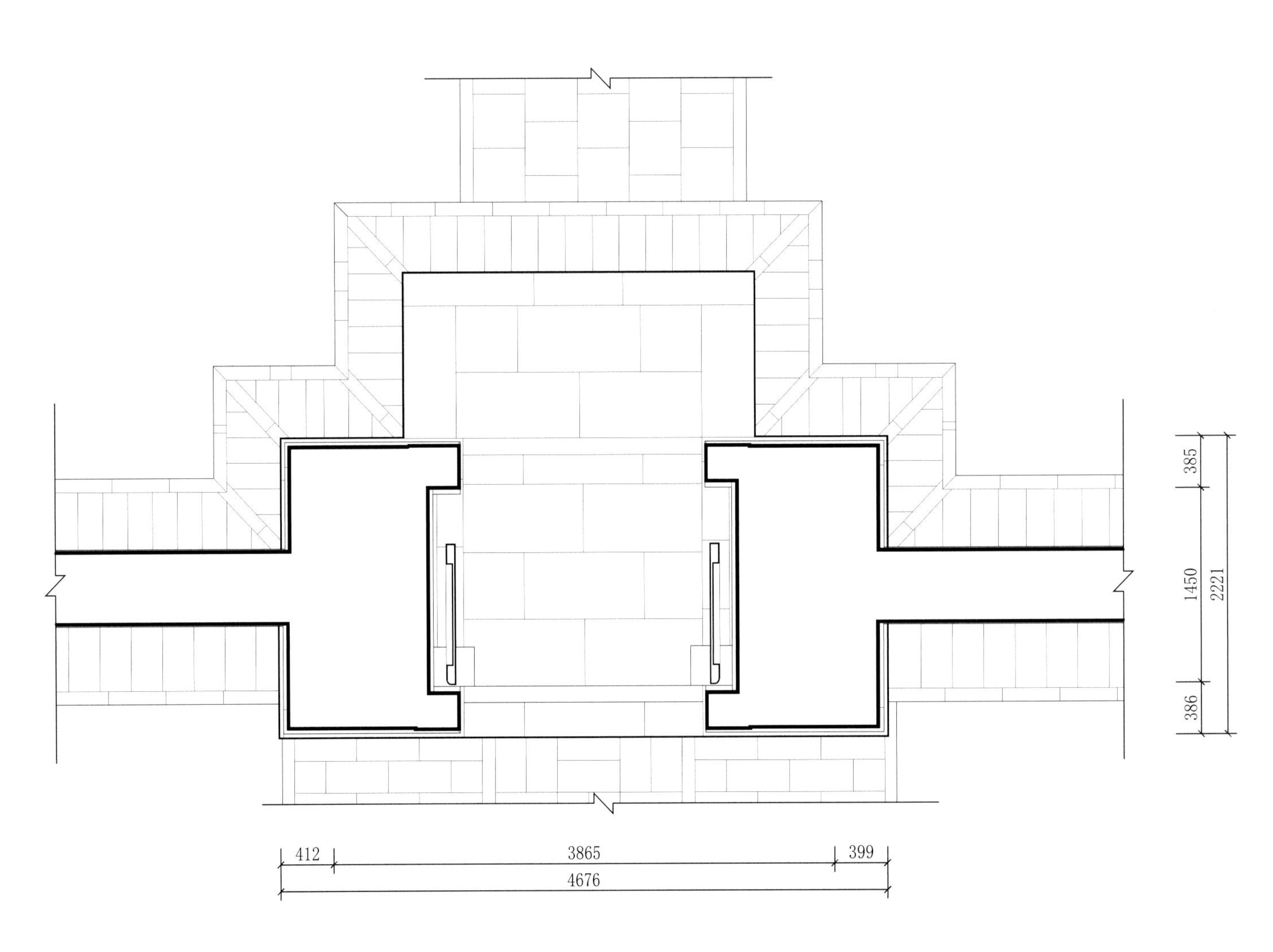

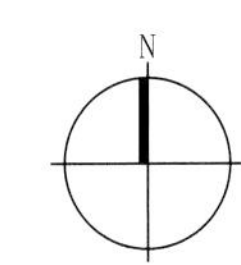

关帝庙琉璃门平面图

Plan of Colored-glaze Gate of Guandi Temple

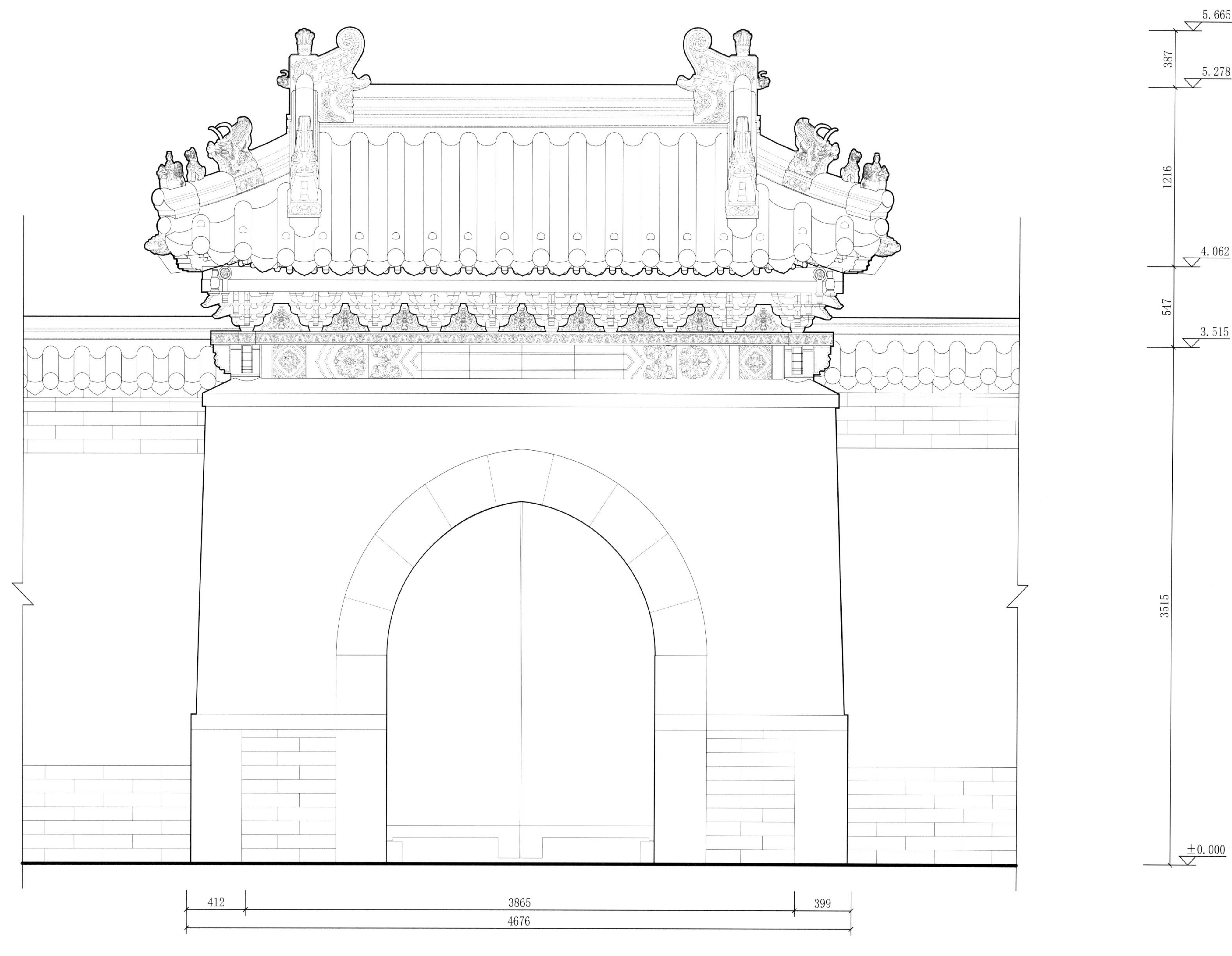

关帝庙琉璃门正立面图
Front elevation of Colored-glaze Gate of Guandi Temple

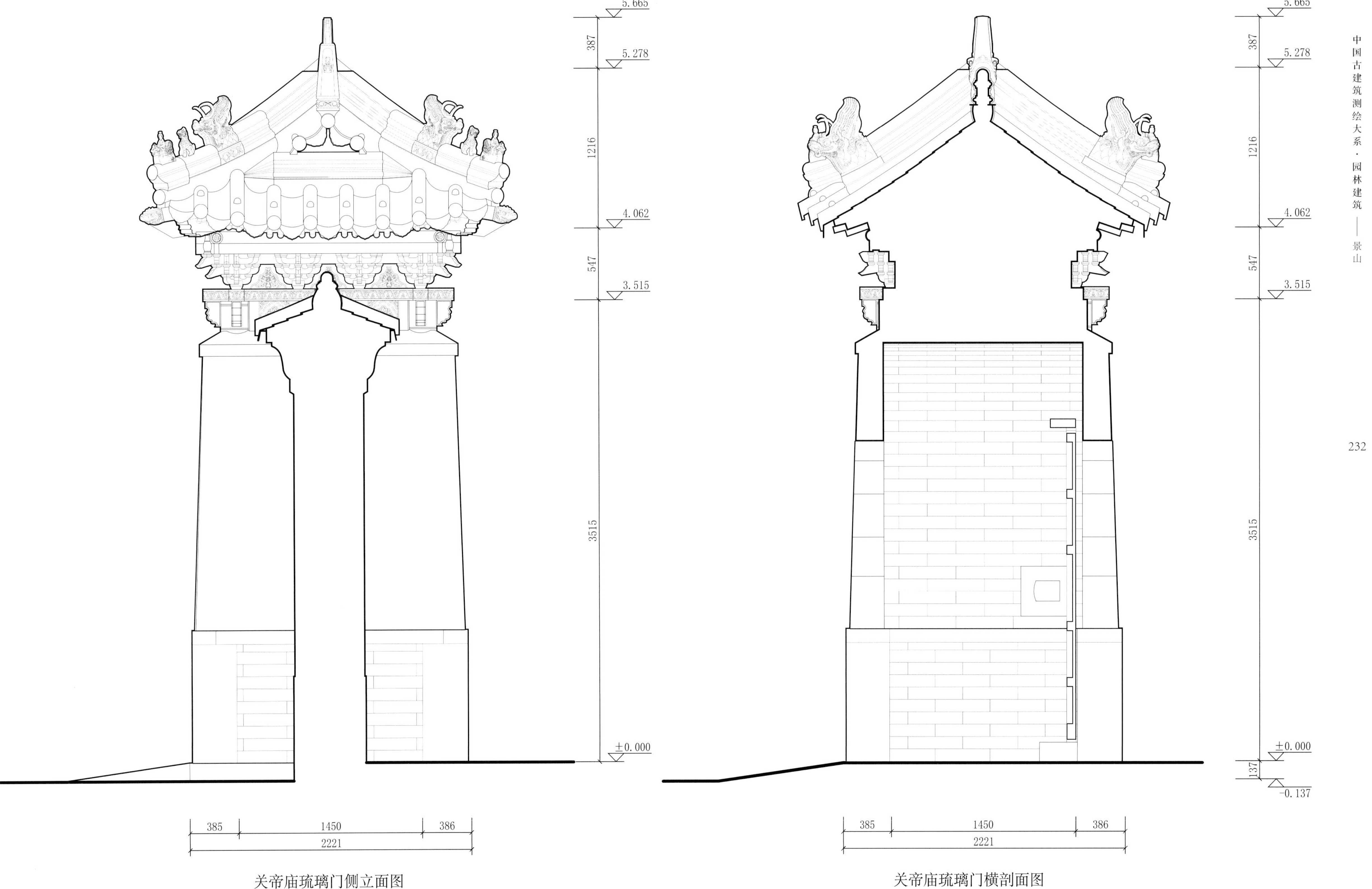

关帝庙琉璃门侧立面图
Side elevation of Colored-glaze Gate of Guandi Temple

关帝庙琉璃门横剖面图
Transverse section of Colored-glaze Gate of Guandi Temple

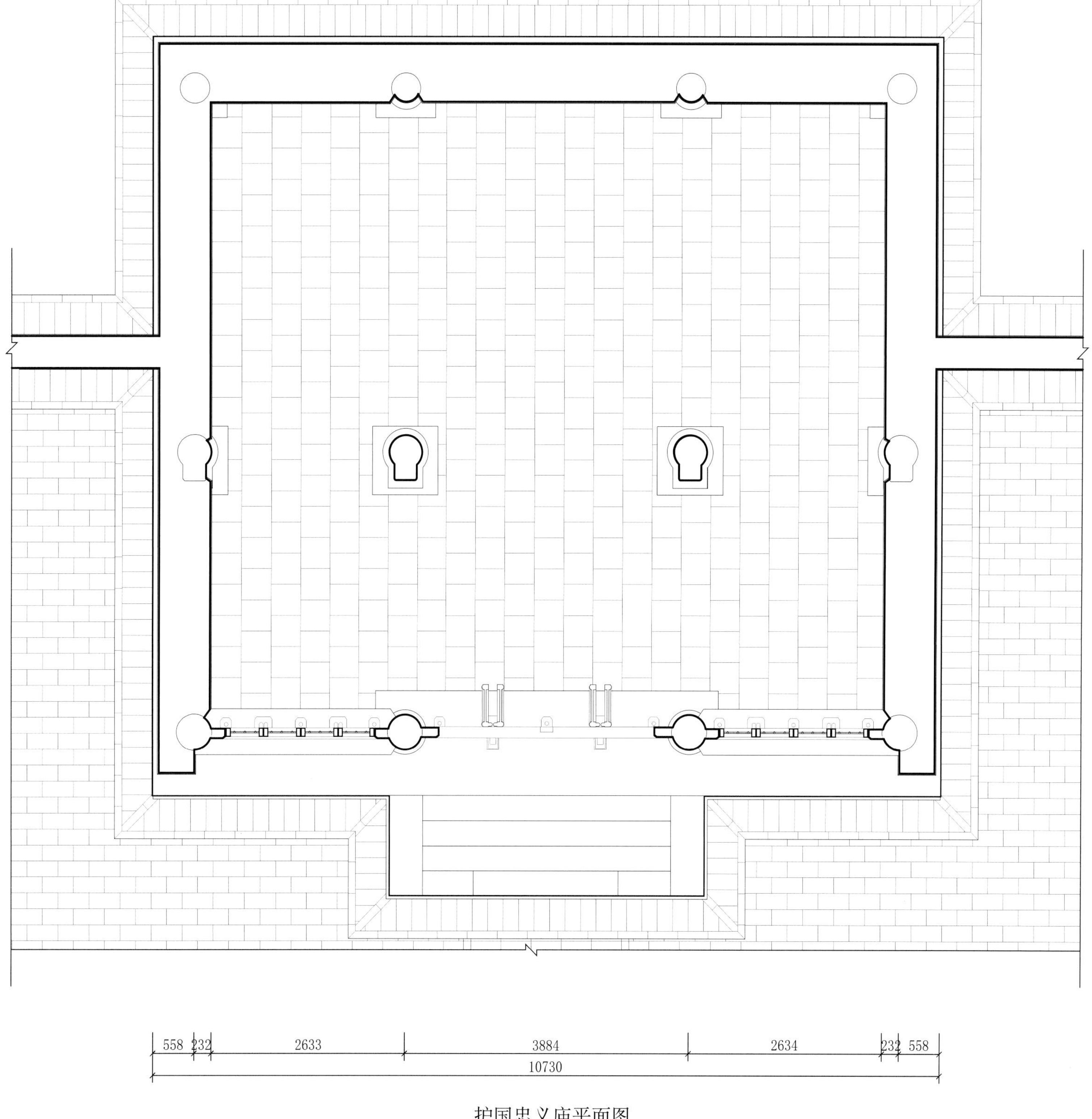

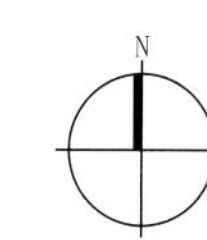

护国忠义庙平面图

Plan of Huguozhongyi Temple

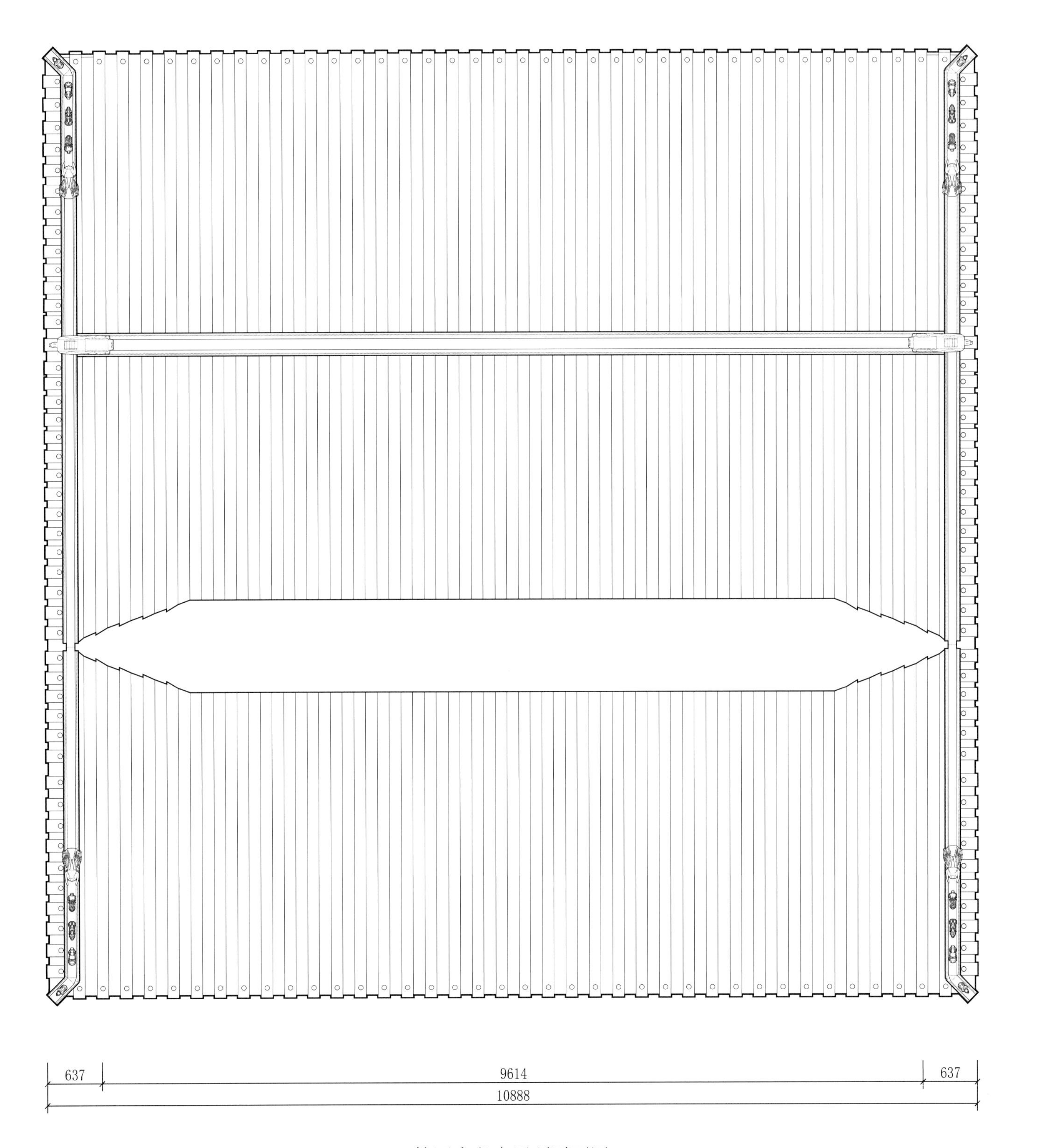

护国忠义庙屋顶平面图
Roof plan of Huguozhongyi Temple

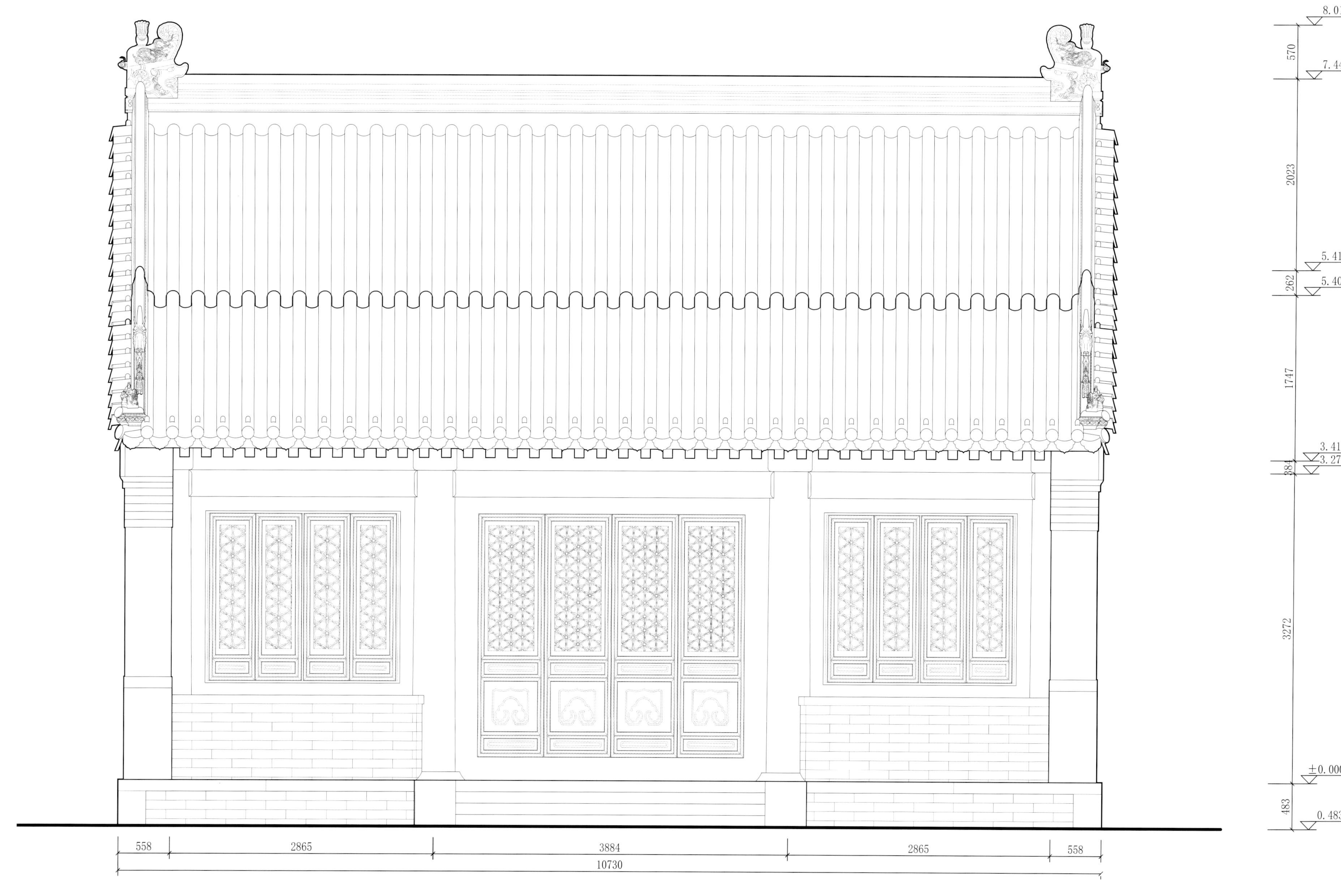

护国忠义庙正立面图

Front elevation of Huguozhongyi Temple

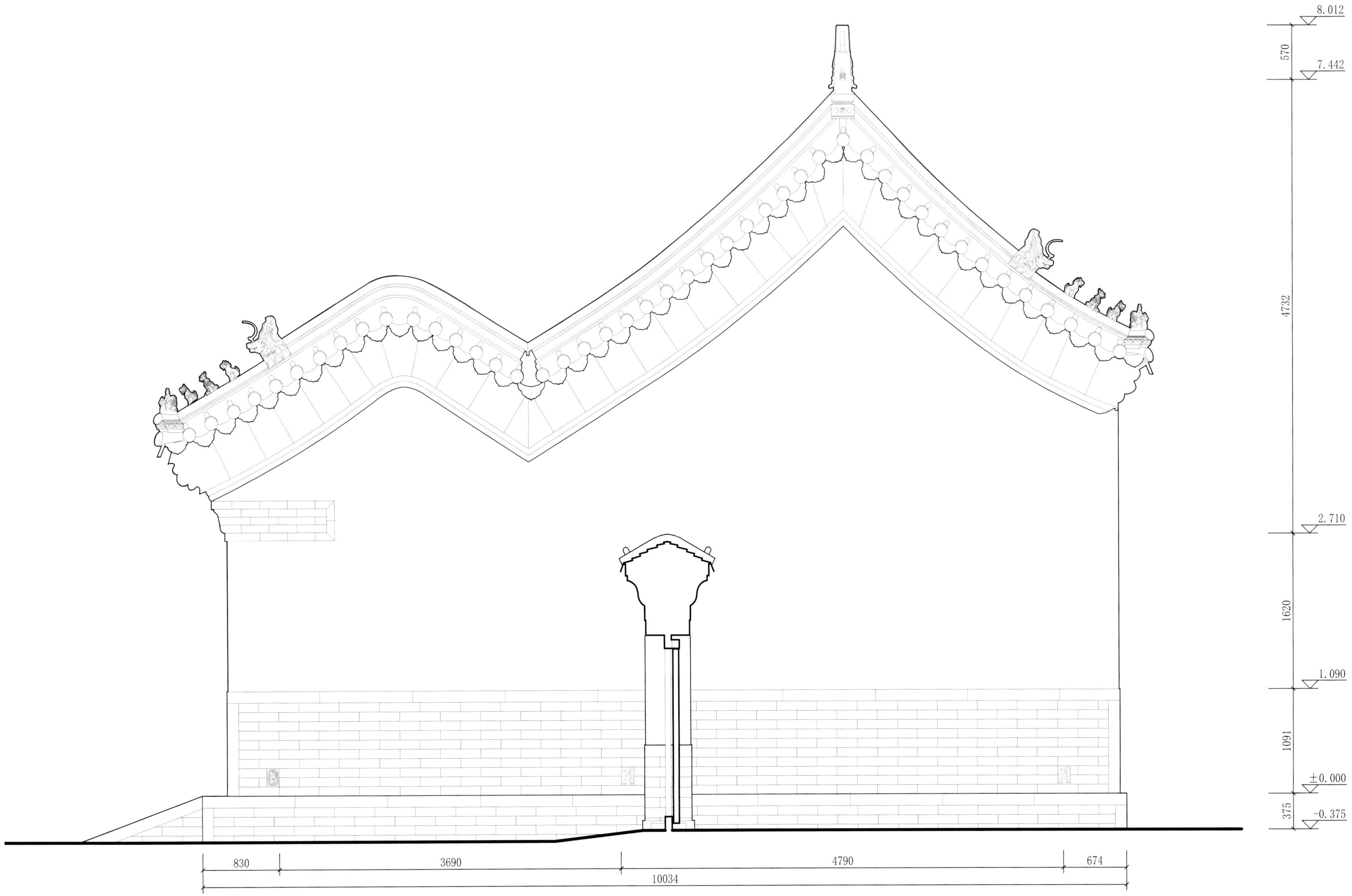

护国忠义庙侧立面图

Side elevation of Huguozhongyi Temple

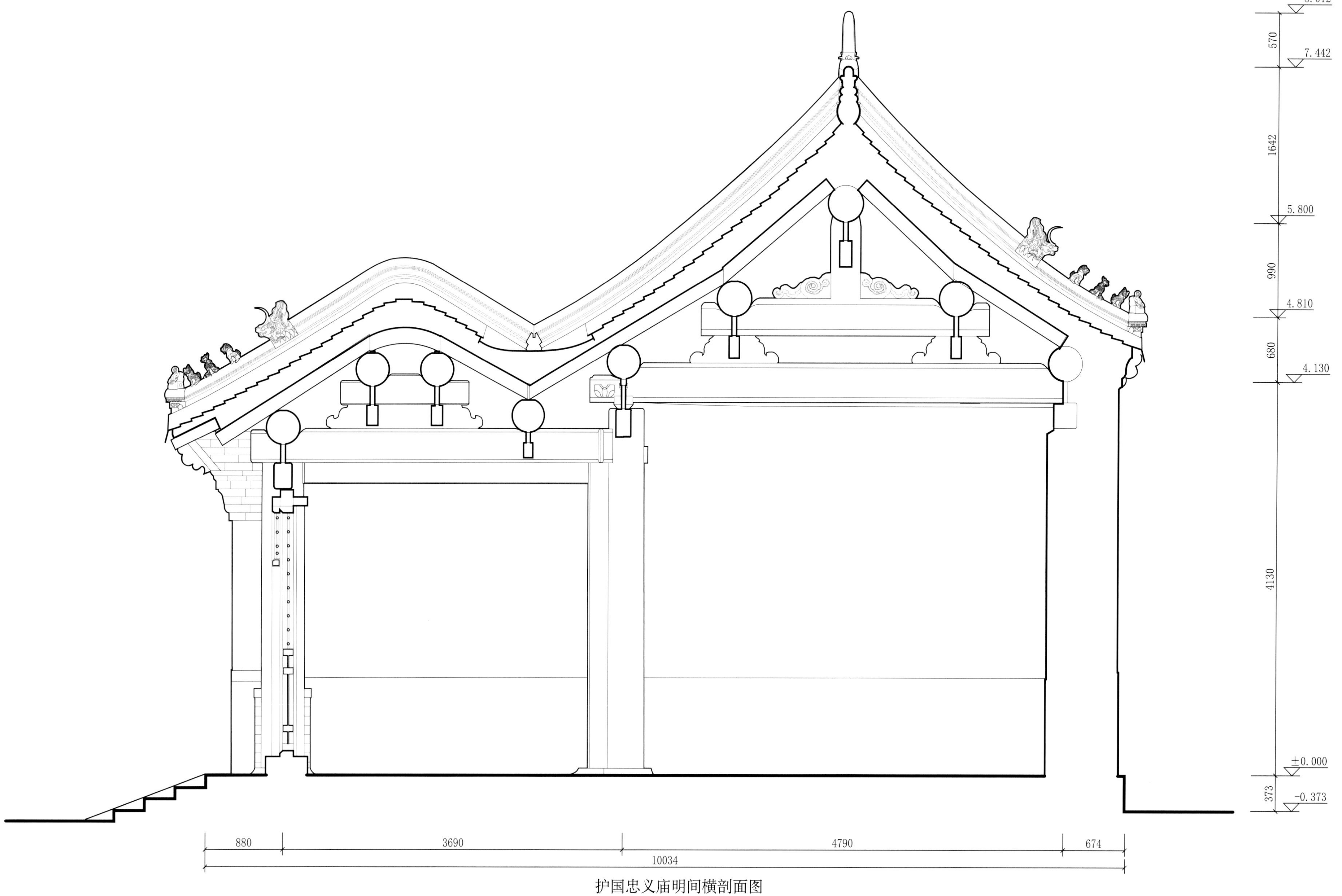

护国忠义庙明间横剖面图

Central bay section of Huguozhongyi Temple

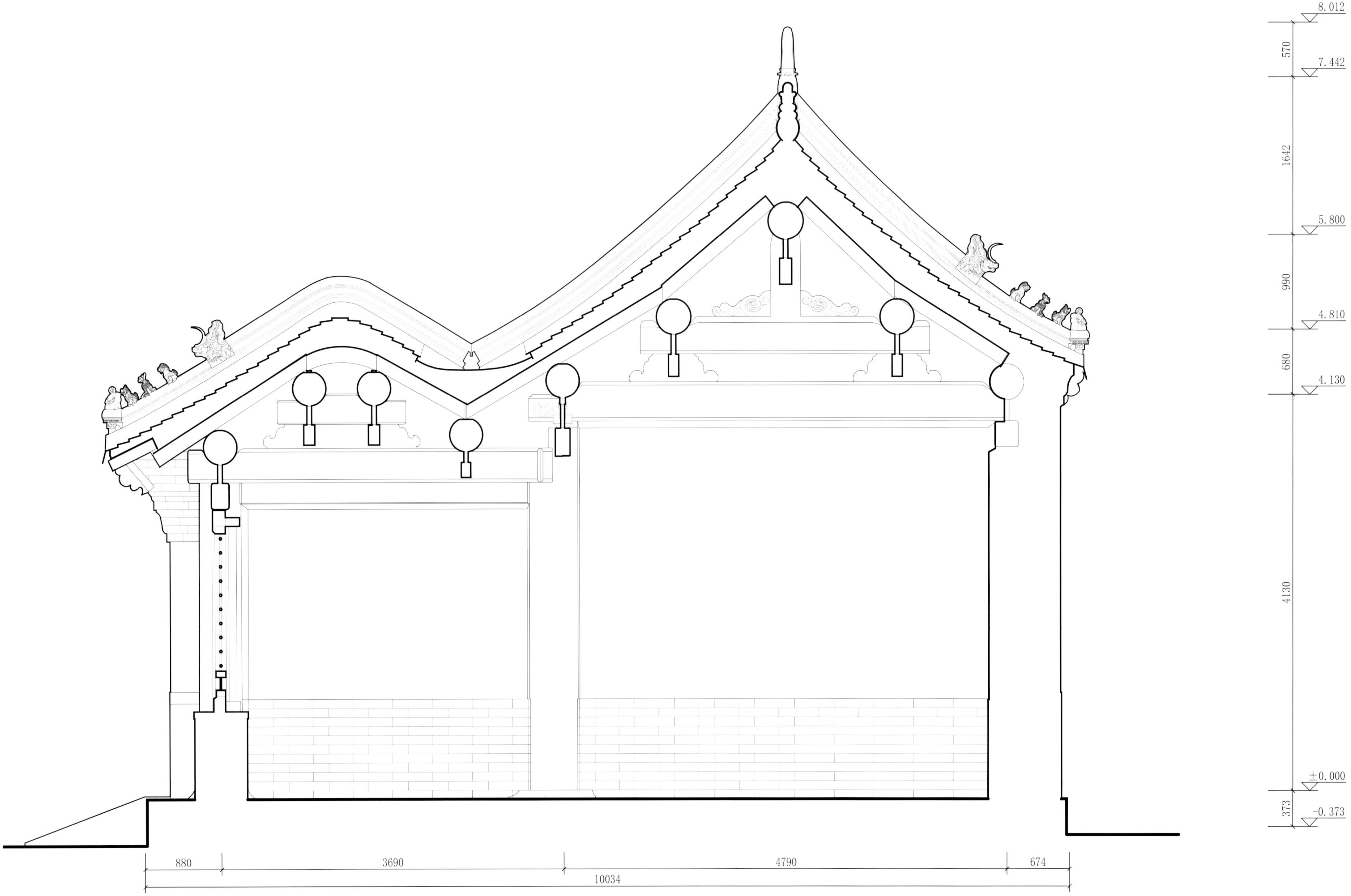

护国忠义庙次间横剖面图
Lateral bay section of Huguozhongyi Temple

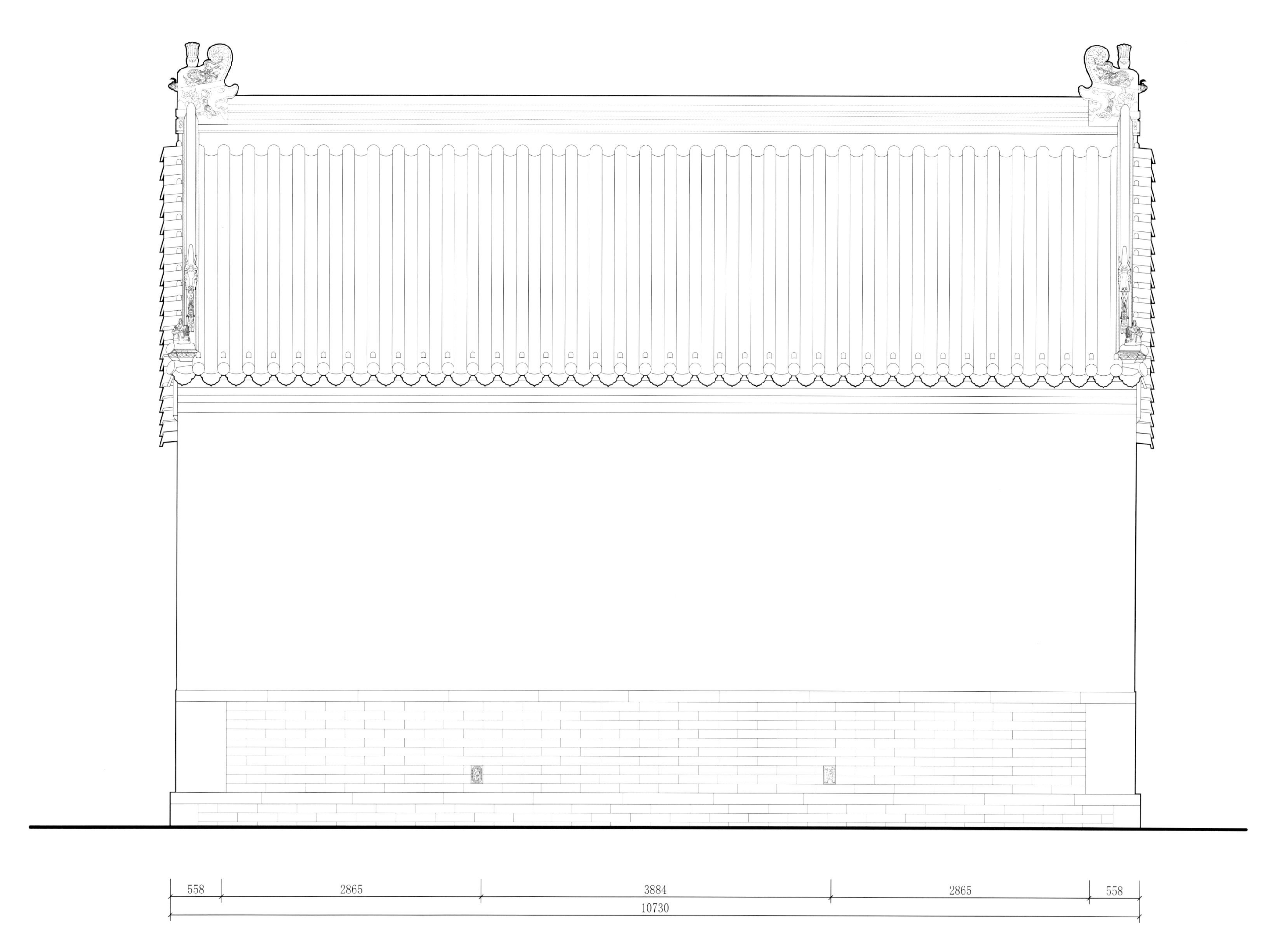

护国忠义庙背立面图

Back elevation of Huguozhongyi Temple

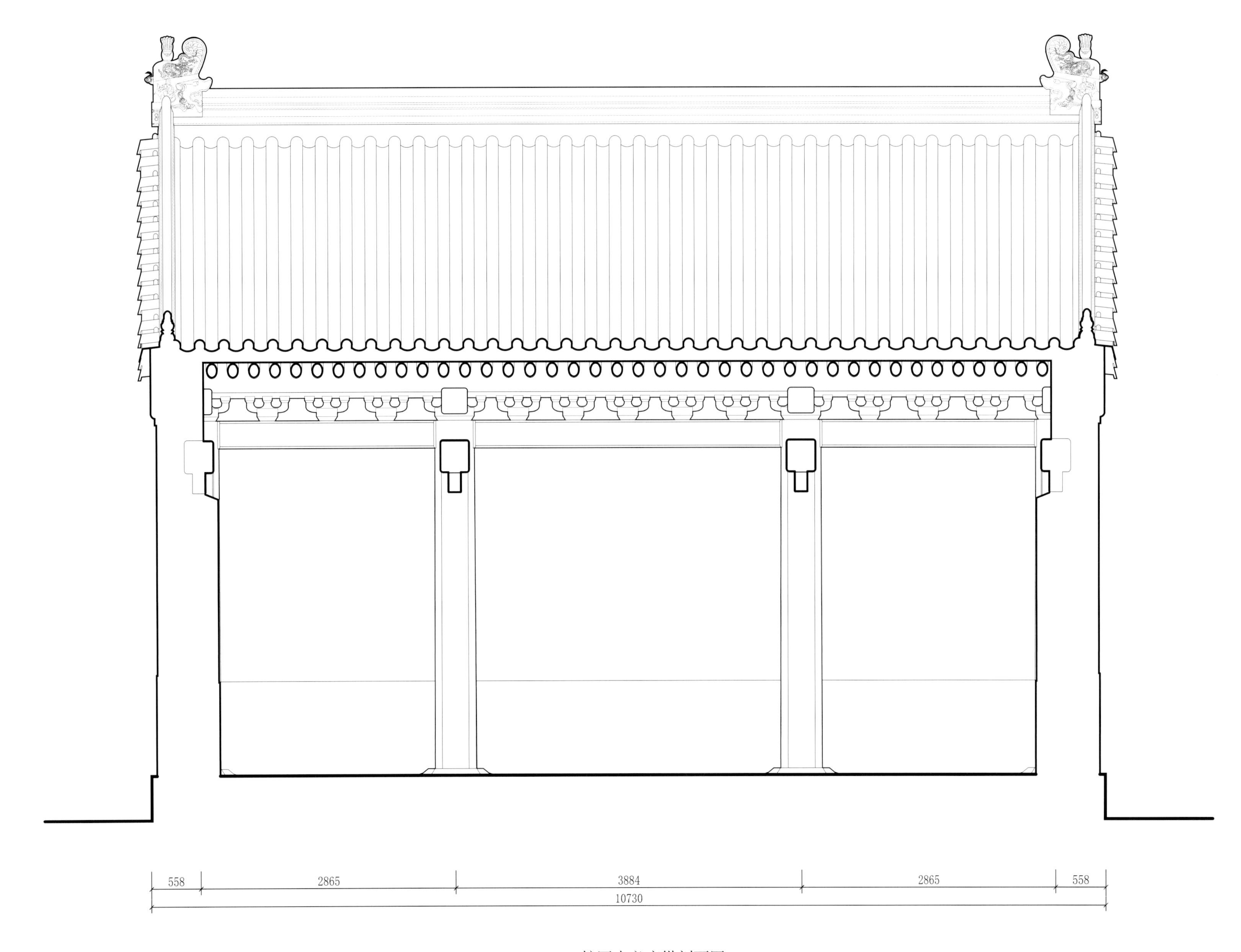

护国忠义庙纵剖面图
Longitudinal section of Huguozhongyi Temple

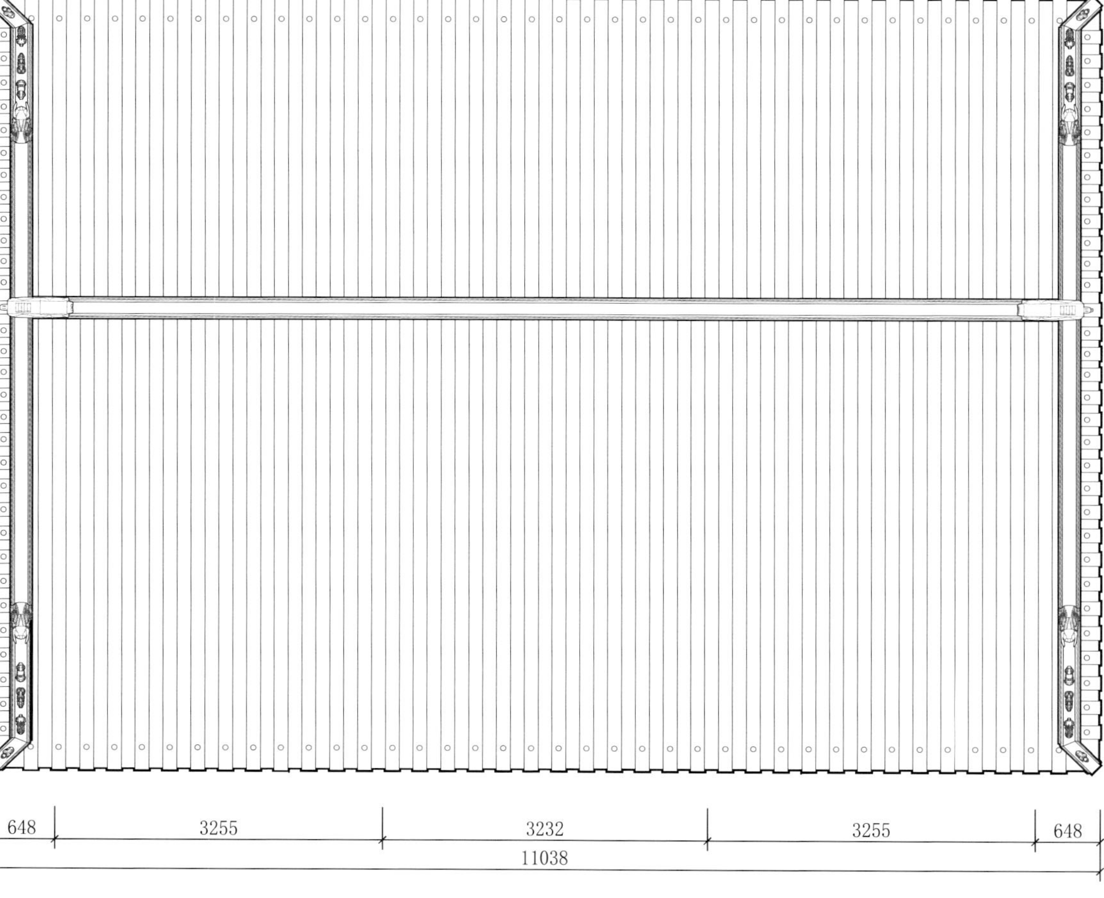

真武殿屋顶平面图
Roof plan of Zhenwu Hall

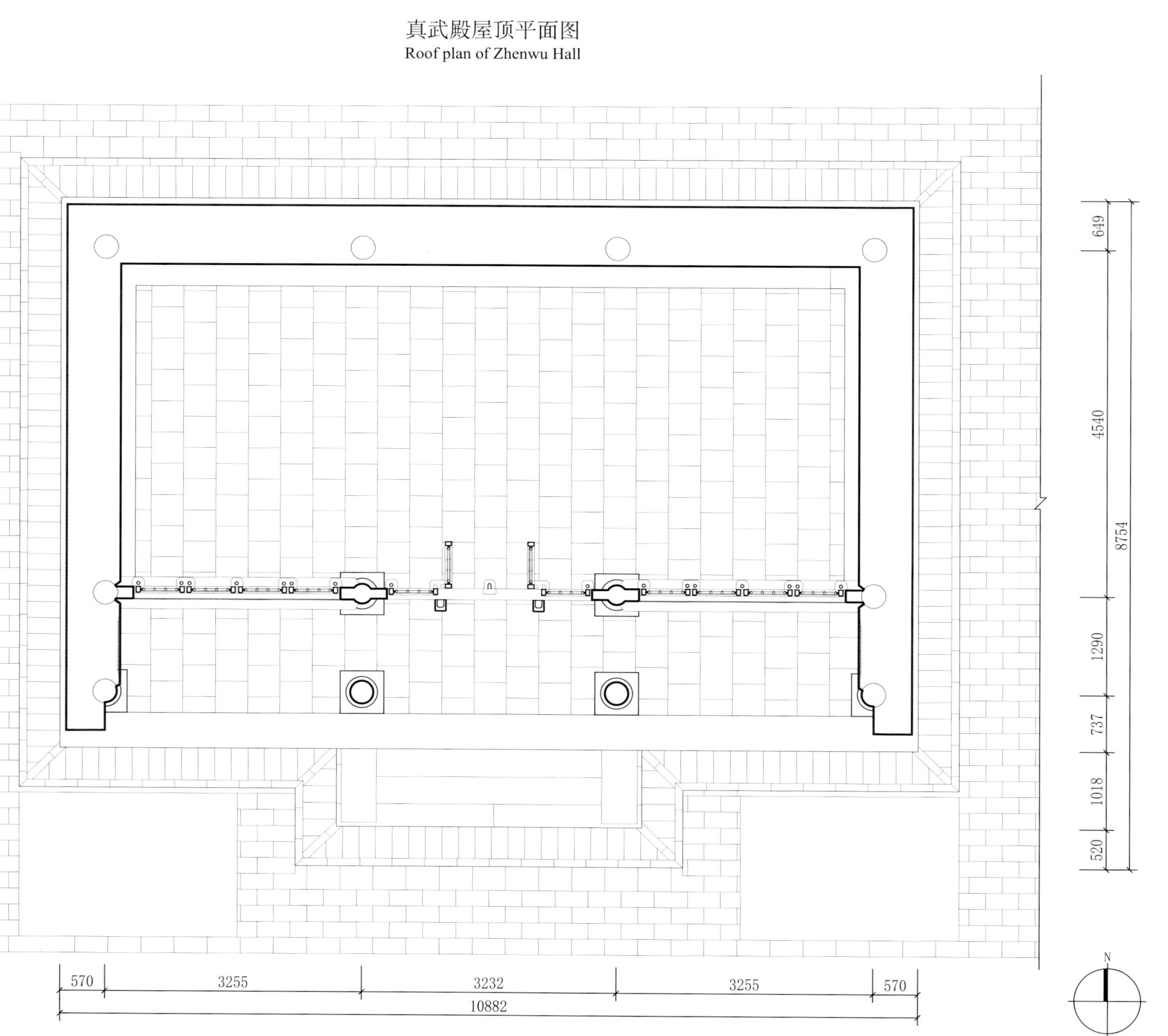

真武殿平面图
Plan of Zhenwu Hall

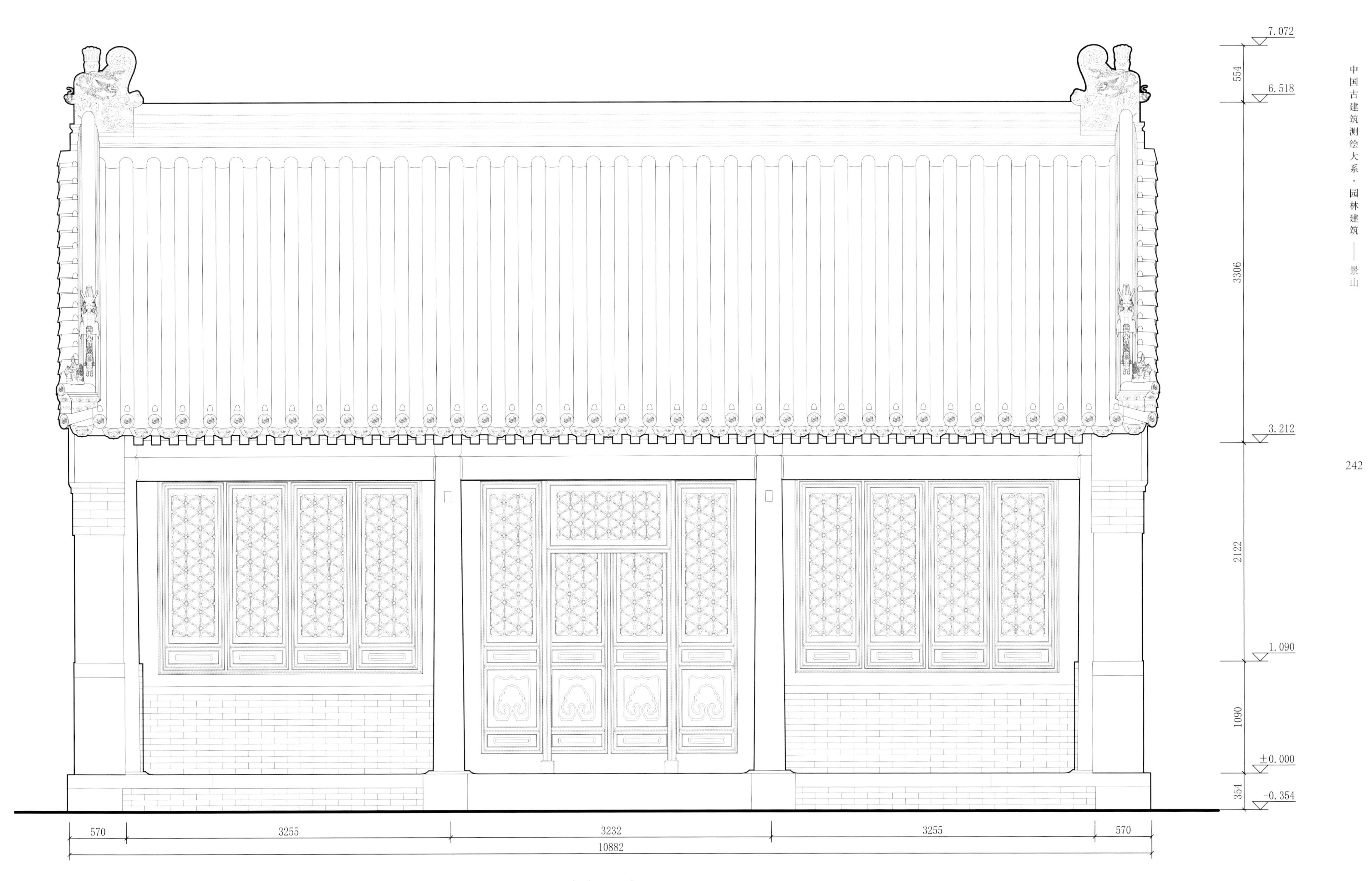

真武殿正立面图
Front elevation of Zhenwu Hall

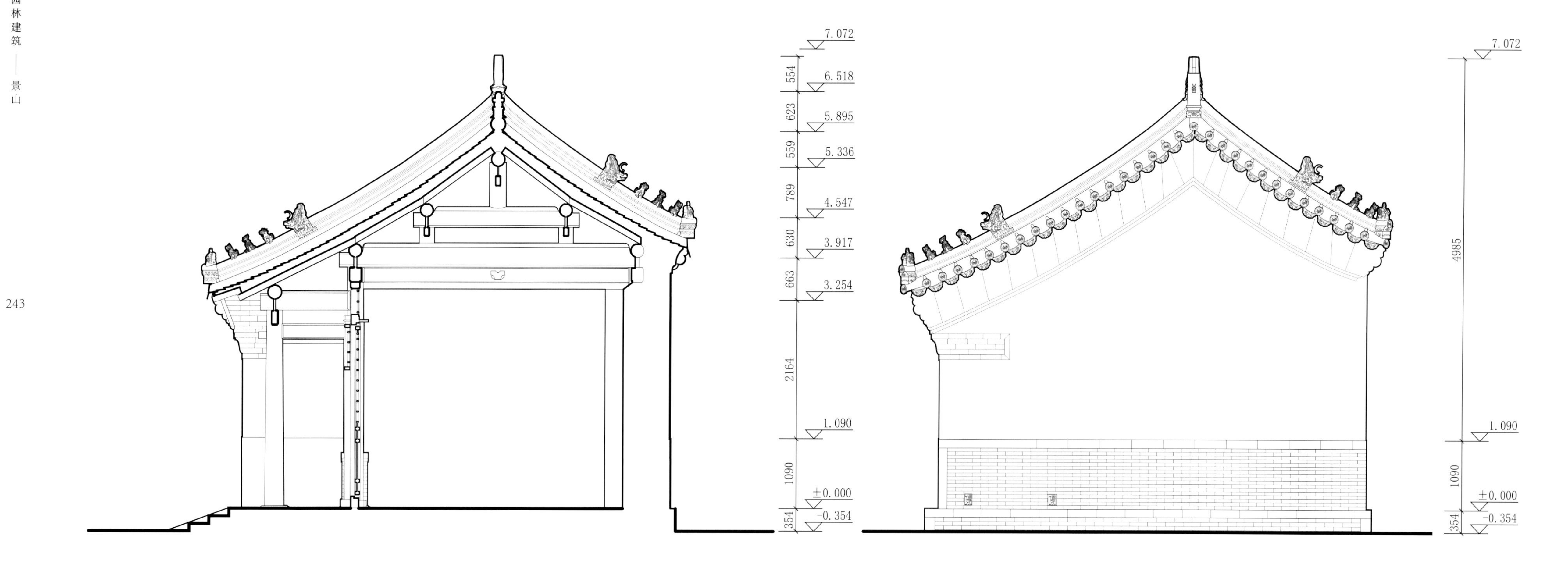

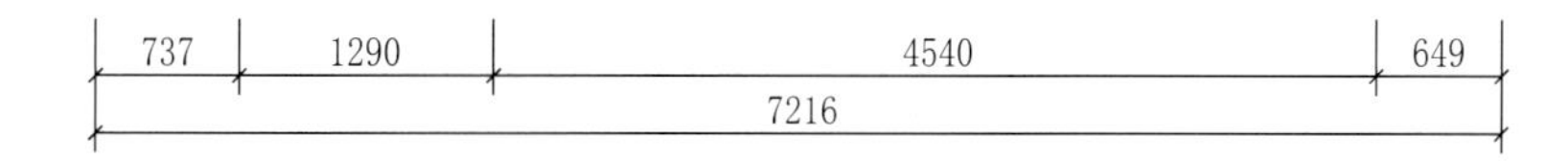

真武殿明间横剖面图

Central bay section of Zhenwu Hall

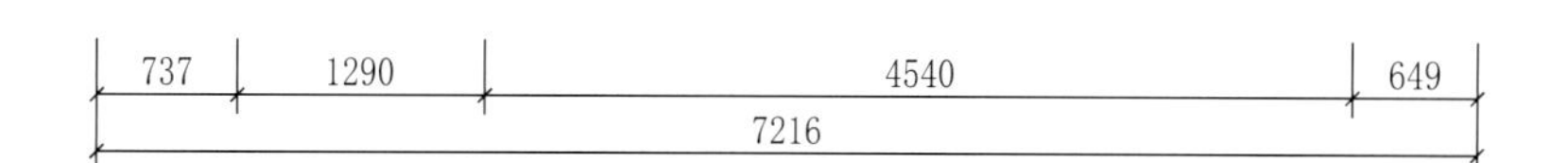

真武殿侧立面图

Side elevation of Zhenwu Hall

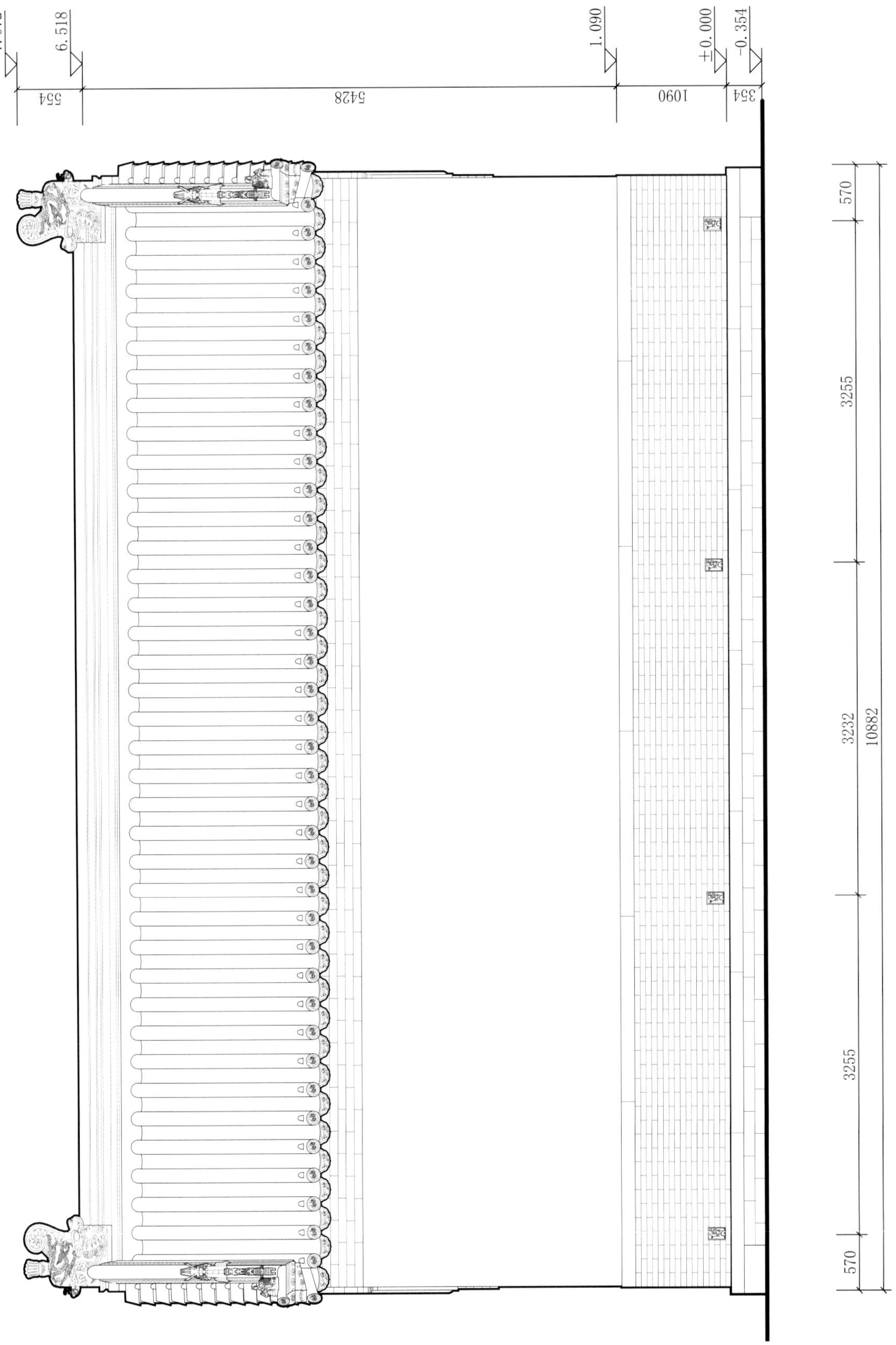

真武殿背立面图
Back elevation of Zhenwu Hall

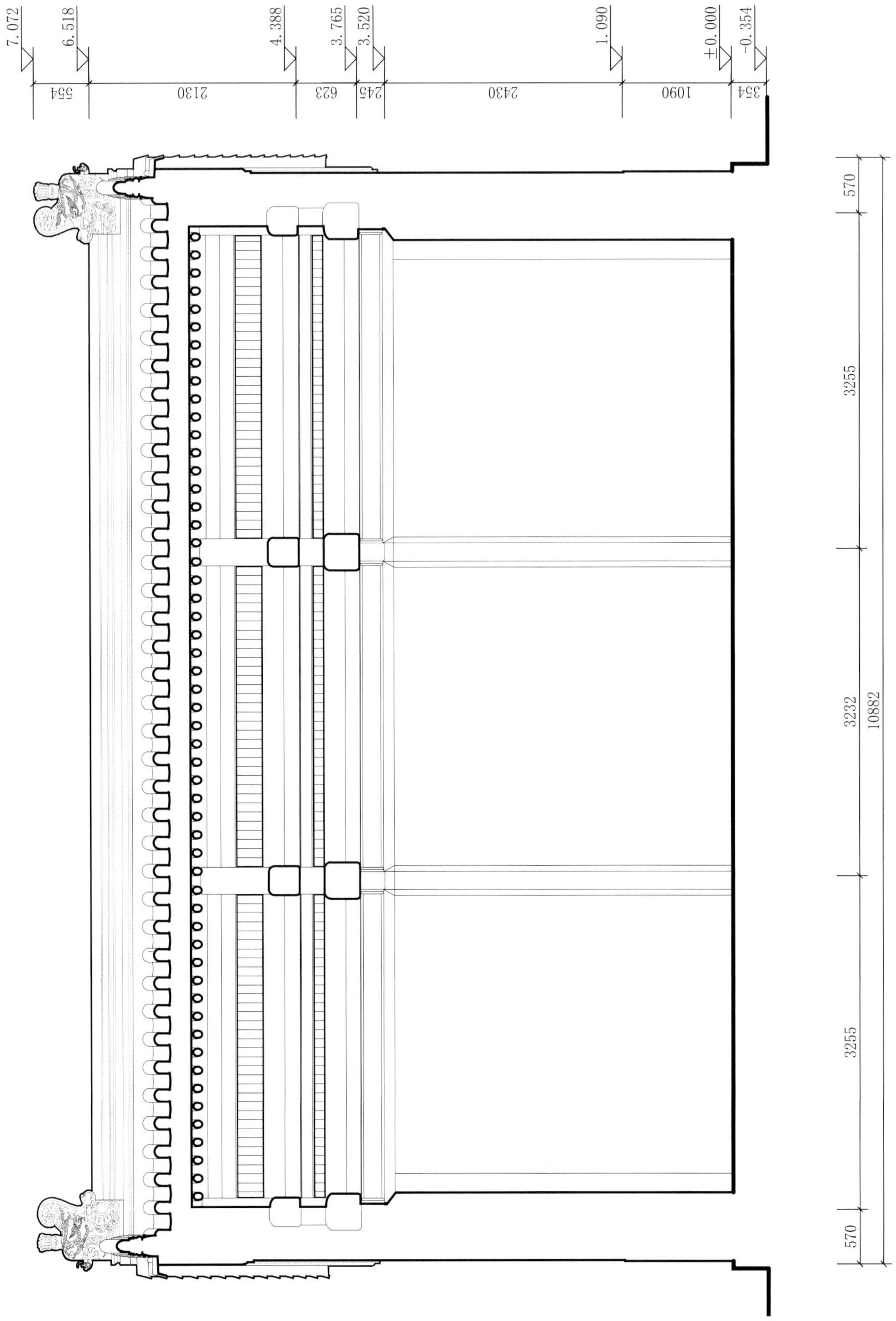

真武殿纵剖面图
Longitudinal section of Zhenwu Hall

主要参考文献
References

[一]（元）李志常. 长春真人西游记[M]. 北京：文物出版社，1988.

[二]（元）陶宗仪. 南村辍耕录[M]. 沈阳：辽宁教育出版社，1998.

[三]（明）宋濂等. 元史. 文渊阁四库全书（电子版）. 上海：上海人民出版社，上海迪志文化出版有限公司.

[四]（明）刘侗，于奕正. 帝京景物略[M]. 北京：北京古籍出版社，1935.

[五]（明）沈德符. 万历野获编[M]. 北京：中华书局，1959.

[六]（明）萧询. 故宫遗录[M]. 北京：北京古籍出版社，1982.

[七]（明）佚名. 北平考[M]. 北京：北京古籍出版社，1982.

[八]（清）孙承泽. 春明梦余录[M]. 扬州：江苏广陵古籍刻印社，1990.

[九]（清）高士奇. 金鳌退食笔记[M]. 北京：北京古籍出版社，1962.

[十]（清）乾隆御制诗文集. 四库全书. 成于乾隆四十九年（1784年）.

[十一]（清）于敏中等. 日下旧闻考[M]. 北京：北京古籍出版社，1982.

[十二]（清）鄂尔泰，张廷玉等. 国朝宫史[M]. 北京：北京古籍出版社，1987.

[十三]（清）励宗万. 京城古籍考[M]. 北京：北京古籍出版社，1981.

[十四]（清）阙名. 日下尊闻录[M]. 北京：北京古籍出版社，1981.

[十五]（清）昆冈. 大清会典事例. 大清会典.

[十六]（清）清实录[M]. 北京：中华书局，1991.

[十七]（清）吴长元. 宸垣识略[M]. 北京：北京古籍出版社，1981.

[十八]（民国）赵尔巽，柯劭忞等. 清史稿[M]. 北京：中华书局，1977.

[十九]北京北海公园管理处. 北海景山公园志[M]. 北京：中国林业出版社，2000.

[二十]周维权. 中国古典园林史[M]. 北京：清华大学出版社，1999.

参与测绘及相关工作的人员名单

指导教师： 王其亨　白成军　张凤梧　杨　菁　张　龙　何蓓洁

辅导员： 来　琳　谢　舒

本科生

2012级：宋　文　王楚瑶　王　尧　白　丹　刘晗之　雷琳馨　李　牧
王　储　冉子嵘　郭永健　杜邦国　张佳欣　曾昱程　黄昱锟
魏万豪　刘琦蕾　邓惠予　薄　珏　吉瀚林　王春艺　刘克嘉
刘　畅　蒋洒洒　田英祯　潘艾婧　沈　季　高翔宇　张　宇
谢成溪　吴　凡　梁　露　黄兰琴　王思琦　张　航　曾　韵
李佳泊　边玉麒　邱　彤　任爱婕　张书涵　张浩然　张　欢
赵星宇　奚雪晴　钟　升　尹波宁　王　茜　仝存平　董韵笛
贺　妍　美　乐

2013级：蔡焱南

2014级：冯兰萌　龙治至　魏逸忱　王东辉　李　利　马培铨　陈　鹏
刘欣佳　李子昂　宋晨阳　杜兴科　董鑫伟　康鑫宇　罗海亮
赵文昊　郭　强

2015级：张栖宁　刘宇珩　王爱嘉　刘雨卿　谭凯家　牛宇豪　王济时
谢靖嵘　余思苇　闫方硕　陈　怡　龚江宇　胡振宇　李栋钰
运乃博　欧士銮

硕士研究生

2013级：梁　璐　李程远　韩　涛　张雨奇　荣　幸

2014级：徐　丹　赵蓬雯　王　齐　肖芳芳　马胜楠　李东遥

2015级：周悦煌　杨　洁　付蜜桥

2016级：张煦康　张净妮

博士研究生

2016级：王笑石

摄影测量：李 哲 张 文 邵浩然 闫 宇 吴晓冬

技术人员：张志强 张志勇 李 港

此次出版图纸整理人员

图纸审阅：王其亨 张凤梧 杨 菁 张 龙

图纸整理修改：周悦煌 张煦康 杨 洁 付蜜桥 林 涛 王宏伟

钱一畅 马晓菡 王涯琪 刘凯旋 庞 磊 曹博雅

刘 洋 祁 爽 赵欣宜 童成蹊 牛嘉城 张 伟

李 峰 周 颖 刘丽子 段文星 於昊臻 席坤杨

王紫晗

景山公园管理处：从一蓬 宋 愷 陈艳红 邹 雯 张平水

都艳辉 李 宇

英文翻译：［奥］荷雅丽

List of Participating Involved in Surveying and Related Works

Instructor: WANG Qiheng, BAI Chengjun, ZHANG Fengwu, YANG Jing, ZHANG Long, HE Beijie

Counsellor: LAI Lin,XIE Shu

Bachelor Student

2012 Year: SONG Wen, WANG Chuyao, WANG Yao, BAI Dan, LIU Hanzhi, LEI Linxin, LI Mu, WANG Chu, RAN Zidie, GUO Yongjian, DU Bangguo, ZHANG Jiaxin, CENG Yucheng, HUANG Yukun, WEI Wanhao, LIU Qilei, DENG Huiyu, BAO Jue, JI Hanlin, WANG Chunyi, LIU Kejia, LIU Chang, JIANG Sasa, TIAN Yingzhen, PAN Aijing, SHEN Ji, GAO Xiangyu, ZHANG Yu, XIE Chengxi, WU Fan, LIANG Lu, HUANG Lanqin, WANG Siqi, ZHANG Hang, CENG Yun, LI Jiapo, BIAN Yuqi, QIU Tong, REN Aijie, ZHANG Shuhan, ZHANG Haoran, ZHANG Huan, ZHAO Xingyu, XI Xueqing, ZHONG Sheng, YIN Boning, WANG Qian, TONG Cunping, DONG Yundi, HE Yan, MEI Le

2013 Year: CAI Yannan

2014 Year: FENG Lanmeng, LONG Zhizhi, WEI Yichen, WANG Donghui, LI Li, MA Peiquan, CHEN Peng, LIU Xinjia, LI Ziang, SONG Chenyang, DU Xingke, DONG Xinwei, KANG Xinyu, LUO Hailiang, ZHAO Wenhao, GUO Qiang

2015 Year: ZHANG Qining, LIU Yuhang, WANG Aijia, LIU Yuqing, TAN Kaijia, NIU Yuhao, WANG Jishi, XIE Jingrong, YU Siwei, YAN Fangshuo, CHEN Yi, GONG Jiangyu, HU Zhenyu, LI Dongyu, YUN Naibo, OU Shiluan

Master Student

2013 Year: LIANG Lu, LI Chengyuan, HAN Tao, ZHANG Yuqi, RONG Xing

2014 Year: XU Dan, ZHAO Pengwen, WANG Qi, XIAO Fangfang, MA Shengnan, LI Dongyao

2015 Year: ZHOU Yuehuang, YANG Jie, FU Miqiao

2016 Year: ZHANG Xukang, ZHANG Jingni

Doctoral Student

2016 Year: WANG Xiaoshi

Potographic Surveying: LI Zhe, ZHANG Wen, SHAO Haoran, YAN Yu, WU Xiaodong

Technician: ZHANG Zhiqiang, ZHANG Zhiyong, LI Gang

The Publication of the Drawing Collation Personnel

Drawing Check and Proof: WANG Qiheng, ZHANG Fengwu, YANG Jing, ZHANG Long

Drawing Arrangement and Modification: ZHOU Yuehuang, ZHANG Xukang, YANG Jie, FU Miqiao, LIN Tao, WANG Hongwei, QIAN Yichang, MA Xiaohan, WANG Yaqi, LIU Kaixuan, PANG Lei, CAO Boya, LIU Yang, QI Shuang, ZHAO Xinyi, TONG Chengqi, NIU Jiacheng, ZHANG Wei, LI Feng, ZHOU Ying, LIU Lizi, DUAN Wenxing, YU Haozhen, XI Kunyang, WANG Zihan

Beijing Municipal Administration Office of Jingshan Park: CONG Yipeng, SONG Kai, CHEN Yanhong, ZOU Wen, ZHANG Pingshui, DOU Yanhui, LI Yu

English Translator: Alexandra Harrer

图书在版编目（CIP）数据

景山 = JINGSHAN PARK：汉英对照 / 王其亨主编；张凤梧，杨菁编著 .—北京：中国建筑工业出版社，2019.12
（中国古建筑测绘大系 · 园林建筑）
ISBN 978-7-112-24546-8

Ⅰ . ①景… Ⅱ . ①王… ②张… ③杨… Ⅲ . ①景山—园林建筑—建筑艺术—图集 Ⅳ . ① TU-882

中国版本图书馆CIP数据核字（2019）第284396号

丛书策划 / 王莉慧
责任编辑 / 李　鸽　刘　川
英文审稿 / [奥] 荷雅丽（Alexandra Harrer）
书籍设计 / 付金红
责任校对 / 王　烨

中国古建筑测绘大系 · 园林建筑
景山
天津大学建筑学院
北京市景山公园管理处
合作编写
王其亨　主编
张凤梧　杨　菁　编著

Traditional Chinese Architecture Surveying and Mapping Series: Garden Architecture
JINGSHAN PARK
Compiled by School of Architecture, Tianjin University &
Beijing Municipal Administration Office of Jingshan Park
Chief Edited by WANG Qiheng
Edited by ZHANG Fengwu, YANG Jing

*
中国建筑工业出版社出版、发行（北京海淀三里河路 9 号）
各地新华书店、建筑书店经销
北京方舟正佳图文设计有限公司制版
北京雅昌艺术印刷有限公司印刷
*
开本：787 毫米 × 1092 毫米　横 1/8　印张：34½　字数：954 千字
2022 年 3 月第一版　2022 年 3 月第一次印刷
定价：**268.00** 元
ISBN 978-7-112-24546-8
（35217）